Empowering Sustainable Industrial 4.0 Systems With Machine Intelligence

School of Electrical Engineering and Computer Science (SEECS), National University of Sciences and Technology (NUST), Islamabad, Pakistan

Noor Zaman

Taylor's University, Malaysia

A volume in the Advances in Logistics,
Operations, and Management Science (ALOMS)
Book Series

Published in the United States of America by
  IGI Global
  Business Science Reference (an imprint of IGI Global)
  701 E. Chocolate Avenue
  Hershey PA, USA 17033
  Tel: 717-533-8845
  Fax: 717-533-8661
  E-mail: cust@igi-global.com
  Web site: http://www.igi-global.com

Library of Congress Cataloging-in-Publication Data

Names: Ahmad, Muneer, 1977- editor. | Jhanjhi, Noor Zaman, 1972- editor.
Title: Empowering sustainable industrial 4.0 systems with machine
  intelligence / Muneer Ahmad, and Noor Zaman Jhanjhi, editor.
Description: Hershey, PA : Business Science Reference, [2022] | Includes
  bibliographical references and index. | Summary: "This book addresses a
  wide range of topics related to industry 4.0 issues, its challenges and
  employment of smart and intelligent solutions to contain the errors and
  their impacts affecting stakeholders of such systems including, health
  industry, pharmaceutical industry, education industry, defense systems,
  manufacturing industry, logistic support systems, supply chain
  management, social media industry, and disaster management systems
  etc"-- Provided by publisher.
Identifiers: LCCN 2021044511 (print) | LCCN 2021044512 (ebook) | ISBN
  9781799892014 (hardcover) | ISBN 9781799892021 (paperback) | ISBN
  9781799892038 (ebook)
Subjects: LCSH: Artificial intelligence--Industrial applications. |
  Management information systems. | Sociotechnical systems.
Classification: LCC TA347.A78 E47 2022 (print) | LCC TA347.A78 (ebook) |
  DDC 006.3--dc23/eng/20211013
LC record available at https://lccn.loc.gov/2021044511
LC ebook record available at https://lccn.loc.gov/2021044512

This book is published in the IGI Global book series Advances in Logistics, Operations, and Management Science (ALOMS) (ISSN: 2327-350X; eISSN: 2327-3518)

British Cataloguing in Publication Data
A Cataloguing in Publication record for this book is available from the British Library.

For electronic access to this publication, please contact: eresources@igi-global.com.

Advances in Logistics, Operations, and Management Science (ALOMS) Book Series

John Wang
Montclair State University, USA

ISSN:2327-350X
EISSN:2327-3518

MISSION

Operations research and management science continue to influence business processes, administration, and management information systems, particularly in covering the application methods for decision-making processes. New case studies and applications on management science, operations management, social sciences, and other behavioral sciences have been incorporated into business and organizations real-world objectives.

The **Advances in Logistics, Operations, and Management Science** (ALOMS) Book Series provides a collection of reference publications on the current trends, applications, theories, and practices in the management science field. Providing relevant and current research, this series and its individual publications would be useful for academics, researchers, scholars, and practitioners interested in improving decision making models and business functions.

COVERAGE

- Marketing engineering
- Operations Management
- Risk Management
- Finance
- Decision analysis and decision support
- Networks
- Information Management
- Services management
- Political Science
- Production Management

IGI Global is currently accepting manuscripts for publication within this series. To submit a proposal for a volume in this series, please contact our Acquisition Editors at Acquisitions@igi-global.com or visit: http://www.igi-global.com/publish/.

Titles in this Series

For a list of additional titles in this series, please visit: http://www.igi-global.com/book-series/advances-logistics-operations-management-science/37170

Management Strategies for Sustainability, New Knowledge Innovation, and Personalized Products and Services
Mirjana Pejic-Bach (University of Zagreb, Croatia) and Çağlar Doğru (Ufuk University, Turkey)
Business Science Reference • © 2022 • 365pp • H/C (ISBN: 9781799877936) • US $215.00

Cases on Supply Chain Management and Lessons Learned From COVID-19
Ana Paula Lopes (Polytechnic of Porto, Portugal)
Business Science Reference • © 2022 • 238pp • H/C (ISBN: 9781799891406) • US $225.00

Key Factors and Use Cases of Servant Leadership Driving Organizational Performance
Maria Pressentin (International School of Management, France)
Business Science Reference • © 2022 • 368pp • H/C (ISBN: 9781799888208) • US $215.00

Handbook of Research on Innovative Management Using AI in Industry 5.0
Vikas Garg (Amity University, Greater Noida, India) and Richa Goel (Amity University, Noida, India)
Business Science Reference • © 2022 • 351pp • H/C (ISBN: 9781799884972) • US $295.00

Contemporary Challenges for Agile Project Management
Vannie Naidoo (University of Kwa-Zulu Natal, South Africa) and Rahul Verma (Delhi University, India)
Business Science Reference • © 2022 • 354pp • H/C (ISBN: 9781799878728) • US $225.00

Cases on Optimizing the Asset Management Process
Vicente González-Prida (University of Seville, Spain & National University of Distance Education, Spain) Carlos Alberto Parra Márquez (University of Seville, Spain) and Adolfo Crespo Márquez (University of Seville, Spain)
Business Science Reference • © 2022 • 457pp • H/C (ISBN: 9781799879435) • US $215.00

Handbook of Research on Current Trends in Asian Economics, Business, and Administration
Bülent Akkaya (Manisa Celal Bayar University, Turkey) Kittisak Jermsittiparsert (Henan University of Economics and Law, China) and Ayse Gunsel (Kocaeli University, Turkey)
Business Science Reference • © 2022 • 497pp • H/C (ISBN: 9781799884866) • US $295.00

Logistics and Supply Chain Management in the Globalized Business Era
Lincoln C. Wood (University of Otago, New Zealand) and Linh N.K. Duong (University of the West of England, Bristol, UK)
Business Science Reference • © 2022 • 413pp • H/C (ISBN: 9781799887096) • US $225.00

IGI Global
PUBLISHER of TIMELY KNOWLEDGE

701 East Chocolate Avenue, Hershey, PA 17033, USA
Tel: 717-533-8845 x100 • Fax: 717-533-8661
E-Mail: cust@igi-global.com • www.igi-global.com

Editorial Advisory Board

Table of Contents

Detailed Table of Contents

Chapter 1

Hafsa Zubair, National University of Sciences and Technology, Pakistan
Rafia Mumtaz, National University of Sciences and Technology, Pakistan
Hassan Kumail Ali, National University of Sciences and Technology, Pakistan
Abdullah Nasir, National University of Sciences and Technology, Pakistan

The periodic time series analysis of different aspects of urban areas is essential owing to rise in population, stressed resources, and lack of technology-based solutions. In this regard, temporal analysis of water quality holds paramount importance, and for this purpose, the data from satellite remote sensing, geographic information system (GIS), and internet of things (IoT) are collected to perform water quality trend analysis. The study area is Rawal Dam, where data are processed to derive water quality parameters (WQPs) and thereafter water quality index (WQI). The monthly, yearly, seasonal, pre- and post-COVID-19 temporal analyses are performed to analyze the trends of different WQPs and overall WQI, using suitable machine learning (ML) models over the last eight years (2013-20). The water quality classification is performed using neural networks (NN) with an accuracy of 80%, and predictions are made using vector auto-regression (VAR) and long short-term memory (LSTM) networks with an average root mean squared error (RMSE) of 25.63 and 2.664, respectively.

Chapter 2

Aymen Bashir, National University of Sciences and Technology, Pakistan
Abdullah Mughal, National University of Sciences and Technology, Pakistan
Rafia Mumtaz, National University of Sciences and Technology, Pakistan
Muhammad Ali Tahir, National University of Sciences and Technology, Pakistan

As of today, increased air pollution has disrupted the air quality levels, deeming the air unsafe to breathe. Traditional systems are hefty, costly, sparsely distributed, and do not provide ubiquitous coverage. The interpolation used to supplement low spatial coverage induces uncertainty especially for pollutants whose concentrations vary significantly over small distances. This chapter proposes a solution that uses satellite images and machine/deep learning models to timely forecast air quality. For this study, Lahore is chosen as a study area. Sentinel 5-Precursor is used to gather data for Sulphur Dioxide (SO2), Nitrogen Dioxide (NO2), and Carbon Monoxide (CO) for years 2018-2021. The data is processed for several AI models, where convolutional neural networks (CNN) performed the best with mean squared error (MSE) 0.0003 for the pollutants. The air quality index (AQI) is calculated and is shown on web portal for data visualization. The trend of air quality during COVID-19 lockdowns is studied as well, which showed reduced levels of NO2 in regions where proper lockdown is observed.

Chapter 3

Abdul Aziz Chaudhry, National University of Sciences and Technology, Pakistan
Rafia Mumtaz, National University of Sciences and Technology, Pakistan
Usman Ahmad Siddiqui, National University of Sciences and Technology, Pakistan
Syed Hassan Muzammil, National University of Sciences and Technology, Pakistan
Muhammad Ali Tahir, National University of Sciences and Technology, Pakistan

Livestock monitoring is one of the most common problems in the current time, and to sustain the lifecycle and support the nature of domesticated animals, the standard checking of animal wellbeing is fundamental. Moreover, many diseases are spread from animals to human beings; hence, an early prognosis in regard to cow wellbeing and illness is essential. This chapter proposed an internet of things (IoT)-based framework for domesticated animal wellbeing checking. The proposed framework comprises of a specially crafted multi-sensor board to record a few physiological boundaries including skin temperature, pulse, and rumination with regards to encompassing temperature, stickiness, and a camera for picture examination to recognize diverse standards of health. The data is collected using LoRa gateway technology, where gathered data is examined and utilized for performing ML models to identify diseased and healthy creatures and foresee cow wellbeing for giving early and convenient clinical consideration. The results obtained are used for careful insights regarding animal health and wellbeing.

Chapter 4

María A. Pérez-Juárez, University of Valladolid, Spain
Javier M. Aguiar-Pérez, University of Valladolid, Spain
Javier Del-Pozo-Velázquez, University of Valladolid, Spain
Miguel Alonso-Felipe, University of Valladolid, Spain
Saúl Rozada-Raneros, University of Valladolid, Spain
Mikel Barrio-Conde, University of Valladolid, Spain

The Fourth Industrial Revolution, under the name of Industry 4.0, focuses on obtaining and using data to facilitate decision-making and thus achieve a competitive advantage. Industry 4.0 is about smart factories. For this, a series of technologies have emerged that communicate the physical and the virtual world, including Internet of Things, Big Data, and Artificial Intelligence. These technologies can be applied in many areas of the industry such as production, manufacturing, quality, logistics, maintenance, or security to improve the optimization of the production capacity or the control and monitoring of the production process. An important area of application is maintenance. Predictive maintenance is focused on monitoring the performance and condition of equipment during normal operation to reduce the likelihood of failures with the help of data-driven techniques. This chapter aims to explore the possibilities of using artificial intelligence to optimize the maintenance of the machinery and equipment components so that product costs are reduced.

Shier Khee Saw, Department of Surgery, School of Medical Sciences, Universiti Sains
Malaysia Health Campus, Kubang Kerian, Malaysia
Syaiful Azzam Sopandi, Department of Surgery, Hospital Raja Perempuan Zainab II, Kota
Bharu, Malaysia
Rosnelifaizur bin Ramely, Department of Surgery, School of Medical Sciences, Universiti
Sains Malaysia Health Campus, Kubang Kerian, Malaysia
Chow Khuen Chan, Department of Biomedical Engineering, Universiti Malaya, Kuala
Lumpur, Malaysia
Michael Pak Kai Wong, Department of Surgery, School of Medical Sciences, Universiti Sains
Malaysia Health Campus, Kubang Kerian, Malaysia
Shier Nee Saw, Department of Artificial Intelligence, Faculty of Computer Science and
Information Technology, Universiti Malaya, Kuala Lumpur, Malaysia

Abdominal aorta aneurysm (AAA) is defined as an abnormal dilatation of the aorta at least 50% more than the adjacent normal vessel diameter. AAA is usually asymptomatic until complications occur such as aorta dissection and ruptured AAA, which has a direct relationship with the size of the aneurysm. Early detection with early intervention of AAA reduces the mortality rate related to rupture. In the era of digitalization, medical data such as electronic medical record, ultrasound images, and physical measurements are available for analysis. Furthermore, with the advancement of artificial intelligence (AI) technologies, numerous AI models have been proposed and shown to improve AAA diagnosis and prognostication. AI technologies, with no doubt, possess an infinite potential to improve the services of healthcare providers. Hence, this chapter targets the audience from all professions: clinicians, radiologists, and computer scientists. This chapter aims to close the gap between the medical profession and computer scientists and thus to design an AI model that can be clinically used.

Aamir Farooq Khan, National University of Sciences and Technology, Pakistan
Rafia Mumtaz, National University of Sciences and Technology, Pakistan
Muhammad Usama, National University of Sciences and Technology, Pakistan
Taimoor Khan Mahsud, National University of Sciences and Technology, Pakistan

Remote sensing through satellites and internet of things (IoT) technology are two widespread techniques to assess inland water quality. However, both these techniques have their limitations. IoT provides point data, which is insufficient to represent entire water body, especially if the water body has complex terrain and hydrology. Through remote sensing, we can sample data of a large area, but data acquisition is constrained by satellite. Revisit time and quality of estimates can be affected by image resolution. Moreover, non-optical properties that might affect water quality cannot be sensed through satellites. To complement this, GIS data from labs can be useful for providing higher resolution and accurate data and can be used as ground truth. Thus, in this chapter, the authors aim to integrate both these data collection techniques followed by estimation and prediction through machine learning models. The accumulated datasets are used to train machine learning (ML) models deployed at a server. The selected ML model is an artificial neural network with train accuracy of 97% and test accuracy of 95%.

The current problem with CRM is weak marketing, as there is no individualization strategy, low productivity, and blurred marketing objectives. Next, companies are failing to completely capitalize on and extract useful information from a vast collection of databases. Organizations were also unable to adequately evaluate consumer behaviors and customer expectations that contribute to poor vendor-customer relationships and decrease customer loyalty. Most companies are seeking innovative ways to develop their CRM as it helps to challenge new ways of marketing and growing income, as customer loyalty and sales rely on one another. Whereas some businesses are incorporating data mining techniques in the management of CRM, there are several disadvantages in the market basket analysis, and one of the main drawbacks is that it is difficult to distinguish interesting patterns, as the number of rules obtained is very high. However, we might assume that it is computationally efficient as a minimum support value of 60% with a minimum confidence value of 80%.

Machine intelligence is a backbone of self-driving automotive (SDA) systems. Presently, ResNet, DenseNet, and ShuffleNet V2 are excellent convolution choices, whereas object detection focuses on YOLO and F-RCNN design. This study discovers the uniqueness of methods and argues the suitability of using each design in SDA technology. Real-time object detection is imperative in SDA technology, for CNN, as well as to object detection algorithms, an architecture that is a balance between speed and accuracy is important. The most favorable architecture in the scope of this case study would be ShuffleNetV2 and YOLO since both are networks that prioritize speed. But the drawback of speed prioritization is that they suffer from slight inaccuracies. One way to overcome this is to replace the neural network with a more accurate (albeit slower) model. The other solution is to use reinforcement learning to find the best architecture, basically using neural networks to create neural networks. Both approaches are resource-intensive in the sense of capital, talent, and computational budget.

Chapter 9

Suja A. Alex, St. Xavier's Catholic College of Engineering, India
Ponkamali S., Cognizant Technology Solutions, USA
Andrew T. R., Intuit Inc., USA
N. Z. Jhanjhi, Taylor's University, Malaysia
Muhammad Tayyab, Taylor's University, Malaysia

The stroke is an important health burden around the world that occurs due to the block of blood supply to the brain. The interruption of blood supply depends on either the sudden blood supply interruption to the brain or a blood vessel leak in tissues. It is tricky to treat stroke-affected patients because the accurate time of stroke is unknown. Internet of things (IoT) is an active field and plays a major role in stroke prediction. Many machines learning (ML) techniques have been used to automate the process and enable many machines to detect the prediction rate of stroke and analyze the risk factor. The ML-based wearable device plays a significant role in making real-time decisions that benefit stroke patients. The parameters such as risk factors associated with stroke and wearable sensors and machine learning techniques for stroke prediction are discussed.

Chapter 10

Tasnim Mohamad Naim, Universiti Malaya, Malaysia
Siti Nurnabila Abdul Rashid, Universiti Malaya, Malaysia
Muneer Ahmad, Universiti Malaya, Malaysia & School of Electrical Engineering and
    Computer Science (SEECS), National University of Sciences and Technology (NUST),
    Islamabad, Pakistan
N. Z. Jhanjhi, Taylor's University, Malaysia

Diabetes has become a growing global public health issue. This illness can become chronic since there are no early symptoms and it can lead to several negative health impacts. This study focuses on the early identification of diabetes based on the symptoms of the disease and finds the relation between the symptoms and the diagnosis. Successful early diagnosis of this disease could boost a person to a better treatment plan before it worsens the health to a critical stage. This study exploits recent classification algorithms including Naïve Bayes, Logistic Regression, REPTree, J48, and Random Forest. The association rules mining using the apriori algorithm is used. Further, the 10-fold cross-validation with split criteria was adopted. The authors adopted several evaluation metrics including accuracy, precision, recall, F-measure, and area under the ROC curve. The research findings revealed the Random Forest to be the best classification algorithm as compared to other classifiers.

Chapter 11

Sadia Aziz, Melbourne Institute of Technology, Australia
Qazi Mudassar Ilyas, King Faisal University, Saudi Arabia
Abid Mehmood, King Faisal University, Saudi Arabia
Ashfaq Ahmad, Jazan University, Saudi Arabia

Since its appearance in late 2019, Severe Acute Respiratory Syndrome Coronavirus 2 (SARS-CoV-2) has become a significant threat to human health and public safety. Machine learning has been extensively exploited in the past to solve a range of problems in everyday life. It has also played its role in virtually all aspects of pandemic management, ranging from early detection and contact tracing to vaccine and drugs development and treatment. This chapter discusses some of the ways in which machine learning-based solutions have helped. In this regard, computer vision approaches have been used for the early detection of disease. Contact tracing has been enhanced by machine learning models to improve distance estimation techniques. Similarly, machine learning techniques have been used to accurately predict mortality rates to optimize resource management. These techniques have also helped in the otherwise tedious processes of vaccine and drugs development in numerous ways, such as providing insights into drug target interactions and possibilities of repurposing the existing drugs.

   Afaq Ahmed, National University of Sciences and Technology, Pakistan
   Ahmad Ali Khan, National University of Sciences and Technology, Pakistan
   Ismail Shah, National University of Sciences and Technology, Pakistan
   Muhammad Ali Tahir, National University of Sciences and Technology, Pakistan
   Rafia Mumtaz, National University of Sciences and Technology, Pakistan

This chapter presents a voice-enabled mobile application that provides preliminary medical diagnosis of non-fatal diseases. It operates in Urdu language, asking questions from the user and responding in the form of a voice-based dialogue. It addresses the problem of lack of hospitals and medical facilities, especially in rural areas. Furthermore, due to the ongoing pandemic of COVID-19, people are wary of visiting the over-crowded government hospitals. The mobile application developed by the authors provides a quick and inexpensive preliminary diagnosis and medical advice. In case of emergency, it recommends a nearby hospital/doctor. As the system is in Urdu language and voice based, it is well-suited to the Pakistani population's low literacy rate. The application user interface is engineered to be intuitive and simple to use.

Preface

It gives us immense pleasure to put forth the book title *Empowering Sustainable Industrial 4.0 Systems With Machine Intelligence*. The recent advancement of industrial computerization has significantly helped in resolving the challenges with conventional industrial systems. The Industry 4.0 quality standards demand smart and intelligent solutions to revolutionize industrial applications. The integration of machine intelligence and internet of things (IoT) technologies can further devise innovative solutions to recent industrial application issues.

Empowering Sustainable Industrial 4.0 Systems With Machine Intelligence assesses the challenges, limitations, and potential solutions for creating more sustainable and agile industrial systems. This publication presents recent intelligent systems for a wide range of industrial applications and smart safety measures toward industrial systems. Covering topics such as geospatial technologies, remote sensing, and temporal analysis, this book is a dynamic resource for health professionals, pharmaceutical professionals, manufacturing professionals, policymakers, engineers, computer scientists, researchers, instructors, students, and academicians.

This book is very appropriate in these exceptional times of COVID-19. It comes at a moment of great global turbulence in the industry sector, where working physically is a challenge due to the current pandemic situations. In this book, we attempted to discuss all the latest approaches, experiments, and evaluate actual and potential contributions with cutting edge technologies, in dealing with challenges primarily related to the sustainable industry 4.0. We strongly believe that this critical reference source is ideal for professional, industries, researchers, academicians, and students.

Chapter 1: Time-Series Analysis and Prediction of Water Quality Through Multisource Data

The periodic time series analysis of different aspects of urban areas is essential owing to rise in population, stressed resources, and lack of technology-based solutions. In this regard, temporal analysis of water quality holds paramount importance and for this purpose, the data from satellite remote sensing, Geographic Information System (GIS), and Internet of Things (IoT) is collected to perform water quality trend analysis. The study area is Rawal dam, where, data is processed to derive Water Quality Parameters (WQPs) and thereafter, Water Quality Index (WQI). The monthly, yearly, seasonal, pre-and post-COVID-19 temporal analysis is performed to analyze the trends of different WQPs and overall WQI, using suitable Machine Learning (ML) models, over the last 8 years (2013 - 20). The water quality classification is performed using neural networks (NN) with an accuracy of 80% and predictions are made using Vector Auto-Regression (VAR) and Long Short Term Memory (LSTM) networks with an average root mean squared error (RMSE) of 25.63 and 2.664 respectively.

Chapter 2: Temporal Analysis and Prediction of Ambient Air Quality Using Remote Sensing, Deep Learning, and Geospatial Technologies

As of today, increased air pollution has disrupted the air quality levels, deeming the air unsafe to breathe in. Traditional systems are hefty, costly, sparsely distributed, and do not provide ubiquitous coverage. The interpolation used to supplement low spatial coverage induces uncertainty especially for pollutants whose concentration vary significantly over small distances. This chapter proposes a solution that uses satellite images, and machine/deep learning models to timely forecast air quality. For this study, Lahore is chosen as a study area. Sentinel 5-Precursor is used to gather data for Sulphur Dioxide (SO2), Nitrogen Dioxide (NO2), and Carbon Monoxide (CO) for years 2018-2021. The data is processed for several AI models, where. Convolutional Neural Networks (CNN) performed the best with mean squared error (MSE) 0.0003, for the pollutants. The air quality index (AQI) is calculated and is shown on web portal for data visualization. The trend of air quality during COVID-19 lockdowns is studied as well, which showed reduced levels of NO2 in regions where proper lockdown is observed.

Chapter 3: Automated Multi-Sensor Board for IoT and ML-Enabled Livestock Monitoring

Livestock monitoring is one of the most common problems in the current time and to sustain the lifecycle and support the nature of domesticated animals, the standard checking of animal's wellbeing is fundamental. Moreover, many diseases are spread from animals to human beings, hence, an early prognosis in regard to cow's wellbeing and illness is essential. This paper proposed an Internet of Things (IoT) based framework for domesticated animals' wellbeing checking. The proposed framework comprises of a specially crafted multi-sensor board to record a few physiological boundaries including skin temperature, pulse, and rumination w.r.t encompassing temperature, stickiness, and a camera for picture examination to recognize diverse standards of health. The data is collected using LoRa gateway technology, where gathered data is examined and utilized for performing ML models to identify diseased and healthy creatures and foresee cow's wellbeing for giving early and convenient clinical consideration. The results obtained are used for careful insights regarding animal health and their well-being.

Chapter 4: How Artificial Intelligence Can Enhance Predictive Maintenance in Smart Factories

The Fourth Industrial Revolution, under the name of Industry 4.0, focuses on obtaining and using data to facilitate decision-making and thus achieve a competitive advantage. Industry 4.0 is about Smart Factories. For this, a series of technologies have emerged that communicate the physical and the virtual world, including Internet of Things, Big Data and Artificial Intelligence. These technologies can be applied in many areas of the industry such as production, manufacturing, quality, logistics, maintenance or security to improve the optimization of the production capacity or the control and monitoring of the production process. An important area of application is maintenance. Predictive Maintenance is focused on monitoring the performance and condition of equipment during normal operation to reduce the likelihood of failures with the help of data-driven techniques. This chapter aims to explore the possibilities of using Artificial Intelligence to optimize the maintenance of the machinery and equipment components so that product costs are reduced.

Chapter 5: A Review of Artificial Intelligence Models in Prognosticating Abdominal Aorta Aneurysm

Abdominal aorta aneurysm (AAA) is defined as an abnormal dilatation of the aorta at least 50% more than the adjacent normal vessel diameter. AAA is usually asymptomatic until complications occurred such as aorta dissection and ruptured AAA, which has a direct relationship with the size of the aneurysm. Early detection with early intervention of AAA reduces the mortality rate related to rupture. In the era of digitalization, medical data such as electronic medical record, ultrasound images and physical measurements are available for analysis. Furthermore, with the advancement of artificial intelligence (AI) technologies, numerous AI models have been proposed and shown to improve AAA diagnosis and prognostication. AI technologies, with no doubt, possess an infinite potential to improve the service of healthcare provider. Hence, this review paper targets the audience from all professions: clinicians, radiologists, and computer scientists. This paper aims to close the gap between medical profession and computer scientists and thus able to design an AI model that can be of clinical used.

Chapter 6: Enhanced Water Quality Monitoring and Estimation Using Multi-Modal Approach

Remote sensing through satellites and Internet of Things (IoT) technology are two widespread techniques to assess inland water quality. However, both these techniques have their limitations. IoT provides point data, which is insufficient to represent entire water body, especially if the water body has complex terrain and hydrology. Through remote sensing we can sample data of a large area, but data acquisition is constrained by satellite revisit time and quality of estimates can be affected by image resolution. Moreover, non-optical properties which might affect water quality, cannot be sensed through satellites. To complement this, GIS data from labs can be useful for providing higher resolution and accurate data and can be used as ground truth. Thus, in this chapter we aim to integrate both these data collection techniques followed by estimation and prediction through machine learning models. The accumulated datasets are used to train machine learning(ML) models, deployed at a server. The selected ML model is an artificial neural network with train accuracy of 97% and test accuracy of 95%.

Chapter 7: Machine Intelligence in Customer Relationship Management in Small and Large Companies

The current problem with CRM is weak marketing, as there is no individualization strategy, low productivity, and blurred marketing objectives. Next, companies are failing to completely capitalize on and extract useful information from a vast collection of databases. Organizations were also unable to adequately evaluate consumer behaviors and customer expectations that contribute to poor vendor-customer relationships and decrease customer loyalty. Most companies are seeking innovative ways to develop their CRM as it helps to challenge new ways of marketing and growing income, as customer loyalty and sales rely on one another. Whereas, some businesses are incorporating data mining techniques in the management of CRM. However, there are several disadvantages in the market basket analysis and one of the main drawbacks is that it is difficult to distinguish interesting patterns, as the number of rules obtained is very high. However, we might assume that it is computationally efficient as a minimum support value of 60% with a minimum confidence value of 80%.

Chapter 8: Machine Intelligence as Foundation of Self-Driving Automotive (SDA) Systems

Machine intelligence is a backbone of Self Driving Automotive (SDA) systems. Presently, ResNet, DenseNet, and ShuffleNet V2 are excellent convolution choices, whereas object detection focuses on YOLO and F-RCNN design. This study discovers the uniqueness of methods and argues the suitability of using each design in SDA technology. Real-time object detection is imperative in SDA technology, for CNN, as well as to object detection algorithms, an architecture that is a balance between speed and accuracy is important. The most favorable architecture in the scope of this case study would be Shuffle-NetV2 and YOLO. Since both are networks that prioritize speed. But the drawback of speed prioritization is that they suffer from slight inaccuracies. One way to overcome this is to replace the neural network with a more accurate (albite slower) model. The other solution is to use reinforcement learning to find the best architecture, basically using neural networks to create neural networks. Both approaches are resource-intensive, in the sense of capital, talent as well as computational budget.

Chapter 9: Machine Learning-Based Wearable Devices for Smart Healthcare Application With Risk Factors Monitoring

Stroke has been known as one of the most active and dangerous attack that has caused high mortality and morbidity rate throughout the world. The Stroke is an important health burden around the world that occurs due to the block of blood supply to the brain. The interruption of blood supply depends on either the sudden blood supply interruption to the brain or a blood vessel leak in tissues. It is tricky to treat the stroke affected patients, because the accurate time of stroke is unknown. Internet of Things (IoT) is an active field and plays a major role stroke prediction. Many machines learning (ML) techniques which has been used to automate the process and enable many machines to detect the prediction rate of stroke and analyze the risk factor. The ML based wearable device plays a significant role in making real-time decisions that benefit the stroke patients. The parameters such as risk factors associated with stroke and wearable sensors and machine learning techniques for stroke prediction are discussed.

Chapter 10: Predicting Early Stage of Diabetes and Finding the Association of the Symptoms

Diabetes has become a growing global public health issue. This illness can become chronic since there are no early symptoms and it can lead to several negative health impacts. This study focuses on the early identification of diabetes based on the symptoms of the disease and finds the relation between the symptoms and the diagnosis. Successful early diagnosis of this disease could boost a person to a better treatment plan before it worsens the health to a critical stage. This study exploits recent classification algorithms including Naïve Bayes, Logistic Regression, REPTree, J48, and Random Forest. Besides, the association rules mining using the apriori algorithm is used. Further, the 10-fold cross-validation with split criteria was adopted. We adopted several evaluation metrics including accuracy, precision, recall, F-measure, and area under the ROC curve. The research findings revealed the Random Forest to be the best classification algorithm as compared to other classifiers.

Chapter 11: Role of Machine Learning in Handling the COVID-19 Pandemic

Since its appearance in late 2019, Severe Acute Respiratory Syndrome Coronavirus 2 (SARS-CoV-2) has become a significant threat to human health and public safety. Machine learning has been extensively exploited in the past to solve a range of problems in everyday life. It has also played its role in virtually all aspects of pandemic management, ranging from early detection and contact tracing to vaccine and drugs development and treatment. This chapter discusses some of the ways in which machine learning-based solutions have helped. In this regard, computer vision approaches have been used for the early detection of disease. Contact tracing has been enhanced by machine learning models to improve distance estimation techniques. Similarly, machine learning techniques have been used to accurately predict mortality rates to optimize resource management. These techniques have also helped in the otherwise tedious processes of vaccine and drugs development in numerous ways, such as providing insights into drug target interactions and possibilities of repurposing the existing drugs.

Chapter 12: Rural Healthcare Mobile App – Urdu Voice-Enabled Mobile App for Disease Diagnosis

This chapter presents a voiced-enabled mobile application that provides preliminary medical diagnosis of non-fatal diseases. It operates in Urdu language, asking questions from the user and responding in the form of a voice-based dialogue. It addresses the problem of lack of hospitals and medical facilities, especially in rural areas. Furthermore, due to the ongoing pandemic of COVID-19, people are wary of visiting the over-crowded government hospitals. The mobile application developed by the authors provides a quick and inexpensive preliminary diagnosis and medical advice. In case of emergency, it recommends a nearby hospital/doctor. As the system is in Urdu language and voice based, it is well-suited to Pakistani population's low literacy rate. The application user interface is engineered to be intuitive and simple to use.

Muneer Ahmad
School of Electrical Engineering and Computer Science (SEECS), National University of Sciences and Technology (NUST), Islamabad, Pakistan

Noor Zaman
Taylor's University, Malaysia

Acknowledgment

We would like to express our thanks to Almighty Allah SWT for his all blessings and then great appreciation to all of those we have had the pleasure to work with during this project. The completion of this project could not have been accomplished without their support. First, the editors would like to express deep and sincere gratitude to all the authors who shared their ideas, expertise, and experience by submitting chapters to this book and adhering to its timeline. Second, the editors wish to acknowledge the extraordinary contributions of the reviewers for their valuable and constructive suggestions and recommendations to improve the quality, coherence, and content presentation of chapters. Most of the authors also served as referees. Their willingness to give time so generously is highly appreciated. Finally, our heartfelt gratitude goes to our family members and friends for their love, prayers, caring, and sacrifices in completing this project well in time.

Muneer Ahmad
University of Malaya, Malaysia

Noor Zaman
Taylor's University, Malaysia

Chapter 1

Time–Series Analysis and Prediction of Water Quality Through Multisource Data

Hafsa Zubair

National University of Sciences and Technology, Pakistan

Rafia Mumtaz

National University of Sciences and Technology, Pakistan

Hassan Kumail Ali

National University of Sciences and Technology, Pakistan

Abdullah Nasir

National University of Sciences and Technology, Pakistan

ABSTRACT

The periodic time series analysis of different aspects of urban areas is essential owing to rise in population, stressed resources, and lack of technology-based solutions. In this regard, temporal analysis of water quality holds paramount importance, and for this purpose, the data from satellite remote sensing, geographic information system (GIS), and internet of things (IoT) are collected to perform water quality trend analysis. The study area is Rawal Dam, where data are processed to derive water quality parameters (WQPs) and thereafter water quality index (WQI). The monthly, yearly, seasonal, pre- and post-COVID-19 temporal analyses are performed to analyze the trends of different WQPs and overall WQI, using suitable machine learning (ML) models over the last eight years (2013-20). The water quality classification is performed using neural networks (NN) with an accuracy of 80%, and predictions are made using vector auto-regression (VAR) and long short-term memory (LSTM) networks with an average root mean squared error (RMSE) of 25.63 and 2.664, respectively.

DOI: 10.4018/978-1-7998-9201-4.ch001

INTRODUCTION

Water is vital to the survival of life on Earth. Pollution and population growth are causing a shortage of water supply and an alarming decline in water quality. The increasing environmental changes and problems have affected the quality of water available for drinking and other purposes. The poor water quality can directly result in the death of aquatic life and can cause long-lasting and deadly effects on wildlife, birds, plants, and soil. It can be a root cause for many fatal diseases like diarrhea, cancer, etc. in human beings. In short, it has impacted the whole ecosystem.

Pakistan, being an agricultural country, suffers from a reduction in both water supply and water quality. Many reasons are contributing to deteriorating water quality in Pakistan, including overpopulation, deforestation, and industrial waste discharges into water bodies. According to a UN study, only 20% of the population in Pakistan has access to clean drinking water (WHO/UNICEF Joint Monitoring Program (JMP) for Water Supply, 2015). A report by Pakistan Medical Association (PMA) shows that in Pakistan, 30% of all diseases and 40% of all deaths are due to poor water quality, with diarrhea being the leading cause of death in infants and children. (M. K. Daud, et al., 2017).

This raises the question of how water quality can be monitored so that the harmful effects of poor water quality can be avoided. This study performs a periodic analysis and prediction of water quality using multisource data to help take preventative measures for deteriorating water quality.

The water quality depends on more than 35 parameters including temperature, precipitation, pH, total dissolved solids (TDS), Total Suspended Solids (TSS), minerals, etc. The measurement of these parameters for a given water body can be used to calculate and analyze its water quality. The World Health Organization (WHO) provides guidelines on the presence of safe amounts of these parameters in water. If the quantity of these parameters differs from the provided guidelines, the water quality is affected.

A Water Quality Index (WQI) is a number or a class that can be derived from the Water Quality Parameters (WQPs). WQI can serve as a quick reference for determining the level of water quality for a given sample. Several WQIs have been developed over the years to measure water quality under various conditions and for different purposes. (R.M., et al., 2007)

Water quality data can be collected from different sources such as i) On-site data collection using IoT sensors, ii) Geographical Information Systems (GIS), and Remote Sensing (RS) satellites. The data collected on-site using IoT sensors is the most accurate and reliable, however, the data collected is point data and does not provide full coverage of the water body. To increase the data points, a greater number of IoT nodes need to be added, which will make the entire study costlier, and sometimes it is impossible to cover the whole area due to the limitation posed by the water body. The manual laboratory analysis of water quality is similarly hectic and slow, requiring skilled personnel and a variety of chemicals. Other sources of data are also needed to scale the process of collecting WQPs data.

GIS and RS provide an easy way to collect data at various scales and resolutions. Images of Earth are readily available from European Space Agency (ESA) and National Aeronautics and Space Administration (NASA)'s Sentinel and Landsat satellites, respectively. Therefore, large areas of data can be processed at once. Although, this does come at the cost of less accurate data, and time-consuming preprocessing of data to derive the WQPs using indirect methods.

In order to provide a scalable, robust, and fairly accurate system, in the proposed study, the data from all of these sources are analyzed. A time series analysis is performed to analyze the trend of water quality. The first part of the study involves collecting and acquiring data from GIS, RS, and IoT. The Rawal Lake data is collected from Landsat 8 (OLI/TIRS) and Sentinel 2 Level 2A satellites spanning the years 2015 to 2020. Rawal Lake Filtration Plant provided GIS data that included WQPs data from 2013 to 2020. Additionally, IoT data is provided, but only for the year 2019.

The second part of the study involves data pre-processing, data cleaning, WQI calculations, and employing effective ML techniques to analyze important trends and predict water quality for the near future. A detailed water quality temporal and comparative analysis is performed to monitor the trend of water quality based on seasonal and monthly data of WQPs. The factors affecting the water quality are highlighted. The comparative analysis of the COVID'19 on the water is also performed to indicate the effect of the pandemic on water. Data collected and acquired is used to train Long Short Term Memory (LSTM) Networks and Vector Auto-Regression (VAR) to predict future water quality. The results of predictions, important trends, and statistics are displayed through a web portal in the form of charts, graphs, and tables. In addition, a water quality prediction analytic model using trained NN is developed and deployed on the web portal. The end users can enter five WQPs and the model returns a categorical water quality value. Although several studies along these lines have been conducted in the past, they considered a single data modality at a time for data analysis such as IoT, GIS, or RS. However, this study has demonstrated that the integration of these three data modalities enables us to understand the water quality holistically and more clearly than ever before.

LITERATURE REVIEW

This section describes the research on the use of conventional methods to determine water quality. In addition, studies on using IoT, RS, and GIS data for water quality analysis are discussed. Further, ML techniques are discussed to analyze the water quality.

Conventional Methods for Water Quality Analysis

Laboratory analysis can be performed to determine the WQPs. It has been one of the most common techniques still in use today. Lab analysis involves the use of different chemical experiments to determine different WQPS. It has been often a costly and time-consuming process.

In 2015, a study is conducted in Perak, Malaysia to calculate physical and chemical WQPs like pH, temperature, turbidity, etc. through the use of IoT sensors and lab analysis. (N.Rahmanian, et al., 2015) Different techniques are employed to measure different water parameters. Quantities of different heavy metals like Copper, Zinc, Magnesium, and Arsenic, etc. are measured using Flame Atomic Absorption Spectrometer. WQPs like TDS and TSS are measured using filtration and evaporation techniques, DO and Chemical Oxygen Demand (COD) are measured using titration methods. These processes yielded highly accurate results, but the data is limited to only a few on-site samples.

Similarly, in 2016 a research is conducted in Cauvery River to study WQPs like pH, temperature, electrical conductivity (EC), Total Solids (TS), TDS, TSS, chlorides, along with other chemical parameters like SO4, PO4 (Appavu, Thangavelu, Muthukannan, Jesudoss, & Pandi., 2016). Total Solids are measured using the process of filtration and evaporation. Biological Oxygen Demand (BOD) and DO are measured by titration using sodium thiosulphate. COD is also measured by titration using potassium dichromate and sodium thiosulphate. Almost all of the physiochemical properties are discussed but again the shortcoming is a low number of samples in a large study area. Important WQPs like Chlorophyll-a, fecal bacteria, etc. are also missing.

IoT for Water Quality Analysis

IoT-based sensors have been developed to determine the real-time data of WQPs. These sensors can readily be used to keep track of the water quality of lakes, rivers, dams, etc. by taking in-situ measurements. Recently, IoT-based sensors have proved to be helpful in accurately and precisely determining the physicochemical parameters.

The study that is conducted in Perak, Malaysia involved the use of IoT-based tools combined with lab analysis to study the physicochemical properties of water. pH is measured using a special pH meter (N.Rahmanian, et al., 2015). Similarly, electrical conductivity and turbidity are measured using their respective sensors. The use of IoT provides quick real-time results, but the process still involves manual handling and effort to cover the middle area of the study area. The process is still time-taking as compared to GIS and RS approaches, which will be discussed later in this section.

A study is conducted in 2018 in which IoT-based embedded sensors are used to measure WQPs by taking on-site samples (Shafi, Mumtaz, Anwar, Qamar, & Khurshid, 2018). Temperature, turbidity, and pH are measured using their respective sensors. The IoT-based prototype is also connected to the cloud, where the data is received and could be processed from anywhere. The data collected is accurate and analysis is performed using ML algorithms like Support Vector Machines (SVM), NN, and k-Nearest Neighbors (KNN). The only shortcoming of the research is that it only considered three WQPs i.e., pH, temperature, and turbidity. Water quality depends on a number of other physiochemical, and biochemical parameters like DO, COD, Fecal Bacteria, metals, etc.

Quite recently in 2020, a study has been conducted, suggesting a low-cost IoT-based setup to measure WQPs (Ahmed, Mumtaz, Anwar, Mumtaz, & Qamar, 2020). The setup has been paired up with state-of-the-art ML techniques to determine water quality and to provide real-time visualization and monitoring of water quality. The system is developed to generate alerts in case of out-of-range values of WQPs or other contaminants. The study provided a new and efficient way to monitor water quality in real-time, but how to conduct such a study on a large scale has been still unaddressed.

GIS and RS for Water Quality Analysis

Several research studies have been conducted to analyze water quality using GIS and RS. With the advancement in satellite technology and improvement of temporal and spatial resolution of satellites, now accurate measurements of pH, TSS, turbidity, Chlorophyll-a, TDS, etc. can be carried out. European Space Agency (ESA) has developed a software named Sentinel Application Platform (SNAP) to analyze and process data gathered from Sentinel Satellites. Data from Landsat satellites can be loaded and visualized using ArcGIS software. More and more such tools are being developed and satellite data is being made available online for the general public. This has made it possible to conduct research on large areas effectively.

Remote sensing data collection has been performed using satellites and spectrometers equipped with multi-spectral image sensors. Specifically, for water bodies, the acquired images are analyzed for color reflectance, radiance, and brightness that changes water properties. The image sensors have different sets of color bands and sensitivity levels.

In 2004, Landsat and Terra MODIS satellites are used to analyze WQPS like pH, DO, planktons, chlorophyll-a, salinity, temperature, Secchi depth, etc. in New York Harbor (Hellwegera, Schlossera, U.Lall, & Weissel, 2004). Secchi disk depth, TSS, and chl-a are the main focus of the research. In-situ measurements are compared with the images to correctly analyze the dataset. Additionally, a lot of factors that can affect the analysis when using RS imagery are discussed.

Similarly, in 2011, images from Landsat Thematic Mapper are used to measure WQPs like TSS and Suspended Particulate Matter (SPM) in Penang, Malaysia (Yusop, Abdullah, San, & Bakar, 2011). The collected images are atmospherically corrected using the software. Using the darkest pixel method, the scatteredness from the image is reduced. In this paper, PCI Geomatica software is described to correct atmospheric effects on images, but the analysis is limited only to TSS and SPM. A study is conducted on Lake Buyukcekmece, Istanbul, which involved using the IKONOS satellite to gather multispectral images for measuring NO3-N and TSS. Images' data are converted to spectral radiance measures using blue, green, red, and near-infrared bands, however, only 12 points are taken for comparison.

In 2012, a Water quality analysis of Lake Bhadravathi Taluk is performed using GIS. The Weighted Arithmetic Index is used as the WQI. Parameters like pH, Total Alkalinity, hardness, TDS, EC, and metals like Ca, Mg, Cl, F, Fe, Na, and K are analyzed (Sneha & K, 2012). The paper showed the importance of pre-monsoon and post-monsoon effects on water quality. Although, no satellite images are used and due to recent waste disposal in the area, some samples are compromised.

To determine the water quality of Tenmile Lake in Oregon, in 2014, Landsat 5 satellite is used to analyze WQPs like DO, Hazardous Algae Blooms (HAB), turbidity, chl-a, and biovolume. The study proved a linear relationship between turbidity and radiance in sedimented areas while an inverse relationship in algal-dominated areas. A relationship between biovolume and turbidity is also established. (Waxter, 2014)

In 2017, NASA also defined its Sustainable Development Goal 6 regarding "Clean Water and Sanitation" and also released a full-fledged course on water quality analysis using machine learning. (NASA, 2017)

Data cleaning and preprocessing is a must when collecting data from RS or GIS. There are factors like cloudiness, sun-glare, bad pixels, duplicated images, removing unnecessary parts of images, etc. that must be removed from the image to effectively use it for any purpose. A research activity (Chang, Ba, & Chen, 2017) showed the use of software like SeaDAS and ArcMap for pre-processing ocean images obtained from different satellites. The study also described the image reconstruction process using Single Mode Intermediate Range (SMIR) by combining images from multiple satellites to fill as many data gaps as possible.

In 2019, Sentinel and Landsat satellites are used to measure water quality for water reservoirs in Italy and Greece using multispectral images from Sentinel-2A. Landsat 7 and 8 satellites (Bresciani, et al., 2019). Images taken from 2013 to 2018 are used to retrieve the WQPs (chl-a, turbidity, SDD, and temperature) and a time-series analysis is performed.

Machine Learning (ML) for Water Quality Trend Analysis and Predictions

Several research studies have been conducted to predict and analyze water quality using multiple ML Regression and classification techniques. Dataset gathered from multiple resources has been pre-processed and cleaned to make it suitable for ML algorithms.

A research study is conducted on the data gathered from the IKONOS satellite, where WQPs analyzed are NO3-N and Suspended Sediments. Regression analysis is performed based upon four spectral radiance bands of the IKONOS satellite. The bands are red, green, blue, and near infrared. Results indicated that the model's accuracy improved when all the IKONOS bands are incorporated for water quality prediction.

In 2017, a research team integrated WQPs data from multiple sensors and trained the Artificial Neural Network (ANN) (Chang, Ba, & Chen, 2017). The parameters tested are Total Nitrogen, Total Phosphorus and Chlorophyll-a, and turbidity. An RS platform is presented by data merging and machine learning techniques. SeaDAS and ArcMap software are used for image preprocessing, and SIASS algorithms combined multiple satellite images. Further, a single spatially complete image is constructed using SMIR. A correlation between parameters and ocean color reflectance is established. The research showed how to extract features from images and also described the importance of image reconstruction techniques. The limitation of this study is that the analysis done is for only those WQPs which reflect light.

ANN and Support Vector Regression (SVR) can be seen to outperform other various machine learning algorithms. One such study is conducted on a 55 years' time-series dataset of Tireh River in Iran (Haghiabi, Ali Heider Nasrolahi, & Parsaie, 2018). Models trained are ANN, Group Method of Data Handling (GMDH), and SVR. The parameters trained are temperature, pH, Specific Conductivity, TDS, and other minerals and salts. For the predictive value of one parameter, all the other parameters are provided as input to the ANN. ANN with different variable hidden layers are trained and it is seen that ANN with 2 hidden layers, having 8 and 3 neurons performed the best. Apart from that, SVR with RBD as kernel function performed the best. Overall, SVR performed better than ANN which performed better than GMDH.

Recently many researchers are inclined to use of LSTMs for time-series analysis and forecasting. One such study employed LSTM networks for water quality prediction (Liu, Wang, Sangaiah, Xie, & Yin., 2019). Time-series data is obtained from an automatic water quality monitoring station. Some of the WQPs trained are temperature, pH, DO, turbidity, COD, etc. Auto-Regressive Integrated Moving Average (ARIMA) and SVR models are also trained but LSTM outperformed the other two in terms of MSE and accuracy.

METHODOLOGY

The proposed research work is conducted in a total of four phases, with each phase concerning the retrieval and processing of key components needed to complete the research goals and to help perform detailed trend analysis. The time series analysis entails a large volume of data, which has been collected from three different sources: namely, RS, GIS, and IoT. The phases are discussed below.

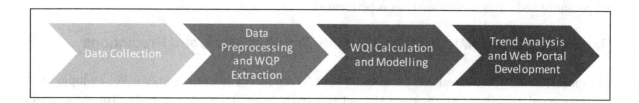

Data Collection

The data for the WQPs has been collected using RS, GIS, and IoT. The WQPs of interest include temperature, pH, turbidity, DO, and TDS. The GIS data is collected from the Rawal lake filtration plant while the IoT data is provided by the NUST-SEECS, IoT lab, Islamabad, Pakistan. The chosen satellites for remote sensing data collection are the Sentinel and Landsat. The research showed that the most used satellites for retrieving WQP information are the Landsat-8 OLI/TIRS (Operational Land Imager/ Thermal Infrared Sensor) and the Sentinel-2 MSI Level 2A and Level 1C. The data of these satellites are downloaded from the Earthdata website and Sentinel's official data hub, the Copernicus Open Access Hub. The acquired data for Landsat-8, spanned a time period of 2015 to 2020, and Sentinel-2 data, spanned a time period of 2016 to 2020 as shown in Table 1.

Table 1. Data collection

Source	Time Period
Landsat-8 OLI/TIRS (Operational Land Imager/Thermal Infrared Sensor)	01-01-2015 to 31-12-2020
Sentinel-2 MSI Level 2A and Level 1C	01-01-2016 to 31-12-2020
IoT Data	20-06-2019 to 04-12-2019
GIS Data	01-01-2013 11-06-2020

Data Preprocessing and WQP Extraction

For the purpose of data pre-processing, Sentinel Application's Platform (SNAP) tool is used, which is developed by the European Space Agency. Each file in the data is called a product and each product underwent the following stages.

Sub Setting

Each product is 800MB to 1GB large and covered a significantly larger area than the study area. Therefore, subsetting allowed to extract a small portion out of the product (the study area) and discard the rest. The precise values of latitudes and longitudes of the area are an important part of the subsetting.

Resampling

Each satellite consists of many bands each with a different resolution. To get a decent amount of pixel values, the resolution of all bands is down sampled to the same resolution and this is where resampling takes place. The data is down sampled instead of up sampling to avoid any extra variability in the value.

Masking

Masking is performed using a shapefile, generated from ArcGIS software. The shapefile outlined the borders of the study area in a polygon shape. Masking is used to ensure that the product only has the pixel values that overlap with the shapefile, thus, removing any unnecessary noise from the final product.

Extracting Pixel Values

After performing the above-mentioned steps, the pixel values of each product are extracted in the form of a CSV file. Each pixel value contains latitude and longitude information with values for each band of the satellite.

Final Cleaning using Scripts

The final cleaning has been done using a python script to remove any erroneous values and fill them with the mean of each product. Further, band ratio formulas are employed that are used by other studies, discussed above, to extract WQP information.

The pre-processing required a lot of computing power as the size of the data is over hundreds of Gigabytes. After the final cleaning, band data is obtained from each satellite band and ratio formulas are used to extract information about WQPs from each pixel value.

WQI Calculation and Modelling

Once all the data is pre-processed, the obtained datasets are combined. The chosen WQPS are temperature, turbidity, pH, dissolved oxygen, and dissolved solids. Three WQI are considered to categorize the water quality and provide labels that would be used by the ML models for prediction. Canadian Council of Ministers of the Environment WQI (CCME WQI), the Weighted Arithmetic WQI (WAWQI), and the National Sanitation Foundation WQI (NSFWQI) are considered for this purpose (see Table 2). After studying these WQIs, it is found that the CCME WQI suited the needs of the proposed research as it is the only WQI that could adapt itself to any number of WQPs. Therefore, CCME WQI is selected for the time series analysis. Three ML models are configured and trained using the prepared dataset. For water quality classification service on the web portal, a NN model is designed. For the time series analysis and future predictions, both the LSTM Networks model and the VAR model are used.

Table 2. Water quality index

Index	Objective	Method
Canadian council of ministers of the environment WQI (CCME WQI)	Drinking water quality	Assesses water quality for drinking purposes using WHO guidelines
Weighted arithmetic WQI (WAWQI)	Lake basin	Assess water quality using calculated weight of each parameter according to its quality rating
National sanitation foundation WQI (NSFWQI)	Inland waters	Assesses water quality using pre-assigned weights of each parameter according to its quality rating

Neural Networks

For the water quality classification service, a NN is developed that could predict the labels set by the CCME WQI. The model used the WQPs as attributes and the WQI labels are chosen as the target class. The model is made using Keras supported by Scikit-Learn, NumPy, and Pandas. The NN model consists of one input layer, two hidden layers with 20 and 6 neurons respectively as shown in Figure 1. The hidden layers used the ReLu activation function and the output layer used the sigmoid function.

The model is trained with over 2050 data points. The train and validation sets are made in 10 splits using repeated kfold to ensure equal representation of each class in each epoch. Every input is scaled to the range of 0 to 1.

The model is used to predict the water quality based on the value of WQPs fed to the model. The inputs accepted by the model are Temperature, Turbidity, pH, Dissolved Oxygen, and Dissolved Solids. Figure 1 shows the structure of neural networks.

Figure 1. Structure of NN

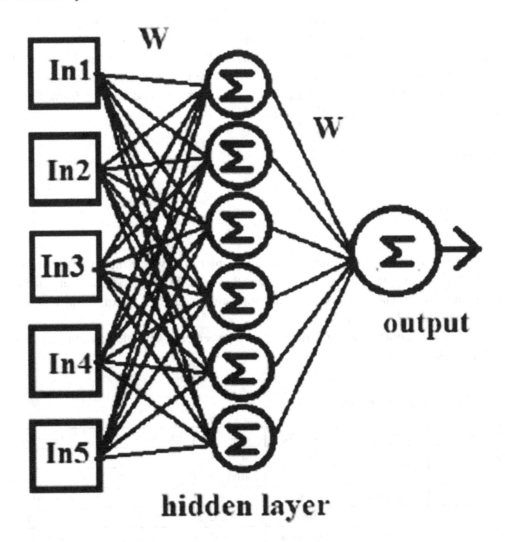

LSTM Networks

For temporal analysis, LSTM networks are used, which are an extension of Recurrent NN, capable of learning long-term dependencies. This means that they are good for sequential data or data based on time. The main intuition behind LSTM Networks is that they are able to learn the context behind the prediction and remember information for a long period of time.

The developed model consists of 3-Stacked Univariate LSTM networks with 50 features in its hidden state. The 3-stacked LSTM networks mean that 3 x LSTM Networks are stacked on top of each other with an output of 1 x LSTM network being used as an input of the next one. The model is developed using PyTorch supported by Sckit-Learn, NumPy, and Pandas.

The model is trained using the same prepared dataset except for this time it is used as a regression model to predict the changes in each WQP to aid in the temporal analysis. An input sequence of the last 30 days is used to predict the WQPs for the next 20 days.

VAR

The VAR model is used due to its multivariate nature and is specifically designed for time series forecasting and analysis. There are two basic requirements when it comes to using VAR and those are that you need at least 2 or more series, and those series should all influence each other. This suited the needs of the proposed research study as well, to see how each WQP influenced the other.

The model is an Autoregressive model because each variable is modeled as a linear function of its past values as well as the past values of the other series.

$$Y_{1,t} = \pm_1 + {}^2_{11,1}Y_{1,t-1} + {}^2_{12,1}Y_{2,t-1} + \in_{1,t} \tag{1}$$

$$Y_{2,t} = \pm_2 + {}^2_{21,1}Y_{1,t-1} + {}^2_{22,1}Y_{2,t-1} + \in_{2,t} \tag{2}$$

The above equations show a VAR model with a lag order of 1, where the lag is the number of past values used. VAR model with a lag order of 7 means that each WQP is modeled as an equation with 7 past values of each WQP. The Statsmodels library supported with Sckit-Learn, NumPy, and pandas are used to develop the model.

VAR is even able to compute a correlation matrix between each variable according to the equations it modeled, as shown in Table 3.

Table 3. Correlation matrix for WQPs

Correlation	Temperature	Turbidity	pH	Dissolved Solids	Dissolved Oxygen
Temperature	1	0.00684	-0.000419	0.012295	0.008117
Turbidity	0.00684	1	-0.000278	-0.002722	0.010346
pH	-0.000419	-0.000278	1	-0.031792	-0.004243
Dissolved Solids	0.012295	-0.002722	-0.031792	1	-0.01358
Dissolved Oxygen	0.008117	0.010346	-0.004243	-0.01358	1

Trend Analysis and Web Portal Development

In this phase, trend analysis is performed by using LSTM networks and VAR. The monthly and seasonal analysis are also performed. In addition, GIS and RS datasets are compared side-by-side. Detailed trend analysis for the water quality during the COVID-19 pandemic duration is also carried out. The comparison of how the models performed for each of the analyses is done by calculating RMSE, MAE, MSE values. Additionally, a brief overview of the statistics of each dataset involved in the analysis is also added.

The web portal is also developed, powered by HTML/CSS, Javascript, Bootstrap for the frontend. The backend has been developed using NodeJS with the models being hosted on Google Cloud. The web portal provides the visualization services such as trend analysis, classification, and predictive analytic. The details of the web portal are given below:

Homepage

The website features an aesthetically pleasing and simple design keeping responsiveness as the core design value. The homepage features a simple introduction to the project coupled with a video, end-user impacts, and a little about the team. Figure 2 shows the homepage of the website.

Figure 2. Homepage of web portal

Dashboard

The dashboard features the accumulation of all of the analysis and research which users can access with just a click of a button. The web portal shows the temporal analysis and future projections using the LSTM model and the VAR model. The monthly, seasonal, and COVID-19 duration-based water quality analysis are carried out to give users a sense of understanding that how the water quality has changed over the years.

Future Predictions

The future predictions of WQPs using LSTM networks and VAR have been displayed on the dashboard through line charts. Figure 3 and 4 shows the predictions of each WQP using LSTM and VAR respectively.

Figure 3. Future predictions using LSTM

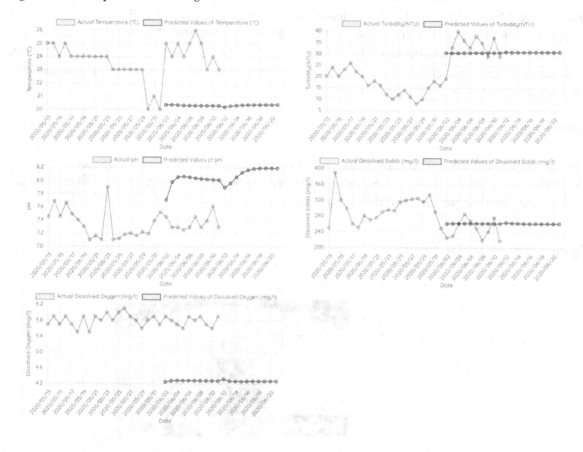

Figure 4. Future predictions using VAR

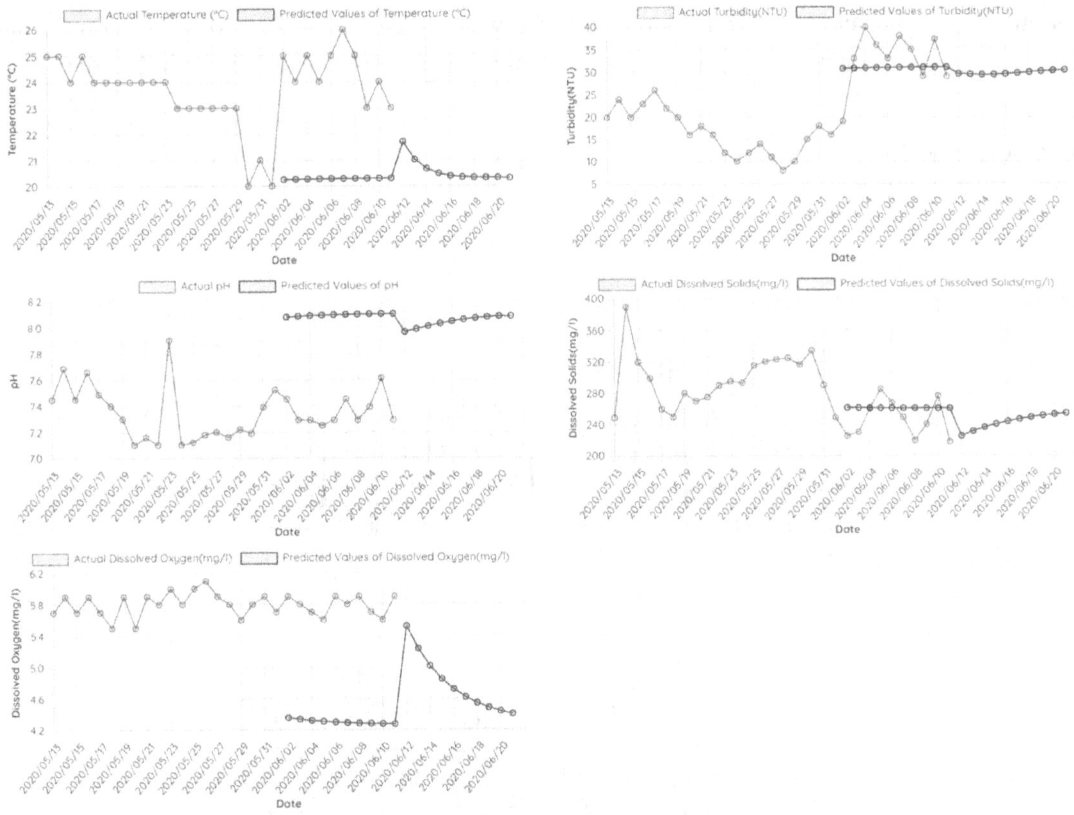

Water Quality Classification Service

The water quality classification service developed using NN is added to the dashboard that predicts the water quality based on the data provided by the user as shown in figure 5.

Figure 5. Water quality classification service on web portal

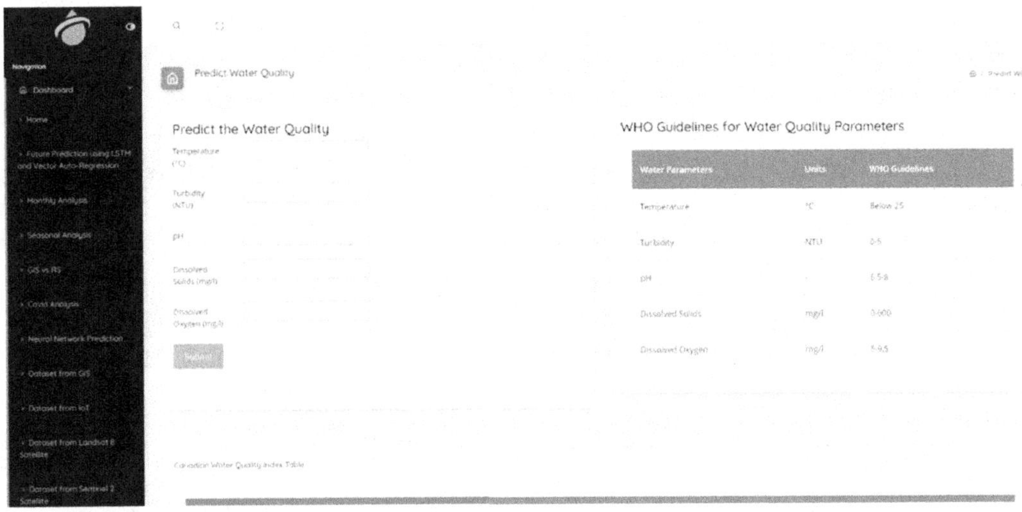

Analysis

The monthly, seasonal, and COVID-19 duration-based water quality analysis is displayed on the dashboard using line and bar charts along with statistical and modeling results as shown in figure 6.

Figure 6. Analysis display on web portal

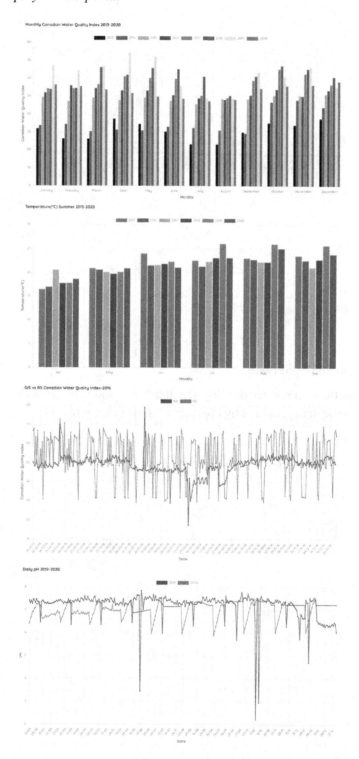

Data Records

All of the data collected through RS, GIS, and IoT have been displayed on the dashboard in tabular form as shown in figure 7.

Figure 7. Data records

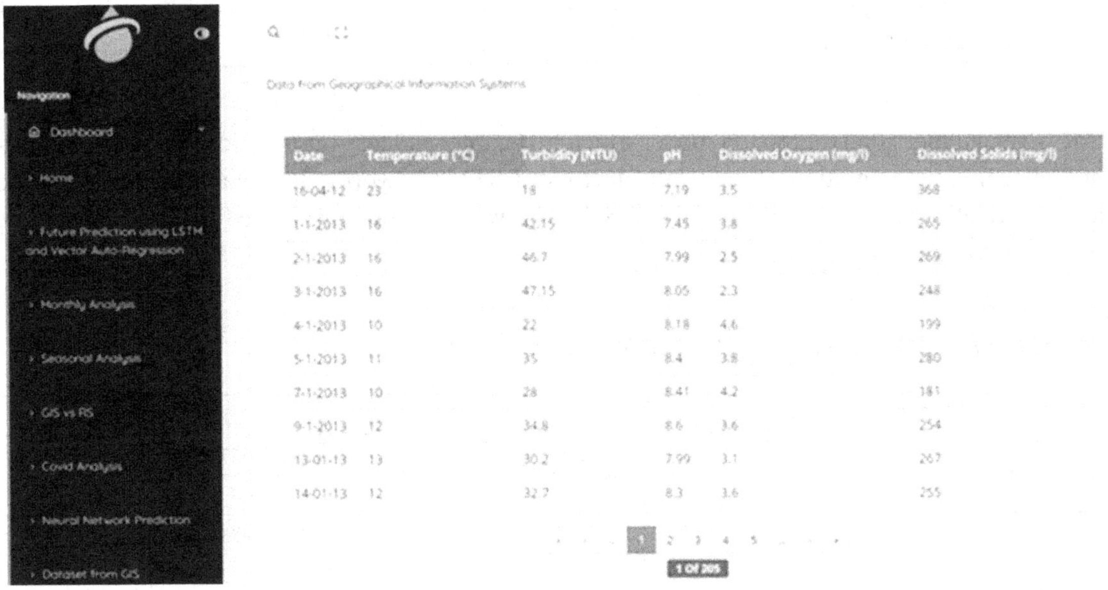

RESULTS AND DISCUSSION

The detailed results have been discussed below.

Water Quality Classification Service powered by NN

For every row of data extracted from GIS, RS, and IoT, a quality class (Poor, Fair, etc.) is assigned to each row of the data based on CCME WQI. All of the data (from 2013 - 2020) is then used to train the ANN model.

The model accuracy comes out to be ~80%. The model is saved and then hosted as a Google Cloud function, from where it loads and returns the output to the website.

Future Predictions using LSTM and VAR

Data taken from GIS has been used to train a separate 3-stacked LSTM Networks model for each WQP. To make the predictions for the next 20 days (02-06-2020 to 21-06-2020), the model is fed data for the last 30 days. Table 4 below shows the Root Mean Squared Error (RMSE) and Mean Squared Error (MSE) for each parameter predicted using LSTM Networks.

Table 4. RMSE and MSE values of WQPs predicted using LSTMs

WQPs	RMSE	MSE
Temperature	1.8361982	3.371624
Turbidity	1.9879257	3.9518487
pH	1.0427667	1.0873624
Dissolved Oxygen	2.062017	4.251914
Dissolved Solids	6.4161096	41.166462

VAR is yet another good regression model which is flexible, and less information and time demanding, and allows easy integration of new data. Just like LSTM Networks, VAR has also been used to predict water quality for each WQP using the GIS data provided by the Rawal lake filtration plant. Table 5 shows the error metrics for VAR on all the WQPs.

Table 5. RMSE and MSE values of WQPs predicted using VAR

WQPs	RMSE	MSE
Temperature	4.7455808	22.520537
Turbidity	75.545238	5707.083
pH	1.1558409	1.3359683
Dissolved Oxygen	2.7903852	7.7862493
Dissolved Solids	44.005115	1936.4501

Water Quality Analysis

Monthly Analysis

For this purpose, the monthly average of each WQP and WQI is taken from 2013-2020 and trends are analyzed. The water quality of the study area exhibited the same trend from 2013-2020 but a relatively noticeable deviation is observed for the years 2018 and 2019. The highest average value of WQI recorded is in April 2019 with a value of 74.46. Over the years, the average temperature is following an overall increasing trend. The average temperature recorded is higher in the months of May-September relative to other months. The turbidity of water is significantly increased over the years especially in 2019 where it hit the peak with an average monthly value of 192.37 NTU in the month of August. Overall, the average monthly value of turbidity of 2019 and 2020 is greater than that of previous years. The average monthly value of pH remained consistent over the span of 8 years. The highest monthly average of pH is recorded in July 2016 with a value of 13.5. The trend in dissolved oxygen in water has significantly increased over the years, where its highest monthly average is recorded for the year 2020 with the maximum value of 7.756 mg/l in December. The amount of monthly average dissolved solids in water increased in the years 2015 and 2016 but decreased in the year 2017 and remained almost consistent afterward. The maximum monthly average value recorded for dissolved solids is 373.994 mg/l in August 2015. Figure 8 shows the monthly trend of WQI from 2013 to 2020.

Figure 8. WQI based on monthly data (2013-2020)

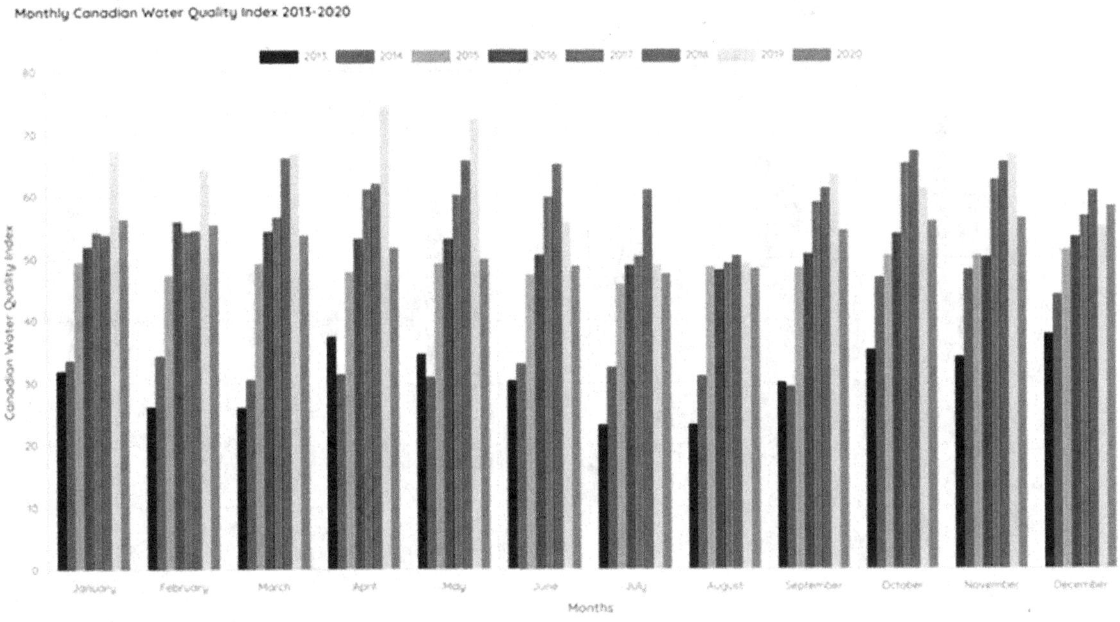

Monthly Canadian Water Quality Index 2013-2020

Seasonal Analysis

For seasonal analysis, the monthly average of each WQP and WQI is taken from 2015-2020. The data is divided into two seasons: ***Summer***: April – September, and ***Winter***: October - March. The temperature hit as low as 9° in summer and 5° in winter while the maximum recorded value is 30° and 29° respectively. The turbidity parameter is observed to be the most affected in summer, with high values of variance and standard deviation of about 2871 and 53 respectively as compared to 951 and 30 in winters. In summers, its highest value recorded is 420 NTU as compared to 353 NTU in winters. This is mainly because of heavy rainfall in the summer season, which causes more soil matter to be dissolved in the water, ultimately increasing the turbidity. No significant change can be seen in the values of pH for both seasons as both have almost the same standard deviation of 0.3 and the average value of about 7.5. Similar is the case with dissolved oxygen. There is a slight increase in the number of dissolved solids in summer, but it is not significant and is under the +5% range. This change is due to rainfall and monsoon season in summer's time period. Figures 9 and 10 show the monthly trend of WQI for the ***Summer*** and ***Winter*** seasons respectively from 2015 to 2020.

Figure 9. Summer WQI based on monthly data (2015-2020)

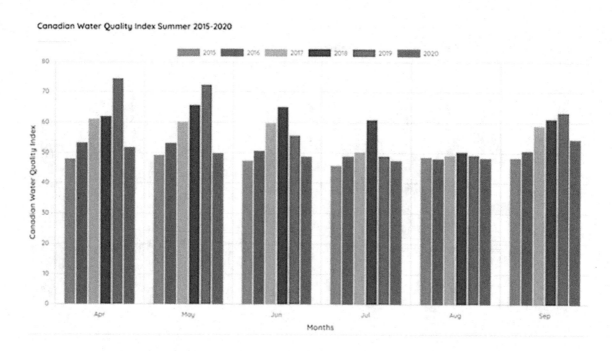

Figure 10. Winter WQI based on monthly data (2015-2020)

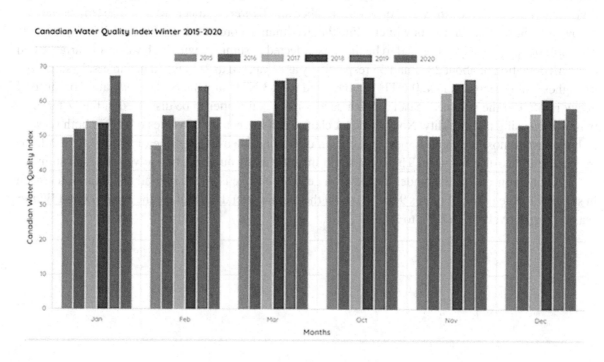

Comparative Analysis

To analyze the results of WQPs collected through RS and GIS, the daily averages of the GIS and the RS dataset for each year are compared from 2016 to 2020 as shown in Figure 5. The WQI calculated from the GIS dataset is quite consistent as compared to the dataset obtained from the RS dataset over the period of January 2016 to June 2020. The highest value recorded is 100 for both datasets and the lowest value is 17.3 and 29.1 respectively. The variance between the WQI obtained from GIS and RS are 127.62 and 143.08 respectively. The major reason for inconsistency in WQI calculated from RS dataset is that the value of WQPs obtained from the RS has a lower temporal resolution. Thus, the data is interpolated to obtain the parameters for each day that affected the results. Secondly, the data obtained from the satellite images depends on the spatial resolution, atmospheric factors, cloudiness, satellite sensors, etc. Therefore, it is relatively less accurate than the data obtained on the ground. For example, if the weather is cloudy, the image is going to get affected which will cause the parameters to be less accurate and eventually end up variating the results. Figure 11 shows the line chart of WQI obtained using GIS and RS for the year 2016.

Figure 11. Daily Average WQI calculated using GIS and RS for the year 2016

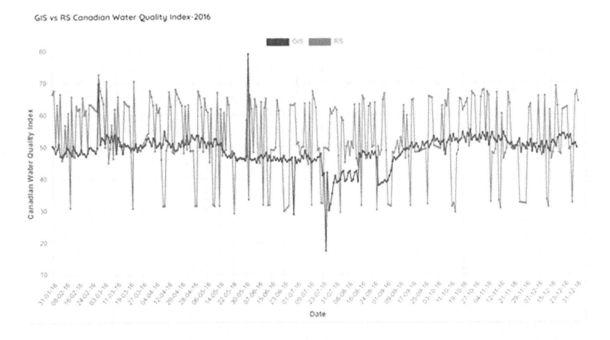

Impact of COVID-19 Pandemic on Water Quality

To analyze the result of the pandemic on water quality, the WQI and WQPs before and during the COVID-19 period are compared from March-Dec 2019 and March-Dec 2020, as shown in Figure 6. The WQI decreased during the COVID-19 period as compared to the period before, with a maximum value of 76.7 and 79.1 respectively. The average WQI recorded before COVID-19 is 61.3 and during pandemic, it is 52.4. The temperature remained almost the same before and during the COVID-19 period with a maximum value of 30.115 and 26.9 ^{0}C and the average value of 22.12 and 22.29 ^{0}C. The turbidity of water is significantly increased during the pandemic time period, where the average turbidity recorded is 81.44 NTU in 2020 and 60.49 NTU in 2019. The value of pH almost remained consistent in both time periods with the average of 7.2 and 7.04 in 2019 and 2020 respectively. There is a minor change in dissolved oxygen when observed in both time periods with the average value of 7.12 and 7.5 mg/l in 2019 and 2020 respectively. The dissolved solids also didn't exhibit any significant change. The average values recorded for dissolved solids are 253 and 246 mg/l in 2019 and 2020 respectively. The reason for decreased water quality in 2020 (during the COVID-19 period) is the increased rainfall. The winter 2020 rainfall across the country (Pakistan) ranked 4th highest seasonal rainfall since 1961. Another reason for low water quality is the lockdown, which led to the increased disposal of domestic waste and contaminants. For example, there is a significant increase in the use of disposable masks and tissues that raised the volume of domestic waste. Figure 12 shows the line charts of WQI for pre and during the COVID'19 period.

Figure 12. Daily Average WQI calculated for Pre and during COVID-19 period

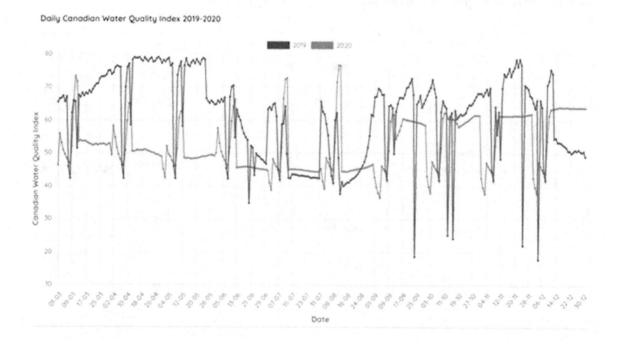

CHALLENGES

There are several issues and challenges related to the collection of data for WQPs. These issues are discussed below:

Sources of Data Collection

To see a true and complete picture of water quality, the data has to be collected from different sources. Each source has its advantages and disadvantages. For instance, the IoT data is more accurate and WQPs can be directly measured using sensors, but the data collected is point data and does not provide ubiquitous coverage of the water body. To increase the data points, a greater number of IoT nodes need to be added, which will make the entire study more expensive. The security of the equipment and the stability of the hardware with respect to the nature of the water body add additional constraints. Therefore, it is difficult to collect the data for the whole area of the water body through IoT instruments. On the other hand, RS and GIS data is easier to collect over a variety of scales and resolutions. The RS data can cover large spatial areas, but it is less accurate due to relatively low resolution. Additionally, the WQPs are measured using indirect means and it lacks contextual information. To overcome all these challenges, the water quality data has to be collected using all three sources and combined to exploit the good characteristics of each data modality.

TEMPORAL RESOLUTION

One of the biggest issues while using remote sensing data is the low temporal resolution. The temporal resolution of Landsat 8 is 16 days while sentinel 2 is 5 days. This means that it's not possible to collect the data of WQPs for each day using satellites. To obtain the daily values, the data collected using satellites is interpolated. This makes the data less accurate and affects the results.

Water Quality Parameter Collection

The water quality depends on more than 40 parameters, and it is very difficult to obtain the daily values of each parameter. To overcome this problem, a correlation is found between different WQPs, and only a few WQPs are used for analysis. This makes the results biased and less accurate.

Seasonal Changes and WQPS

The values of some of the WQPs like total dissolved solids, turbidity, suspended solids, pH, etc., change before and after a certain season, especially during the monsoon season. This seasonal variation greatly affects the accuracy of the data obtained. This demands the need for a complex machine learning model that also considers the seasonal changes in the values of the parameters affecting water quality.

Use of Suitable WQI

There is an assortment of WQIs developed until today. While it is quite advantageous to have several different WQIs to choose from, it becomes quite a hard task to determine which WQI is aligned with the WQPs you choose and would be most suitable to label the data.

CONSISTENT DATA COLLECTION

Imagery data obtained from most of the satellites, be it Landsat or Sentinel, does contain some artifacts and cloudy pixels. These pixels need to be corrected before the data obtained from the image is used for training the models. Most of the time these can easily be removed using image correction techniques often built-in the GIS software, but sometimes it becomes hard to correct the image because it comes at the cost of losing other relevant and important information embedded in the image.

To avoid the issues faced during the temporal analysis and data collection, the data should be collected from all the available sources. More and more satellites can be used to fill the gap of temporal resolution to obtain more accurate results. This will also assist in collecting the data of more WQPs for better results. The techniques like atmospheric correction and the use of tools like the SNAP tool and ArcGIS can minimize the discrepancies of the data. The use of deep learning algorithms is recommended for advanced analysis.

The majority of the research studies are based on GIS and RS data, therefore, before indulging in water quality analysis, it is recommended to gain sufficient domain knowledge of GIS and RS for better understanding. It is very important to understand different tools and products that are being used in the water quality research to avoid any potential errors. The background knowledge of machine learning algorithms, WQI equations, WQP, and band correlations is important to analyze the results. It is also recommended to properly document everything in parallel for future consultation.

CONCLUSION AND FUTURE WORK

In a developing country like Pakistan, the water quality is continuously decreasing due to numerous reasons including population growth, increasing pollution, urbanization, lack of proper waste management, and water quality regulatory and monitoring bodies. There are no advanced and technology-based systems to monitor the water quality deterioration. The country relies on traditional and expensive approaches like lab analysis that makes it difficult to see a bigger picture and monitor water quality on a larger scale.

To analyze the water quality trend, the time series analysis of water quality is performed by collecting the data of the past 8 years; 2013-2020. The historical data is obtained from the Rawal Lake filtration plant, the IoT devices, and the RS. The RS data is derived by obtaining and processing images of Landsat-8 and Sentinel-2 satellites. The Canadian WQI is calculated for each set of WQPs to check the water quality against the WHO Standards.

The collection of data from different sources i.e., RS, GIS, and IoT has proved to be helpful in gathering large amount of data over the longer time period with extended area covered. The multisource approach of data collection has helped in producing accurate and clear results for trend analysis and prediction of water quality. The periodic analysis has helped in identifying the decrease in water quality, and the trends of water quality over time. The temporal analysis shows that the pandemic has further caused a decrease in water quality due to increased pollution. Immediate measurements need to be taken to handle the continuously declining water quality.

LSTM and VAR models are trained for predicting the WQPs for the near future. The LSTM and VAR performed with the average RMSE 2.664 and 25.63 respectively. In addition, a water quality classification service is developed powered by a trained NN model with an accuracy of 80%. The service can be used to classify water quality by providing the values of WQPs.

The future work of this research may include the use of more advanced ML and deep learning models for water quality predictions. The data for this purpose can be collected using more resources like Moderate Resolution Imaging Spectroradiometer Satellite and covering a bigger area for the analysis. The current models developed can be used to perform predictive analysis on a bigger scale to analyze the water quality trends across Pakistan.

The data collected for the present study can be used for future research work regarding water quality. Another useful application of this work is the projection of water quality to provide timely remedial measures in curbing the health hazards caused by poor water quality. The predictions may also indicate the need to take measures to protect marine life and other ecosystem elements which are dependent on water. The models and data collected can also play a part in developing a water drone that will be useful to monitor and analyze the live water quality at any place.

This research work is aligned with United Nations Sustainable Goal 6 regarding Water and Sanitation.

REFERENCES

Ahmed, U., Mumtaz, R., Anwar, H., Mumtaz, S., & Qamar, A. M. (2020). Water quality monitoring: from conventional to emerging technologies. *Water Supply, 20*(1), 28-45.

Appavu, A., Thangavelu, S., Muthukannan, S., Jesudoss, J. S., & Pandi, B. (2016). Study of water quality parameters of Cauvery River water in erode region. *Journal of Global Biosciences*, *5*(9), 4556–4567.

Bresciani, M., Giardino, C., Stroppiana, D., Dessena, M. A., Buscarinu, P., Loretta Cabras, K. S., & Tzimas, A. (2019). Monitoring water quality in two dammed reservoirs from multispectral satellite data. *European Journal of Remote Sensing*, *52*(sup4), 113–122. doi:10.1080/22797254.2019.1686956

Chang, N.-B., Ba, K., & Chen, C.-F. (2017). Integrating multisensor satellite data merging and image reconstruction in support of machine learning for better water quality management. *Journal of Environmental Management*, *201*, 227–240. doi:10.1016/j.jenvman.2017.06.045 PMID:28667841

Daud, M. K., Nafees, M., Ali, S., Rizwan, M., Bajwa, R., Shakoor, M., ... Shui, Z. J. (2017). Drinking Water Quality Status and Contamination in Pakistan. *BioMed Research International*, *2017*, 18. doi:10.1155/2017/7908183 PMID:28884130

Haghiabi, A. H., Nasrolahi, A. H., & Parsaie, A. (2018). Water quality prediction using machine learning methods. *Water Quality Research Journal*, *53*(1), 3–13. doi:10.2166/wqrj.2018.025

Hamid, Jhanjhi, & Humayun. (2020). Digital Governance for Developing Countries Opportunities. *Employing Recent Technologies for Improved Digital Governance*, 36-58.

Hellwegera, F., Schlossera, P., Lall, U., & Weissel, J. (2004). Use of satellite imagery for water quality studies in New York Harbor. *Coastal and Shelf Science*, *61*(3), 437–448. doi:10.1016/j.ecss.2004.06.019

Liu, P., Wang, J., Sangaiah, A. K., Xie, Y., & Yin, X. (2019). Analysis and Prediction of Water Quality Using LSTM Deep Neural Networks in IoT Environment. *Sustainability*, *11*(7), 2058. doi:10.3390u11072058

NASA. (2017). Satellite Observations of Water Quality for Sustainable Development Goal 6. *GEO Week 2017*.

Rahmanian, N., Ali, S. H., Homayoonfard, M., Ali, N. J., Rehan, M., Sadef, Y., & Nizami, A. S. (2015). Analysis of Physiochemical Parameters to Evaluate the Drinking Water Quality in the State of Perak, Malaysia. *Journal of Chemistry*, *2015*, 10. doi:10.1155/2015/716125

R.M., B., Salif, D., Oscar E., N., Niekerk, H. v., Sherbinin, A. d., Vijselaar, L., . . . Al-Lami, A. A.-Z. (2007). *Global Drinking Water Quality Index Development and Sensitivity Analysis Report*. United Nations Environment Programme Global Environment Monitoring System/Water Programme.

Saeed, Soobia, N. Z. Jhanjhi, Naqvi, M., Ponnusamy, V., & Humayun, M. (2020). Analysis of Climate Prediction and Climate Change in Pakistan Using Data Mining. *Industrial Internet of Things and Cyber-Physical Systems: Transforming the Conventional to Digital*, 321-338.

Shafi, U., Mumtaz, R., Anwar, H., Qamar, A. M., & Khurshid, H. (2018). Surface Water Pollution Detection using Internet of Things. *15th International Conference on Smart Cities: Improving Quality of Life Using ICT & IoT (HONET-ICT)*, 92-96. 10.1109/HONET.2018.8551341

Sneha, R. V., & K, M. (2012). Water quality analysis of Bhadravathi taluk using GIS. *International Journal of Environmental Sciences*, 2443–2453.

Waxter, T. M. (2014). *Analysis of Landsat Satellite Data to Monitor Water Quality Parameters in Tenmile Lake*. Civil and Environmental Engineering Master's Project Reports. doi:10.15760/CEEMP.35

WHO/UNICEF Joint Monitoring Program (JMP) for Water Supply. (2015). Progress on sanitation and drinking water: 2015 update and MDG assessment. World Health Organization.

Yusop, S. M., Abdullah, K., San, L. H., & Bakar, M. N. (2011). Monitoring Water Quality from Landsat TM Imagery in Penang, Malaysia. *Proceeding of the 2011 IEEE International Conference on Space Science and Communication (IconSpace)*, 249-253. 10.1109/IConSpace.2011.6015893

Chapter 2
Temporal Analysis and Prediction of Ambient Air Quality Using Remote Sensing, Deep Learning, and Geospatial Technologies

Aymen Bashir

National University of Sciences and Technology, Pakistan

Abdullah Mughal

National University of Sciences and Technology, Pakistan

Rafia Mumtaz

National University of Sciences and Technology, Pakistan

Muhammad Ali Tahir

National University of Sciences and Technology, Pakistan

ABSTRACT

As of today, increased air pollution has disrupted the air quality levels, deeming the air unsafe to breathe. Traditional systems are hefty, costly, sparsely distributed, and do not provide ubiquitous coverage. The interpolation used to supplement low spatial coverage induces uncertainty especially for pollutants whose concentrations vary significantly over small distances. This chapter proposes a solution that uses satellite images and machine/deep learning models to timely forecast air quality. For this study, Lahore is chosen as a study area. Sentinel 5-Precursor is used to gather data for Sulphur Dioxide (SO2), Nitrogen Dioxide (NO2), and Carbon Monoxide (CO) for years 2018-2021. The data is processed for several AI models, where convolutional neural networks (CNN) performed the best with mean squared error (MSE) 0.0003 for the pollutants. The air quality index (AQI) is calculated and is shown on web portal for data visualization. The trend of air quality during COVID-19 lockdowns is studied as well, which showed reduced levels of NO2 in regions where proper lockdown is observed.

DOI: 10.4018/978-1-7998-9201-4.ch002

INTRODUCTION

This chapter introduces an air quality monitoring system after providing some context to the pertinent problems that have been faced in terms of air quality monitoring and forecasting. It discusses the application of statistical, machine learning (ML), and deep learning (DL) models for timely forecasting of air quality values and spreading awareness by the implementation of the optimum model on the web portal. Research involving air quality monitoring through Internet of Things (IoT) systems, ML techniques, and remote sensing has been explored in this chapter. The shortcomings of the existing technologies are also discussed in detail. The research that involved modelling time series air quality data by Statistical, ML, and DL techniques contain models that are either overfitted or under fitted due to the involved dataset size, while those involving IoT frameworks were not applicable on a large scale and held the shortcoming of low spatial coverage.

Conventional air quality monitoring systems are heavy, costly, and, therefore, very sparsely distributed. Though accurate, due to the inconvenience caused by their heavy weight and hence, low spatial coverage, these monitoring stations provide measurements for only a point in space which leads to inaccurate estimations since major pollutants like SO_2 and NO_2 vary over small distances and time. Mobile air quality monitoring equipment is available in various forms, e.g., a monitoring system mounted over a vehicle or bicycles that measure air pollutant levels at different times and locations, but such devices, though better than fixed monitoring devices, still lack certainty and have limited spatial coverage.

Keeping in view all the shortcomings of the on-ground monitoring systems, a solution based on remotely sensed satellite data had been devised to monitor and predict air quality levels. The solution addresses low spatial coverage by using air quality data from remotely sensed satellite data. For this research, the study area is Lahore, which has been constantly listed as one of the world's most polluted cities. Three major pollutants, SO_2, NO_2, and CO have been chosen for this study. The data has been downloaded from Copernicus Sentinel-5 Precursor. After data analysis, various time series forecasting models have been applied to predict air quality levels for the future and issue warnings in case of poor air quality.

LITERATURE REVIEW

This section describes the conventional ways of monitoring air quality and discusses the research on the use of these methods. In addition, studies on using IoT, ML, and GIS data for air quality analysis are discussed. The section also discusses the importance of air quality as a parameter in COVID-19 pandemic related research.

Air Quality Monitoring Methods

This section discusses some of the standard methods used to observe air pollution:

Active Sampling of Gases

In method active sampling, samples of active gases and vapors are collected in a sampling pump. A combination of absorption materials like activated charcoal, silica gel, and Tenex is used inside a sorbent tube to collect these vapors and gases. Another method involves using an annular denuder in which inner walls are coated with a particular material to absorb the agent of interest. Undesired gases and particles will pass through it, and the target particle/gases get absorbed by the absorbent. This method depends upon diffusion to collect gas or vapor particles. (Cleartheair.scottishairquality.scot, 2020))

Passive Sampling of Gases

This method does not require a pump and is developed for gaseous pollutants, including NO_2, SO_2, O_3, and Volatile Organic Compounds (VOCs). A physical process is required to control the sampling rate such as diffusion through a statis air layer or permeation through a membrane. The factors like temperature, humidity, and face velocity have a significant effect on the process. (Cleartheair.scottishairquality.scot, 2020))

Automatic Point Monitoring

A chamber sucks air through an analyzer tube in automatic point monitoring, which automatically identifies the chosen gas and calculates its concentration. It is capable of monitoring air pollutants 24 hours a day. Data is transferred directly into the computer from the site, and it can be observed instantly. (Cleartheair.scottishairquality.scot, 2020))

Photochemical and Optical Sensor System

It is a mobile monitoring system consisting of compact monitoring devices that can continuously record an extensive range of air pollution parameters. The sensitivity of sensors is not sufficient due to which they are implemented at roadsides and a localized source of population. Data can be analyzed once downloaded. (Cleartheair.scottishairquality.scot, 2020))

Remote Optical/Long-Path Monitoring

In this method, light is used to detect pollutants. It detects a sample of air between source and detector, which are placed separately from each other. Real-time measurement can be made using this method. Data is transferred directly into the computer from the site, and it can be seen instantly. (Cleartheair.scottishairquality.scot, 2020))

The Image-Based Method from Space

It uses aerosol optical thickness for accessing air pollutants. Aerosol optical thickness is measured using image-based techniques. Atmospheric correction can be done through the radiative transfer equation. Methods to measure aerosol optical thickness are:

1. The ocean method (used over clear water) (Griggs, 1975)
2. The brightness method (applied over visible land) (Chahine et al., 1983)
3. The contrast-reduction method (applied across the land (Tanre et al., 1979) or on both land and water) (Kaufman et al., 1990)
4. The dark vegetation method (utilizes long-wavelength visible data) (Sifakis, 1998)

5. The temperature attenuation method (Khwaja et al., 2012)
6. The differential texture analysis method (Kanaroglou et al., 2002), (Sifakis & Deschamps, 1992)

Air Quality Monitoring Through IoT Sensors

Traditional ways of monitoring air quality involve the use of IoT sensors. One such system is the Atmospheric Air Surveil System (AASS) that consists of a transportable prototype for distant web-based observation of air quality parameters of outdoor CO and CO_2 gases. It employs microcontrollers, gas sensors, and Global Positioning System (GPS). The sensor is used to measure the number of gases from nearby air, which are further processed via a microcontroller and transmitted using cloud services and IoT devices. MQTT (Message Queuing Telemetry Transport) protocol is used to transfer data to a server stored in a Data Acquisition (DAQ) unit.

The system, however, is limited to CO and CO_2, and it is not designed to predict air quality. Moreover, deploying such monitoring devices is expensive. (Munsadwala et al., 2019). Another proposed method (Zhang & Woo, 2020) to monitor air quality is by using a hybrid approach by integrating fixed and moving IoT sensors installed on cars to collect data. The sensors include temperature and humidity sensors, micro dust sensors, and CO sensors. Microcontrollers were used to process the data. The acquired data is preprocessed by doing outlier detection, interpolation, and data normalization. A visualization tool is also created. Moreover, prediction is made using ML models that included Support Vector Regressor, Random Forest Regressor, and Gradient Boosting. Models were tested using different interpolation and normalization techniques. The results implied that simple classical models performed better than the models mentioned above. However, the dataset used was not large enough for diverse measurements. (Zhang & Woo, 2020)

An additional approach for air quality monitoring proposes Unmanned Aerial Vehicle (UAV). An inexpensive UAV equipped with a limited number of sensors is employed. This is done using Arduino Mini and sensors to measure methane, alcohol, smoke, and propane. Data is transmitted to the base station that is uploaded to a cloud platform. The approach is limited by the battery life of the drone. Moreover, problems like taking permission from authorities for flying drone also introduces limitation to the large-scale implementation of this method. Moreover, sensors from drones may constantly need to be validated. (Vijayakumar et al., 2020)

Another air pollution monitoring system is based on a network of shared bicycles. This bicycling system is built using a single processor, a tiny GPS receiver, a flue gas sensor, a micro-SD card, a Particulate Matter detector, and a Bluetooth module. The sensor starts to collect air pollution data when someone rents a bicycle—the roadway area where the bicycle travels, stores that data into the SD card. Gathered data is uploaded to the cloud storage platform using Bluetooth technology when a user returns its bicycle to a dock station. (Liu et al., 2015)

Air Quality Monitoring Through Satellite Imagery

There are many studies conducted which solely used satellite images to monitor and predict air quality. One of these studies used satellite imagery from planet.com and applied deep neural networks on that data to predict variations of fine particulate matter ($PM_{2.5}$) across northern California. For validation, the ground measured AQI data was retrieved from the United States Environment Protection Agency (EPA). As the satellite's images may vary even at the same levels of AQI due to changing altitude, weather, and time, both the models tested in this article could not bring down both training and validation loss to usable level except conditions. (Cs229.stanford.edu, 2021)

Another study conducted in this domain used satellite-retrieved Aerosol Optical Depth (AOD) $PM_{2.5}$ data from Moderate Resolution Imaging Spectroradiometer (MODIS) instrument and combined it with numerical model simulations obtained from Chemical Transport Model (CTM) using Bayesian Approach to predict $PM_{2.5}$ values in Southeastern US. Overall, this approach resulted in better prediction accuracy than using AOD or CTM separately. However, the CTM simulations can be computationally and economically expensive, making this approach economically challenging.(Murray et al., 2019)

An alternative study in Indonesia used Ozone Measurement Instrument (OMI) mounted on an EOS-Aura satellite launched by NASA to characterize the NO_2 values and find the correlation between satellite's retrieved NO_2 values and ground measurements using Linear Regression. The ground measurements for this purpose were taken from monitoring stations located in Jakarta and Surabaya. Data validation in this study showed a weak correlation between results obtained from satellite imagery and ground monitoring stations. (Darmawan & Syafei, 2019)

An additional study was conducted on determining how accurately $PM_{2.5}$ can be predicted using Moderate Resolution Imaging Spectroradiometer (MODIS) satellite imagery with global coverage since 1999 at 1–2-day intervals. The ground particulate matter data were retrieved from the EPA's pre-generated CSV files. Linear regression models were graphed to observe clustering and the value of a single explanatory variable model for predicting particulate matter. The summer months' linear regression model proved to show the highest percentage of particulate matter variation. The overall spatial coverage in the summer was also proved to be most consistent across the contiguous country due to less cloud coverage. On the other hand, none of the linear regression tests proved to show significantly high R2 values for estimating particulate matter. The tested model provided very sporadic and low predictability results, most likely due to the limited test data and less spatial coverage. (Curtis, 2020)

Air Quality and COVID-19 Pandemic

Recently, the COVID-19 pandemic brought the whole world to a halt. With the huge impact of the pandemic on our health, economy, and lifestyles, research and development to accelerate the world towards a normal have taken pace. Due to the unpredictability of the pandemic, Artificial intelligence is being employed in vast areas to foresee the next move of the pandemic. One such example is using DL models to differentiate COVID infected chest CT scans from CT scans of other diseases. (Humayun & Alsayat, 2022). Moreover, studies are also being conducted to determine the strategies resulting in a low COVID-19 per-capita mortality rate so that pandemic management and response can be optimized. (Girum et al., 2021) Studies to predict the number of cases are also being conducted.(Painuli et al., 2021) Various parameters affected by the pandemic are involved in these studies. Due to the observation of strict lockdowns, the road traffic had significantly reduced and factories producing a huge amount of atmospheric pollutants also remained closed during the time of peaks in the COVID-19 positive cases. Due to the observation of lockdown in the majority of the countries, it is hypothesized that the COVID-19 pandemic might have a significant correlation with the air quality. If any such relationship exists, then air quality during pandemic times can be used as a parameter in a variety of research and development applications concerning the COVID-19 pandemic. Our study aims to determine the air quality pattern in the COVID-19 pandemic and find whether the air quality index can be used as a parameter in COVID-related research.

METHODOLOGY

The methodology followed by this study involved gathering data from an appropriate satellite, choosing a study area where our results could be tested conveniently and considering pollutants that contribute significantly to the increasing levels of air pollution. The data collection step was followed by data pre-processing which involved masking the region of interest, interpolating for faulty values in the data and making data ready to be processed by the time series forecasting models. Various ML, DL and statistical models were then tested and tuned to get the optimum prediction. Then, the best performing model is implemented on the web portal. The detailed steps are discussed below:

Data Collection

Study Area

As per IQAir December 2020, the air quality of Lahore, Pakistan, has fallen to the second worst in the world. For several years now, Smog has been paralyzing life in Punjab, Pakistan; triggered by torching crop wastes, burning residues, ancient technological brick furnaces, the smoke emitting of plants and vehicles alongside a century-old fog. Due to all these adverse factors, Lahore, Pakistan still gives pollution capitals in the world a hard time, often taking the lead. In Lahore alone, 500 brick ovens and almost 2800 plants emit smoke, but the environmental welfare departments fail to tackle the issue by ensuring that environmental policies and prohibitions are fully complied with. This environmental disaster is also affecting the nearby districts in Punjab, Pakistan.

Sentinel-5 Precursor

The Copernicus Sentinel-5 precursor mission is the first to monitor the atmosphere at Copernicus. The result of close cooperation between ESA, the European Commission, the Space Services of the Netherlands, the industry, the data users, and the scientists is the Copernicus Sentinel-5-Precursor. The mission consists of one satellite that is fitted with the TROPO spherical Surveillance Instrument also known as TROPOMI. ESA and the Netherland Space Office financed the TROPOMI instrument. The principal objective of Sentinel-5P is to perform air quality, ozone, UV radiation, climatic monitoring and forecasting atmosphere measurements with a very high spatio-temporal resolution. Table 1 table shows the properties of Sentinel-5 Precursor. Table 2 products are available through Sentinel-5 Precursor.

Table 1. Properties of Sentinel-5 Precursor

Property	Information
Spatial Resolution	Up to 5.5 km x 3.5 km.
Sensor	Tropospheric Monitoring Instrument (TROPOMI), a spectrometer measuring ultraviolet and visible (270-495 nm), near-infrared (675-775 nm) and shortwave infrared (2305-2385)
Revisit time	Less than one day.
Spatial coverage	Global coverage
Data availability	Since April 2018.
Common usage/purpose	To provide global information on a multitude of atmospheric trace gases, aerosols and cloud distributions affecting air quality and climate.

Table 2. Products of Sentinel-5 Precursor

Name	Description
CO	Carbon Monoxide
HCHO	Formaldehyde.
NO_2	Nitrogen Dioxide
O_3	Ozone
SO_2	Sulphur Dioxide
CH_4	Methane
AER_AI340_380	UV (Ultraviolet) Aerosol Index calculated based on wavelengths of 340 nm and 380 nm
AER_AI354_388	UV (Ultraviolet) Aerosol Index calculated based on wavelengths of 354 nm and 388 nm

Selected Parameters

The data is available in the form of NetCDF file format from Sentinel 5-Precursor. The most relevant information in the file is the georeferenced total vertical column of a pollutant for the region of Lahore, Pakistan. One file can cover up to 2400 km, due to which the size of the file can go as large as 2 GB. Table 3 table shows the available data for various air pollutants. All the relevant available data has been downloaded and used for this research project.

Table 3. Data availability for the chosen air pollutants

Pollutant	Time Period			
	2018	**2019**	**2020**	**2021**
SO_2	Available	Available	Available	Available
NO_2	Available	Available	Available	Available
CO	Not Available	Available	Available	Available

Data Preprocessing

Masking

Masking is a process of subsetting a specific region from a file based on some criteria. In data files, the only required region was the study area of Lahore, Pakistan, which forms a very small portion of the original satellite file. Geographical coordinates i.e., longitude and latitude boundary of Lahore was used for region extraction. While masking, NaN values were also removed. Four coordinates were chosen and treated as four corners of a region of interest. A special Python library Xarray was used for this purpose. The original file contained about 350x400 values, while the masked file contained only about 8x9 values, decreasing the file size considerably.

Interpolation

The spatial resolution of the data is 5x7 km^2. Bicubic interpolation was performed to increase the spatial resolution by two times. Spatial interpolation was performed using the Python library Pyinterp, which is used to perform interpolation in NetCDF format files. The data has been then organized to consist of a total 64 locations in and around Lahore. For each location, there is a time series for each pollutant. The 64 locations are chosen in such a way that they form a grid or image of size 8x8. Each location is treated as one of the features of the data and each sample corresponds to one date.

The data contained negative values in between due to cloud cover or dysfunctional spectrometer sensor units. These values were removed and replaced by interpolated values. Bicubic interpolation was performed for this purpose. Each row represents values of an air pollutant on a specific date. From each row, data is converted into a 2D NumPy array of size 8x8 and negative values are replaced by NaN. This 2D array is passed to SciPy's interp2d method. Interpolated array is converted back to a one-dimensional array and inserted into a new '.csv' file.

The data of SO_2, NO_2 and CO as described above came in the form of sequential 2D - arrays for each day with each point in the 2D array corresponding to a longitude - latitude value of different regions in Lahore. For the problem of forecasting, each 2D array had to be considered one sample of the time series. These 2D arrays could be treated as an image and the whole forecasting problem could be modelled as a problem of next image prediction in the time series where each image feature is one longitude-latitude location.

Modelling Multivariate Time Series

A multivariate time series has multiple time-dependent variables. Each variable depends on its past values along with past values of other time series variables. This dependency is used for forecasting future values. For this purpose, 8-by-8 image size or 64 locations were modelled in such a way that the upcoming air quality value for a particular location was dependent not only on the past value of the location itself, but also on the past values of the neighboring locations. The existence of spatio-temporal relationships between different locations creates a way to model the forecasting problem as a multivariate time series problem.

The data is divided into three datasets for each pollutant, and each dataset consists of continuous values of a particular air pollutant for 64 locations in and around Lahore. The shift functionality was used to model the problem of forecasting multivariate time series in the form of input and output. The output is the present values, and the input is the past values of the time series. Multistep output can also be utilized, which result in forecasting for more than one day. An optimum lag is the one with maximum autocorrelation. Figure 1 shows the average of autocorrelation for different lags values for 64 locations. The maximum autocorrelation is observed for lag of about 30. Therefore, past 30 values as input for predictions of the next values were chosen.

Figure 1. Autocorrelation of pollutants with various lags

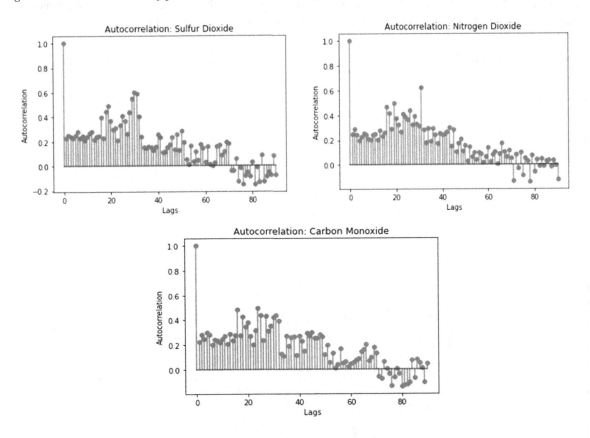

Data Augmentation and Normalization

The dataset consisted of air quality parameters values from 2018 to 2021 daily, which was deemed insufficient for reducing the loss. Therefore, to increase the data size, data augmentation was performed. Data augmentation in time series data does not allow us to mix the data to an extent where time-series patterns in the data are lost. Therefore, it is important to keep in mind that the data augmentation technique should not disturb the time series pattern and at the same time increases the data size.

For data augmentation, data size was increased up to 3 times by randomly swapping the immediate and third time-series values. This introduced a good amount of noise in the data to avoid overfitting and did not affect the overall pattern while also increasing bias to an unacceptable amount. For data normalization, the data was brought in the range of 0 and 1 using min-max normalization. This reduces the training time and normalizes the error metrics. Final data consisted of input of past 30 values and output of next 15 days. Each sample consisted of 30 past values for all 64 locations and as input and next 15 values for all 64 locations as output. For all the air pollutants, the training data consists of 80% of the total data and test data makes the rest of the 20% of the data.

Forecasting Models

Various forecasting models can be used to forecast the multivariate air quality values at the 64 locations in Lahore. Some of the most well-known forecasting models have been tested to model the problem for each air pollutant. The following algorithms have been tested for the problem:

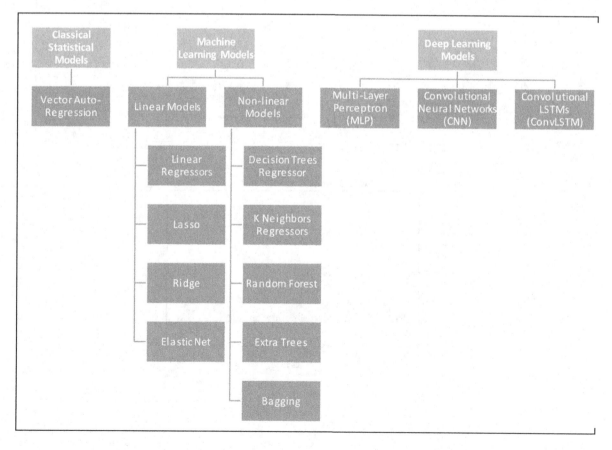

Models Architecture

The model architecture of the employed forecasting model is explained below:

1.   Vector Autoregression (VAR)

VAR takes input series and predicts the upcoming values by using the past values. The model predicts outputs for all 64 locations. The prediction made by the model is used to predict the next value. As the actual data comes, the model is fed the actual data to keep up with the new trend.

2.   Machine Learning

  a.   Linear Models

Four linear models have been tested for forecasting. Linear Regression fits a linear model with coefficients $w = (w1, ..., wp)$ to minimize the sum of squares between the observed targets in the dataset and the targets predicted by the linear regression approximation. Ridge regression model solves a regression model where the loss function is the linear least square's function and regularization is given by the L2-norm. The Lasso is a linear model that estimates sparse coefficients with L1 regularization. Elastic-Net is a linear regression model trained with both L1 and L2 norm regularization of the coefficients.

  b.   Non-Linear Models

The decision tree regressor used the mean square error (MSE) as the criteria to fit the data. The maximum depth of the tree was chosen to be 16. Minimum sample split was kept at 2 for all three pollutants. Ten estimators were used in decision trees. MSE was used as a criteria function. Ten estimators with a maximum depth of 15 were used. MSE was used as a loss function. All pollutants were modelled using the mentioned specifications. Twelve estimators were used for bagging, and MSE was used as the loss function. All pollutants were modelled using the mentioned specifications. KNN regression is a non-parametric method that, in an intuitive manner, approximates the association between independent variables and the continuous outcome by averaging the observations in the same neighborhood. The size of the neighborhood needs to be set by the analyst or can be chosen using cross-validation to select the size that minimizes the mean-squared error. K, in this case, is 20. All pollutants were modelled using the mentioned specifications.

3.   Deep Learning

  a.   Nitrogen Dioxide (NO_2)
    i.   Convolutional Neural Network (CNN)

Figure 2 shows the architecture of the CNN used to achieve optimum MSE. The model uses Adam optimizer with default learning rate of 0.001. One convolutional layer is used with 64 filters. LeakyReLU function is used as an activation function followed by a max pooling layer with filter size 2. After that, the data goes into a fully connected layer with 50 neurons with LeakyReLU as activation function and an output layer with size $64*15 = 960$.

Figure 2. CNN model summary for NO$_2$

```
Model: "sequential"

_____
Layer (type)                 Output Shape              Param #
=================================================================
conv1d (Conv1D)              (None, 28, 64)            12352

leaky_re_lu (LeakyReLU)      (None, 28, 64)            0

max_pooling1d (MaxPooling1D) (None, 14, 64)            0

flatten (Flatten)            (None, 896)               0

dense (Dense)                (None, 50)                44850

leaky_re_lu_1 (LeakyReLU)    (None, 50)                0

dense_1 (Dense)              (None, 960)               48960
=================================================================
Total params: 106,162
Trainable params: 106,162
Non-trainable params: 0
_____
```

ii. Multi-Layer Perceptron (MLP)

MLP for NO$_2$ uses 3 dense hidden layers and one output layer with sizes 150, 100, 100 and 960, respectively. LeakyReLU is used as an activation function and MSE is used as a loss. Adam optimizer with learning rate of 0.001 is employed. The summary of the MLP model for NO$_2$ is given in the Figure 3.

Figure 3. Multi-layer Perceptron model summary for NO_2

```
Model: "sequential_1"
```

Layer (type)	Output Shape	Param #
dense_2 (Dense)	(None, 150)	288150
leaky_re_lu_3 (LeakyReLU)	(None, 150)	0
dense_3 (Dense)	(None, 100)	15100
leaky_re_lu_4 (LeakyReLU)	(None, 100)	0
dense_4 (Dense)	(None, 100)	10100
leaky_re_lu_5 (LeakyReLU)	(None, 100)	0
dense_5 (Dense)	(None, 960)	96960

```
Total params: 410,310
Trainable params: 410,310
Non-trainable params: 0
```

b.   Sulphur Dioxide (SO_2)

i.   Convolutional Neural Networks (CNN)

Figure 4 shows the architecture of the CNN used to achieve optimum MSE. The model uses Adam optimizer with default learning rate of 0.001. One convolutional layer is used with 64 filters. LeakyReLU function is used as an activation function and lastly there is an output layer with size 64*15 = 960. The model performs the best with only one Convolutional layer.

Figure 4. CNN model summary for SO$_2$

```
Model: "sequential_17"

_____
Layer (type)                 Output Shape              Param #
=================================================================
conv1d_10 (Conv1D)           (None, 26, 64)            20544

_____
leaky_re_lu_19 (LeakyReLU)   (None, 26, 64)            0

_____
flatten_6 (Flatten)          (None, 1664)              0

_____
dense_30 (Dense)             (None, 960)               1598400
=================================================================
Total params: 1,618,944
Trainable params: 1,618,944
Non-trainable params: 0
_____
```

ii.   Multi-Layer Perceptron

The Multi-Layer Perceptron involved only one hidden dense layer with 150 neurons and 1 output layer LeakyReLU was used as an activation function and MSE as the loss function. Adam optimizer was employed as a criterion with a learning rate of 0.001.

c.   Carbon Monoxide (CO)

i.   Convolutional Neural Networks (CNN)

Figure 5 shows the architecture of the CNN used to achieve optimum MSE. The model uses Adam optimizer with default learning rate of 0.001. One convolutional layer is used with 64 filters. Max Pooling with filter size 2 has also been used LeakyReLU function is used as an activation function and lastly there is an output layer with size $64*15 = 960$. The model performs the best with only one Convolutional layer.

Figure 5. Convolutional Neural Network model summary for CO

```
Model: "sequential_9"

Layer (type)                    Output Shape          Param #
=================================================================
conv1d_8 (Conv1D)               (None, 26, 64)        20544

leaky_re_lu_9 (LeakyReLU)       (None, 26, 64)        0

max_pooling1d (MaxPooling1D)    (None, 13, 64)        0

flatten_5 (Flatten)             (None, 832)           0

dense_15 (Dense)                (None, 960)           799680
=================================================================
Total params: 820,224
Trainable params: 820,224
Non-trainable params: 0
```

ii.   Multi-Layer Perceptron (MLP)

The MLP involved only one hidden dense layer with 150 neurons and 1 output layer. LeakyReLU was used as an activation function and MSE as the loss function.

Convolutional Neural Networks (CNN)

Table 4. CNN model specifications for the pollutants under study.

	Loss	Optimizer	Activation	Number of Parameters
NO_2	MAE	Adam	LeakyReLU	106162
SO_2	MSE	Adam	LeakyReLU	1618944
CO	MSE	Adam	LeakyReLU	820224

Multi-Layer Perceptron (MLP)

Table 5. MLP model specifications for the pollutants under study

	Loss	Optimizer	Activation	Number of Parameters
NO_2	MAE	Adam	LeakyReLU	410310
SO_2	MSE	Adam	LeakyReLU	433110
CO	MSE	Adam	LeakyReLU	433110

Convolutional Long Short Term Memory Networks (ConvLSTMs)

Figure 6 shows the summary of Convolutional LSTMs used for SO2, NO_2 and CO. The model consists of two 2-Dimensional Convolutional layers.

Figure 6. Convolutional Long Short Term Memory Networks summary for all pollutants

```
Model: "sequential_5"

_____
Layer (type)                  Output Shape               Param #
=================================================================
module_wrapper_29 (ModuleWra  (None, 30, 15, 8, 8)       24060

module_wrapper_30 (ModuleWra  (None, 30, 15, 8, 8)       0

module_wrapper_31 (ModuleWra  (None, 30, 15, 8, 8)       32

module_wrapper_32 (ModuleWra  (None, 30, 1, 8, 8)        260

module_wrapper_33 (ModuleWra  (None, 30, 1, 8, 8)        0

module_wrapper_34 (ModuleWra  (None, 30, 1, 8, 8)        32

module_wrapper_35 (ModuleWra  (None, 1, 1, 8, 8)         811
=================================================================
Total params: 25,195
Trainable params: 25,163
Non-trainable params: 32
_____
```

RESULTS AND DISCUSSION

Nitrogen Dioxide (NO₂)

Table 6, Figure 7, and Figure 8 shows the comparison of various error metrics and results in the form of images for NO_2. The error metrics used to evaluate performance on the test set includes MSE, RMSE and MAE.

Table 6. Comparison of error metrics for the forecasting models for NO_2

	Mean Squared Error	Root Mean Squared Error	Mean Absolute Error
CNN	0.00381	0.06169	0.04072
MLP	0.00477	0.06908	0.04721
Bagging	0.00552	0.07432	0.05596
Extra Trees	0.00662	0.08135	0.05877
Random Forest	0.00778	0.08819	0.06261
K Neighbour	0.00799	0.08938	0.06429
Decision Trees	0.01014	0.10067	0.07225
Elastic Net	0.01270	0.11269	0.09187
Lasso	0.01270	0.11269	0.09187
Ridge	0.02375	0.15410	0.11176
ConvLSTMs	0.02376	0.15413	0.11185
VAR	0.02666	0.16329	0.11831
Linear Regressor	0.04771	0.21842	0.16944

Observations

The best performance is shown by CNN with MSE at 0.0038 on the test set. MLP also shows good performance along with CNN as shown in Table 6. The worst performance is shown by Linear Regression which concludes that the input and output has a more complex relationship than linear relationship. Other linear models, i.e., Lasso, Elastic Net and Ridge also do not show good performances. The ensemble methods take the lead after CNN and MLP and perform relatively better than Decision Trees and K Neighbors. Graphs in Figure 7 show actual NO2 levels over Allama Iqbal International Airport, Lahore, Pakistan, from April 2021 to mid-May 2021 and the predictions up to mid-June 2021.

Figure 7. Graph comparing predictions and actual values of NO$_2$ over Allama Iqbal International Airport, Lahore, Pakistan, from April 2021 to mid-May 2021

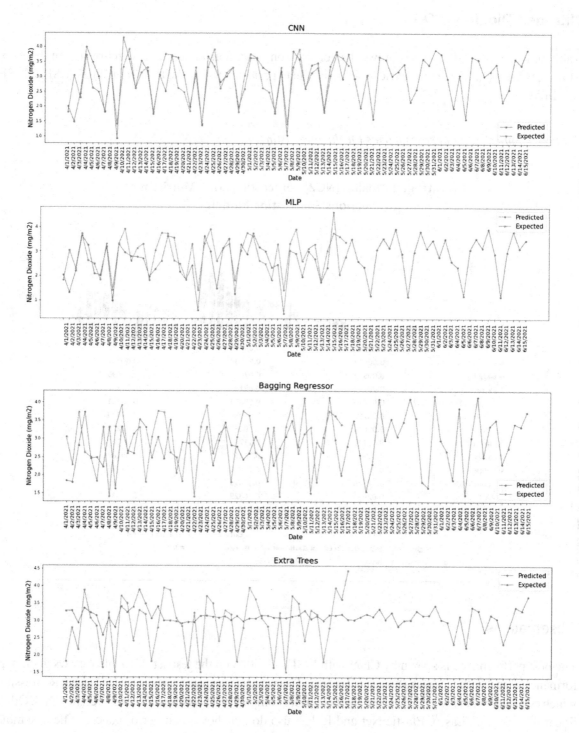

The comparison in Figure 7 clearly shows that CNN followed by MLP shows the best performance. The performance gets bad with Random Forest and K Neighbor Regressor, which makes us conclude that the DL Models perform the best at picking the spatio-temporal patterns in the time series data for the region of Lahore, which can also be observed in Figure 8.

Figure 8. Results comparing top four models for NO$_2$ for all 64 locations

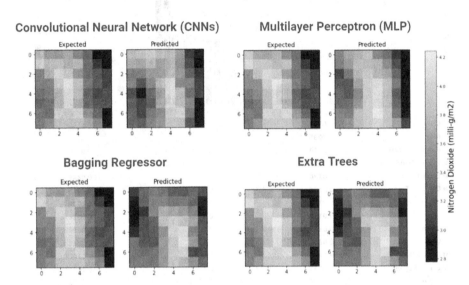

Sulphur Dioxide (SO$_2$)

Table 7, Figure 9, and Figure 10 show the comparison of errors metrics for the forecasting models performance on the test set for SO2:

Table 7. Comparison of error metrics for the forecasting models for SO2

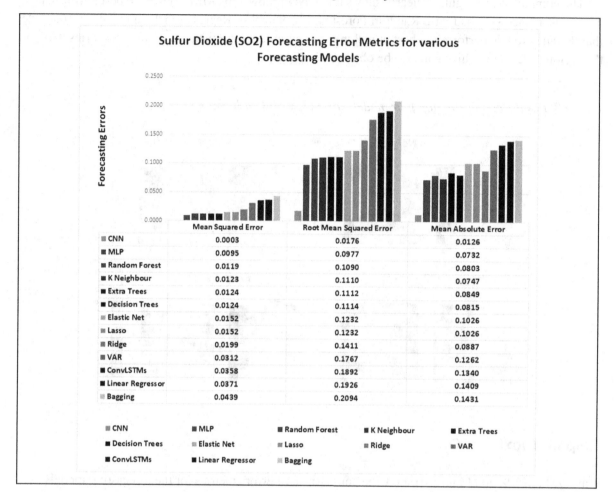

	Mean Squared Error	Root Mean Squared Error	Mean Absolute Error
CNN	0.0003	0.0176	0.0126
MLP	0.0095	0.0977	0.0732
Random Forest	0.0119	0.1090	0.0803
K Neighbour	0.0123	0.1110	0.0747
Extra Trees	0.0124	0.1112	0.0849
Decision Trees	0.0124	0.1114	0.0815
Elastic Net	0.0152	0.1232	0.1026
Lasso	0.0152	0.1232	0.1026
Ridge	0.0199	0.1411	0.0887
VAR	0.0312	0.1767	0.1262
ConvLSTMs	0.0358	0.1892	0.1340
Linear Regressor	0.0371	0.1926	0.1409
Bagging	0.0439	0.2094	0.1431

Observations

The best performance is shown by CNN with MSE at 3e-4 on the test set. The worst performance is shown by Bagging and Linear Regressor, as shown in Table 7. Non-linear algorithms show a relatively better performance than statistical model Vector Autoregression and linear regression models of Lasso, Ridge, and Elastic Net. Convolutional Long Short Memory Networks also show bad performance. Graphs in Figure 9 show actual SO_2 levels over Allama Iqbal International Airport, Lahore, Pakistan, from April 2021 to mid-May 2021 and predictions up to mid-June 2021.

Figure 9. Graph comparing predictions and actual values of SO$_2$ over Allama Iqbal International Airport, Lahore, Pakistan, from April 2021 to mid-May 2021

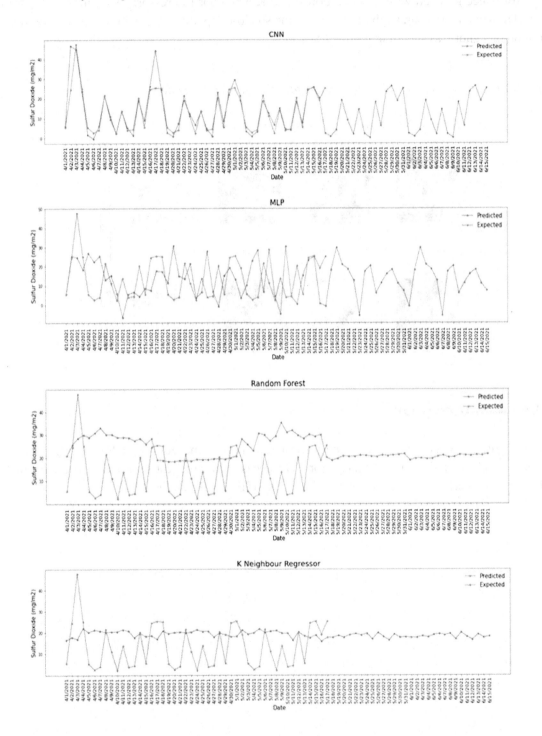

The comparison in Figure 9 clearly shows that CNN followed by MLP shows the best performance. The performance gets bad with Random Forest and K Neighbor Regressor, which makes us conclude that the DL Models perform the best at picking the spatio-temporal patterns in the time series data for the region of Lahore, as shown in Figure 10 as well.

Figure 10. Results comparing top four models for SO$_2$ for all 64 locations

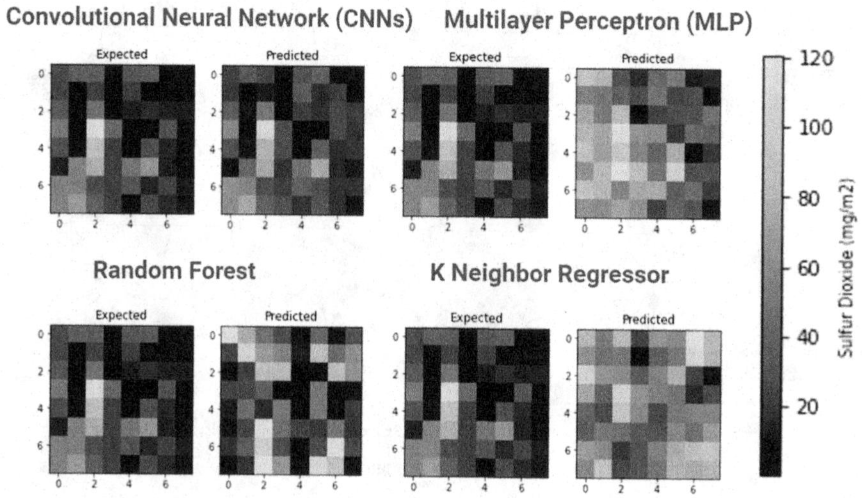

Carbon Monoxide (CO):

Table 8, Figure 11, and Figure 12 show the comparison of errors metrics for the forecasting models and comparative results in the form of images for CO forecasting on the test set for CO. The error metrics used to evaluate performance on the test set includes MSE, RMSE and MAE.

Table 8. Comparison of error metrics for the forecasting models for CO

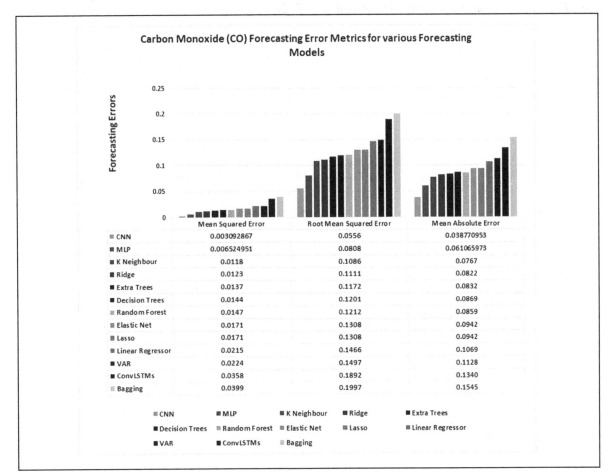

Observations

The best performance is shown by CNN with a MSE at 3e-2 on the test set. MLP also shows good performance with CNN. The worst performance is shown by Bagging and ConvLSTMs, as shown in Table 8. Non-linear algorithms show a relatively better performance than statistical model Vector Auto Regression. Linear regression models of Ridge also show a good performance proving that CO may have a linear relationship between past 30 values and future 15 values.

Graphs in Figure 11 show actual CO levels over Allama Iqbal International Airport, Lahore, Pakistan, from April 2021 to mid-May 2021 and predictions up to mid-June 2021.

Figure 11. Graph comparing predictions and actual values of CO over Allama Iqbal International Airport, Lahore, Pakistan, from April 2021 to mid-May 2021

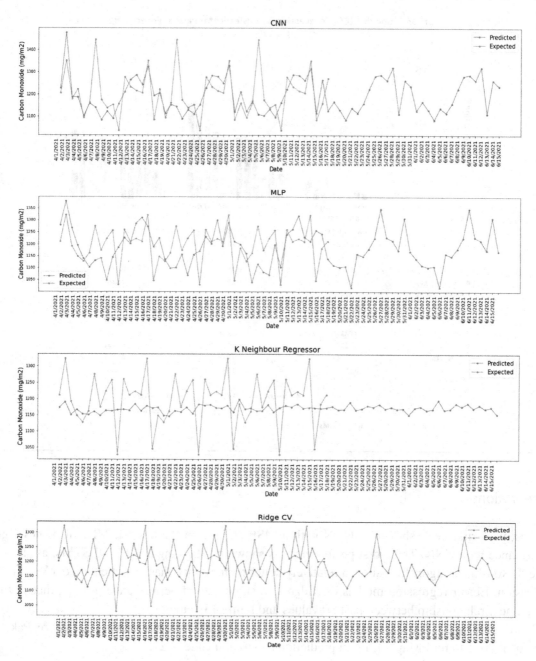

It can be observed that CNN shows the best performance, and the error increases from MLP to Ridge CV and can be observed in Figure 12.

Figure 12. Results comparing top four models for CO for all 64 locations

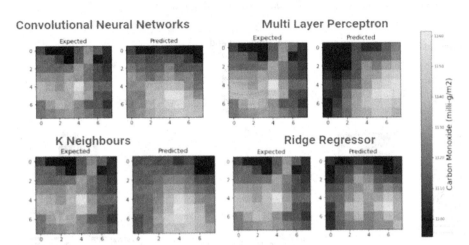

Web Portal

A web portal is implemented for public viewing and environment health regulatory units to take precautionary measures. The web portal is integrated with a prediction model which is deployed on a Virtual Machine provided by SEECS (School of Electrical Engineering and Computer Science, NUST). All the data is downloaded from https://s5phub.copernicus.eu/dhus/#/home and stored in the virtual machine. An automated code applies masking and interpolation on that data and then converts it into '.csv' format. Weights of the prediction model are saved and loaded as per requirement. The model can predict data for one month. Data from the last 15 days is appended at the end of previous data, and the models are retrained every 15 days. The model generates a new data file in '.csv' format. All data files containing actual and predicted values are uploaded to the web server. Actual and predicted data of each air pollutant is stored in different files. Another file is generated, which contains predicted AQI (calculated from predicted air pollutant values). The backend of the web portal is developed using Node.js, while the frontend is developed using HTML, CSS, and JavaScript, as shown in Figure 13. The web portal consists of five tabs, and each tab represents data in a different format. It is deployed on Heroku: Cloud Application Platform.

Figure 13. Web portal architecture
The URL of the web portal is: https://air-quality-monitoring-pk.herokuapp.com/

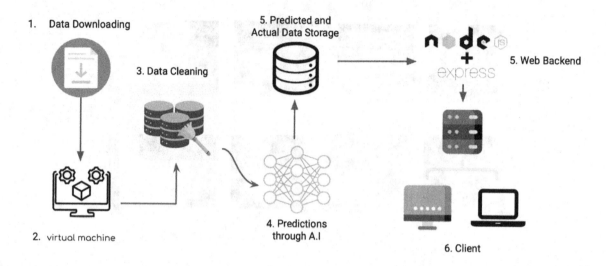

Home Page

The screenshot of the homepage is given in Figure 14. Users can view the current air pollution levels as well as the AQI index of Lahore, Pakistan. An air quality prediction table is also developed where a user can view predicted air pollutant levels for the next 15 days. AQI values are presented using six colors. Each color is associated with a specific range of AQI. Instructions related to that specific AQI range are also given. Values between 0 and 50 are considered good, while values between 301 to 500 range are considered hazardous. AQI meter is implemented using a JavaScript library gauge.js.

Figure 14. Screenshot of the Home page displaying the predicted Air Quality Index for Lahore, Pakistan, with a table showing the predicted Air Quality Index for the upcoming 14 days

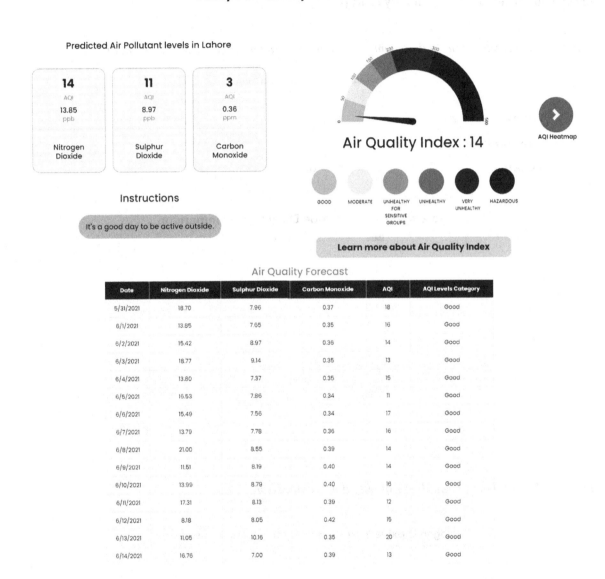

Graphical Data Tab

To get a visual understanding of air pollutant values, they are implemented in graphical form, as shown in Figure 15. Users can select only one air pollutant at a time. Users can also view the air pollutant level of a specific location in Lahore and the average air pollutant level of Lahore. A time period of one month is selected by default but a user can change it as per their requirements. Three graphs are available for each air pollutant. One graph shows the daily comparison of actual and predicted values of an air pollutant, while the other two shows the weekly and monthly data comparison based on years, as shown in Figure 16. The x-axis of the chart represents the date, while the y-axis represents the concentration of air pollutants. Users can also enable and disable certain parameters of the chart as well. These charts are developed using JavaScript library chart.js.

Figure 15. Screenshot of daily data comparison between actual and predicted air quality

Figure 16. Screenshot of monthly air quality data comparison

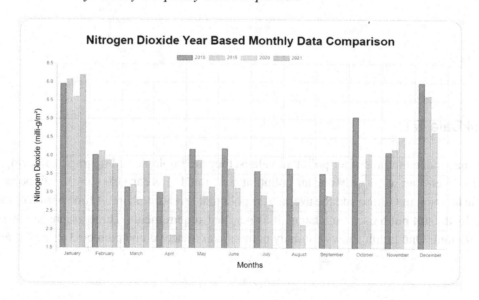

Tabular Data Tab

Data is also represented using the tabular form. In this tab, users can select only one air pollutant at a time. Users can also define the period by selecting the starting and ending dates. Each row of the table represents the concentration of an air pollutant on a specific date. The leftmost column represents the date, while the other column represents the locations, as shown in Figure 17. The table is implemented using JavaScript and jQuery.

Figure 17. Screenshot of the Tabular Data tab displaying the NO$_2$ levels

Choose an air pollutant:	Nitrogen Dioxide ⌄
Choose your desired location/Data Type:	Lahore ⌄
Start Date	05/07/2021 ▢	End Date	05/15/2021 ▢

Nitrogen Dioxide values in milli-g/m²

Date	Bahria Town Marquee	Barkat Ali Market, Riwind Rd	Bakra Mandi, Defence Road	Lahore Ring Road, Kahna Interchange	Model Baraz, Chung	Bilal Town, LDA Avenue	Wapda Town, Phase 1	Pak-Arab Housing Scheme
5/15/2021	6.92343728	5.21467845	5.4046484	5.28093838	3.35159046	4.41779881	4.92412225	5.5037019
5/14/2021	5.49564987	9.08826359	3.93445185	2.26833646	5.1041208	5.42695776	3.31290692	2.11566391
5/13/2021	2.29486916	2.22204508	2.27191995	2.00852475	2.53401924	2.28239462	2.18486751	2.12433085
5/12/2021	6.21002794	8.01169255	3.12876737	2.82446197	5.42782237	4.47656271	3.42323493	2.98189845
5/11/2021	1.68989272	3.84819523	4.91623256	3.36315988	1.70136724	3.53701004	5.70413898	3.7586137
5/10/2021	2.6570238	3.62606838	3.29444801	2.87302117	2.55554548	3.27796169	3.40988072	3.25030931
5/9/2021	3.24649702	3.24649702	4.05683732	4.86717729	3.06080798	3.04529975	3.9235258	4.80175217
5/8/2021	5.48085462	10.37997357	6.54264648	3.70233754	4.21990168	7.09298477	5.39740676	4.5822955

Maps Tab

For better visual understanding, classification heatmaps are also implemented on the web portal. Predicted values from the Deep Learning models are interpolated ten times to generate a dense heatmap. The red color represents the higher concentration, while the blue color represents the lower concentration of an air pollutant, as shown in Figure 18. The Heatmap layer is developed using a JavaScript library called leaflet.js, while the map layer is added using OpenStreetMap contributors. Marker points are also added to make it convenient for users to see pollutant levels at a specific location. A gradient bar is also implemented for better understanding and visualization. Users can select any air pollutant and view forecasted heatmaps by selecting future dates.

Figure 18. Screenshot of the Maps tab showing a color-coded map for NO$_2$ over Lahore, Pakistan

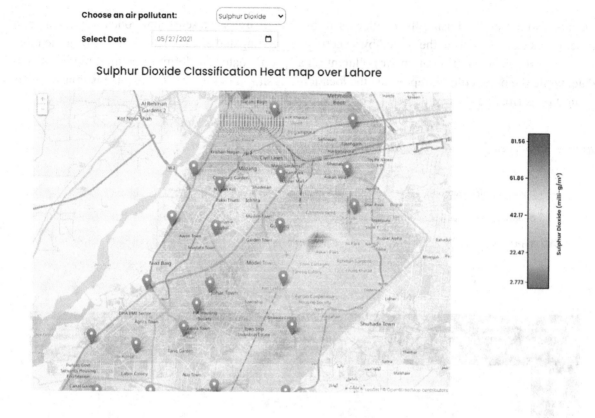

COVID-19 Analysis

The COVID-19 pandemic generated a global response followed by exceptional reductions in economic and industrial affairs. Due to COVID-19, people stopped driving and flying in large numbers and completely staying at home caused a drastic change in the air pollution pattern. Combustion of fossil fuels in industries and transportations are major sources of NO$_2$, both of which were considerably reduced during strict lockdown to prevent the outspread of COVID-19. In this study, it can be observed that these strict lockdowns caused the reduction of NO$_2$ concentration by almost 30%, as shown in Figure 19.

Figure 19. Reduced levels of NO$_2$ during lockdown March 2021 - May 2021

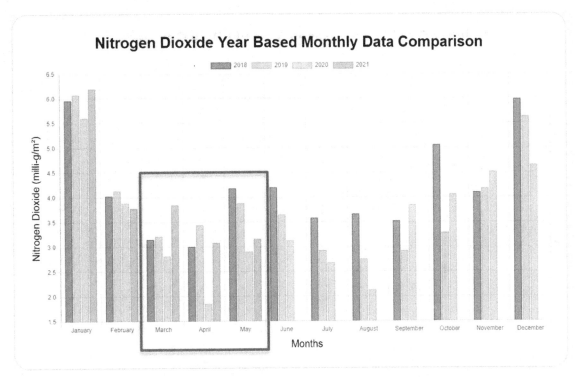

In 2020, Pakistan's Government enforced two major lockdowns to limit the spread of Coronavirus. The first lockdown was announced in the last week of March 2020 and lasted till 9th May 2020, while the second lockdown was announced in July and Ended in August 2020. In both lockdowns, there were very few economic activities. Transportation was heavily affected, which led to better air quality. A massive drop in NO$_2$ concentration in April 2020 and August 2020 compared to 2018 and 2019 can be observed in Figure 19.

Figure 20. Comparison of SO$_2$ levels from 2018-2021

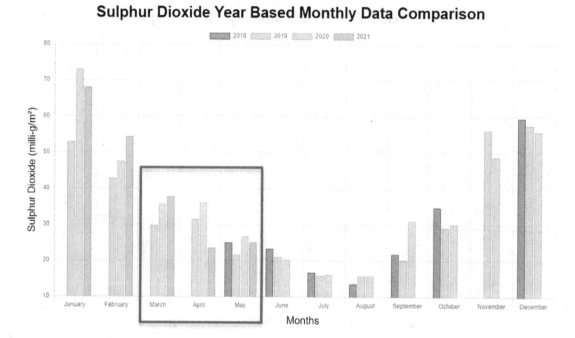

Lockdown did not have a considerable impact on the concentration of SO$_2$, as shown in Figure 20. The concentration of the SO$_2$ decreased by a minimal amount. Sources of SO$_2$ is combustion of furnace oil which is mainly used by the cement industry, sugar industry and power plants. Those sectors were working at their maximum capacity during the lockdown period.

CONCLUSION AND FUTURE WORK

According to World Health Organization, air pollution is a growing concern for the whole world since ambient air pollution causes nearly 4.2 million deaths per year. Pakistan has two cities, Lahore and Karachi, listed as two of the top ten most polluted cities in the world. The air pollution in Lahore exceeds the safe limits and is classified as hazardous by experts. These unsafe levels of pollutants in the atmosphere pose serious health risks for adults and children. Air pollution is directly linked with cardiovascular diseases, including diabetes, high blood pressure, increased risk of heart attacks, strokes, and aortic dissections. In children, air pollution affects brain development. Studies have shown that air pollution also affects mental health negatively.

With such risks to physical and mental health, it is concerning that the first step towards combating air pollution, i.e., air pollution monitoring, is inefficient. This is because the conventional air pollution monitoring techniques use heavy and costly on-ground monitoring systems that only provide point measurements in the atmosphere. Due to inconvenience in their use, these monitoring stations are distributed sparsely and only estimate air quality levels at a certain point in the atmosphere which introduces a lot of uncertainty since major pollutants like SO_2 and Particulate Matter 2.5 vary significantly over small distances and time periods.

The proposed solution can help monitor and forecast air pollution with better spatial coverage than on-ground monitoring stations in a relatively cost-effective way. Monitoring air quality through satellite spectral data is the way forward to monitor air quality more conveniently and cost-effectively since the solution is software-based and does not involve heavy machinery. The method involves the analysis of SO_2, NO_2 and CO levels over the city of Lahore. This data is obtained from a state-of-the-art satellite, Sentinel-5 Precursor, which has a temporal resolution of 24 hours and spatial resolution of around 2.5x7 km^2.

For forecasting purposes, various models were tested, and CNN performed the best with only a few Convolutional layers. Adam optimizer gave the most optimum results. LeakyReLU improved the test errors as compared to other activation functions. The air quality index is calculated from the predicted pollutant concentrations and has been displayed on the web portal in graphs, charts, and heatmaps.

The future work for this research project includes incorporating data of other cities of the world and including other major air pollutants from more sources. The data from IoT devices will also be incorporated in the future to validate the current data. With time, the latest and better ML / DL models will be implemented to improve the prediction errors and efficiency of the project. This project will also be a way forward for the analysis of heavy Smog and its patterns in Lahore caused by air pollution. The project can be extended through the development of Android and iOS applications to provide an effortless, faster, and more responsive user experience. The current user interface of the website can also be improved to enhance the user experience and make the website more robust. Another way forward is to make APIs for the interpolated and forecasted data and make it available for broad usage.

REFERENCES

Cs229.stanford.edu. (2021). Available at: http://cs229.stanford.edu/proj2019spr/report/22.pdf

Chahine, M. T., McCleese, D. J., Rosenkranz, P. W., & Staelin, D. H. (1983). Interaction mechanisms within the atmosphere. *Manual of Remote Sensing, 1*(5).

Cleartheair.scottishairquality.scot. (2020). *How do we monitor air pollution? Clear the Air, a learning resource in Scotland*. Available at: http://cleartheair.scottishairquality.scot/about/how-do-we-monitor

Curtis, A. (2020). *Using MODIS Satellite Imagery to Estimate Particulate Matter*. Academic Press.

Darmawan, T. B., & Syafei, A. D. (2019, May). Characterizing NO_2 in Indonesia Using Satellite Ozone Monitoring Instruments. *IOP Conference Series. Earth and Environmental Science, 284*(1), 012011. doi:10.1088/1755-1315/284/1/012011

Girum, T., Lentiro, K., Geremew, M., Migora, B., Shewamare, S., & Shimbre, M. (2021). Optimal strategies for COVID-19 prevention from global evidence achieved through social distancing, stay at home, travel restriction and lockdown: A systematic review. *Archives of Public Health, 79*(1), 150. Advance online publication. doi:10.118613690-021-00663-8 PMID:34419145

Griggs, M. (1975). Measurements of Atmospheric Aerosol Optical Thickness over Water Using ERTS-1 Data. *Journal of the Air Pollution Control Association, 25*(6), 622–626. doi:10.1080/00022470.1975.10470118 PMID:1141544

Humayun, M., & Alsayat, A. (2022). Prediction Model for Coronavirus Pandemic Using Deep Learning. *Computer Systems Science and Engineering, 40*(3), 947–961. doi:10.32604/csse.2022.019288

Kanaroglou, P. S., Soulakellis, N. A., & Sifakis, N. I. (2002). Improvement of satellite derived pollution maps with the use of a geostatistical interpolation method. *Journal of Geographical Systems, 4*(2), 193–208. doi:10.1007101090100080

Kaufman, Y. J., Fraser, R. S., & Ferrare, R. A. (1990). Satellite measurements of large-scale air pollution: Methods. *Journal of Geophysical Research, D, Atmospheres, 95*(D7), 9895–9909. doi:10.1029/JD095iD07p09895

Khwaja, H., Fatmi, Z., Malashock, D., Aminov, Z., Kazi, A., Siddique, A., Qureshi, J., & Carpenter, D. (2012). Effect of air pollution on daily morbidity in Karachi, Pakistan. *Journal of Local and Global Health Science, 2012*(1). Available at: https://www.qscience.com/content/journals/10.5339/jlghs.2012.3

Liu, X., Li, B., Jiang, A., Qi, S., Xiang, C., & Xu, N. (2015, June). A bicycle-borne sensor for monitoring air pollution near roadways. In *2015 IEEE International Conference on Consumer Electronics-Taiwan* (pp. 166-167). IEEE. 10.1109/ICCE-TW.2015.7216835

Munsadwala, Y., Joshi, P., Patel, P., & Rana, K. (2019, April). Identification and visualization of hazardous gases using IoT. In *2019 4th International Conference on Internet of Things: Smart Innovation and Usages (IoT-SIU)* (pp. 1-6). IEEE. 10.1109/IoT-SIU.2019.8777481

Murray, N. L., Holmes, H. A., Liu, Y., & Chang, H. H. (2019). A Bayesian ensemble approach to combine PM2. 5 estimates from statistical models using satellite imagery and numerical model simulation. *Environmental Research, 178*, 108601. doi:10.1016/j.envres.2019.108601 PMID:31465992

Painuli, D., Mishra, D., Bhardwaj, S., & Aggarwal, M. (2021). Forecast and prediction of COVID-19 using machine learning. *Data Science for COVID-19*, 381–397. doi:10.1016/B978-0-12-824536-1.00027-7

Sifakis, N., & Deschamps, P. Y. (1992). Mapping of air pollution using SPOT satellite data. *Photogrammetric Engineering and Remote Sensing, 58*, 1433–1433.

Sifakis, N. I. (1998). Quantitative mapping of air pollution density using Earth observations: A new processing method and application to an urban area. *International Journal of Remote Sensing, 19*(17), 3289–3300. doi:10.1080/014311698213975

Tanre, D., Herman, M., Deschamps, P., & de Leffe, A. (1979). Atmospheric modeling for space measurements of ground reflectances, including bidirectional properties. *Applied Optics, 18*(21), 3587. doi:10.1364/AO.18.003587 PMID:20216655

Vijayakumar, P., Khokhar, A., Pal, A., & Dhawan, M. (2020, July). Air Quality Index Monitoring and Mapping Using UAV. In *2020 International Conference on Communication and Signal Processing (ICCSP)* (pp. 1176-1179). IEEE. 10.1109/ICCSP48568.2020.9182374

Zhang, D., & Woo, S. S. (2020). Real time localized air quality monitoring and prediction through mobile and fixed IoT sensing network. *IEEE Access: Practical Innovations, Open Solutions, 8*, 89584–89594. doi:10.1109/ACCESS.2020.2993547

KEY TERMS AND DEFINITIONS

Aerosal Optical Depth (AOD): It is the measure of aerosols (e.g., urban haze, smoke. particles, desert dust, sea salt) distributed within a column of air from the instrument.

Air Quality Index (AQI): An air quality index is used to communicate to the public on how polluted the air currently is or how polluted it's forecast to become.

Cloud Virtual Machine: A virtual machine or operating system that runs over a cloud.

ConvLSTM: It is a type of recurrent neural network for spatio-temporal prediction with convolutional structures in both the input-to-state and state-to-state transitions.

Express.js: It is a backend web application framework for Node.js. It is designed for building web applications and APIs.

JavaScript: A web-scripting language that creates animations and adds dynamic functionality for websites.

Keras: Keras is an open-source software library that provides a Python interface for artificial neural networks.

Multivariate Time Series: Series that consist of more than one time-dependent variable and each variable depends on the past values of other variables.

MySQL: MySQL is an open-source relational database management system.

NETcdf: Network Common Data Form is a set of software libraries and machine-independent data formats to support creation, access, and sharing of array-oriented scientific data.

Node.js: Node.js is an open-source, cross-platform, backend JavaScript runtime environment that runs on the V8 engine and executes JavaScript code outside a web browser.

Orbiting Carbon Observatory-2: It is an American environmental science satellite launched by NASA as a replacement for the Orbiting Carbon Observatory, which was lost in a launch failure in 2009.

React.js: React is an open-source, front-end JavaScript library for building user interfaces or UI components.

Satellite: It is an artificial body placed in orbit around the earth or moon or another planet to collect information or for communication.

Sentinel-5 Precursor: Sentinel-5 Precursor is an Earth observation satellite developed by ESA as part of the Copernicus Program to close the gap in the continuity of observations between Envisat and Sentinel-5.

Smog: A mixture of smoke and fog is called smog.

Spatial Resolution: It refers to the size of the smallest feature that can be resolved by the satellite sensor.

Spatio-Temporal Data: Data collected across space and time.

Temporal Resolution: It refers the smallest possible time difference between two readings of a sensor.

TensorFlow: Tensorflow is a symbolic math library based on dataflow and differentiable programming. It can be used across a range of tasks but has a particular focus on the training and inference of deep neural networks.

TROPOMI: The TROPOspheric Monitoring Instrument is a satellite instrument on board the Copernicus Sentinel-5 Precursor satellite. It was launched in October 13, 2017.

Chapter 3
Automated Multi–Sensor Board for IoT and ML–Enabled Livestock Monitoring

Abdul Aziz Chaudhry

National University of Sciences and Technology, Pakistan

Rafia Mumtaz

National University of Sciences and Technology, Pakistan

Usman Ahmad Siddiqui

National University of Sciences and Technology, Pakistan

Syed Hassan Muzammil

National University of Sciences and Technology, Pakistan

Muhammad Ali Tahir

National University of Sciences and Technology, Pakistan

ABSTRACT

Livestock monitoring is one of the most common problems in the current time, and to sustain the lifecycle and support the nature of domesticated animals, the standard checking of animal wellbeing is fundamental. Moreover, many diseases are spread from animals to human beings; hence, an early prognosis in regard to cow wellbeing and illness is essential. This chapter proposed an internet of things (IoT)-based framework for domesticated animal wellbeing checking. The proposed framework comprises of a specially crafted multi-sensor board to record a few physiological boundaries including skin temperature, pulse, and rumination with regards to encompassing temperature, stickiness, and a camera for picture examination to recognize diverse standards of health. The data is collected using LoRa gateway technology, where gathered data is examined and utilized for performing ML models to identify diseased and healthy creatures and foresee cow wellbeing for giving early and convenient clinical consideration. The results obtained are used for careful insights regarding animal health and wellbeing.

DOI: 10.4018/978-1-7998-9201-4.ch003

INTRODUCTION

Livestock is a trillion-dollar global asset and according to UNDP, the demand for animal products by 2050 will increase by 70% owing to the escalating world population. Today, dairy farmers face the problems of infrastructure, connectivity monitoring and management. These problems can decrease milk production up to 35L/day. According to UN report, approximately $21 billion losses are incurred each year due to animal diseases. Moreover, traditional livestock farming contributes to 14.5% in Greenhouse emissions, and also causes soil erosion, deforestation, and pollution.

Livestock production is an important component of agriculture and performs a very supportive economic role in the country. It is considered a sub-sector of Pakistan agriculture and contributes to about 56% of the value addition in agriculture and nearly 11% to the GDP. Hence, the market size is large enough to deal with the problem both quantitatively and qualitatively.

The main challenge for veterinarians and animal owners is the establishment of an accurate and timely diagnosis of ill and dying animals to implement intervention strategies to minimize and control of diseases. Despite such worries, technological advancements that make it possible for farmers to manage the health status of animals with fewer resources are a current source of research and development. Such technologies include biosensors, wearable technologies, and non-invasive approaches using artificial intelligence tools for disease surveillance. In this regard, IoT based sensors can play a key role in monitoring the health of 12 livestock that can sense and record the physiological parameters and biomarkers such as pulse, respiratory rate, body temperature, activity levels, and calories burned, and body posture.

More than half of Pakistan's population is dependent on livestock and agriculture as a source of income and livestock in itself contributes more than fifty percent to Pakistan's Agricultural Gross Domestic Product (AGDP). Furthermore, conventional farming methods are responsible for deforestation and soil erosion as well as producing greenhouse gases many times higher than all the worlds' automobiles and aircrafts combined so it is essential to root out inefficiencies in the sector so that farmers can increase their income stream and reduce unpredictability in animal health and yield.

The top consumed animal products include milk, meat, and dairy products. Their qualitative and quantitative productivity greatly depends upon the livestock health, physical activity, physiological parameters, food intake, environmental conditions, breed and many more. A constant checking on animal health is important as timely diagnosis of ill-animals can help us implement intervention strategies and prevent the spread of diseases in a herd by closely observing the well-being of livestock through automatic sensing and visual monitoring. With the limited resources available to the farmers, it is a substantial challenge for them to meet these ever-increasing dairy demands and at the same time deal with the far-reaching challenge of global warming.

So, we propose a sustainable idea of LiveIoT, a tech-based Smart-livestock system for managing and real-time monitoring. Today, there exists a dire need for an automated real time farm management system whereby the health status and yield of different animals can be monitored, so that prominent actions can be taken against problematic factors.

The objectives for which are defined as under:

- Help farmers to monitor the health of their animals.
- Early detection of any animal disease using predictive model and encourage them to take necessary advance actions for controlling the disease.
- Increase the dairy products production by carefully analyzing animal's health.

- Make the monitoring of the cattle farms easy by technological devices monitoring.

LITERATURE REVIEW

For this purpose, various papers were studied thoroughly to see how the system was implemented in past and how we can add upon it to give our innovation. Although not widely implemented commercially, much research has been done in this field which involve the use of sensors like pulse rate sensor, temperature and humidity sensor, blood pressure sensor etc. have been used for measurement of different factors affecting the health of cattle animals. The readings from these sensors are transmitted through a wireless module like Xbee in case there is an internet facility nearby. Otherwise, GSM and Bluetooth modules have been used for communication. With the advancement in technology and advent of machine learning and artificial intelligence, much more sophisticated techniques have been developed.

Now-a-days, cameras are being used for observing behavior of animals and collect different types of data from it. Multi sensor boards allow for the use of different sensors in concurrence. Currently, these boards exist in various types, each having specialized characteristics and distinct capacity of sensors. These include the boards like Turnkey Sensor Board (Ahmad, 2017), Thunder Board IoT development kit, ROHM sensor evaluation kit etc. where different sensors (Humidity, Temperature etc.) are mounted as SMD, providing plug and play capabilities. When powered the data is sent to their developed applications through Cloud.

However, on the customized implementation of our multi-sensor board, different sensors can be connected to the microcontroller board by the user. Our requirement was to monitor livestock health parameters; thus, relevant sensor functionalities exist on the board. The data collection through sensor board technology serves as the basis of our project. The sensor board is deployed for real-time monitoring of the livestock. The data collected on the app/server is utilized for the Machine learning enabled Livestock monitoring. Different behavioral patterns and physiological parameters including 16 pulse rate, cow body temperature, neck movements, and surrounding conditions are monitored to detect sick animals and welfare index.

The parameters are measured through two different technologies for comparison and a better predictive model for livestock health. The first technique of data collection is through the sensors where pulse rate sensor, temperature sensor and humidity sensor are used for the data collection. The second technique involves the image processing analysis on the collected image and videos containing animal behaviors, postures, motion and rest time and states to estimate animal welfare. (Costa, 2018)

Physiological parameters of animals are one of the significant factors that determine livestock health and contribute towards increase production. The parameters used for determining animal welfare include temperature, humidity, heart rate or pulse rate for normal monitoring of the animals. Lastly, observation of process of rumination also determines the animal's health and production. The environmental conditions also greatly influence the livestock especially the grazing conditions, resting time, and surroundings. Its body condition and the fat levels on its body can also analyze animal health status. Physical changes in body structure give detailed information about declining conditions. Thus, visual analysis of behavioral and physical parameters is equally important.

Use of IoT Based Sensor Unit to Detect Skin Temperature, Heart Rate and its Variability

In IoT based sensor innovation, temperature, and heartbeat rate estimated through wearable collar gadgets (Costa, 2018), (D.Aswini, 2017). The test gadget for estimating comprises of a sensor-based portable unit base mounted with respect to the creature understudy and a decent base unit. Remote information correspondence between the two was completed by the Xbee module. The microcontroller was utilized to deal with every one of the activities of the portable unit base including information obtaining, stockpiling, and transmission. The fixed unit spoke with the data set and HyperTerminal programming was utilized for ongoing information representation and investigation. For pulse, the test hardware readings were analyzed against the control ECG gear (Smart ECG convenient electrocardiograph model SE-1) estimations. In view of the aftereffects of the pulse esteems, there were no significant contrasts seen between the information securing gadgets.

For skin temperature, infra-red cameras and thermometer were utilized as the control hardware. It was seen that the temperature readings acquired from the test gadget were somewhat higher than the test readings. Contact sensors and air temperature were named as the primary obstructions. Two-units sent to gauge the temperature incorporate a fixed-base unit and a portable based unit. A fixed-base unit was utilized to gauge the surrounding temperature while the portable base unit was utilized to quantify the temperature and heartbeat pace of the creature.

Use of Machine Learning Models (ANN, RF, SVM) to Detect Lameness in Cattle

To identify the weakness potential in dairy cattle, the LP2 computational model best suits the necessities and prerequisites dependent on the outcomes (Liakos, 2017). This computational model records four elements: number of steps, strolling distance in a day (m), lying each day (min), eating each day (min). For forecast and preparing, six cases for both positive and negative cases were utilized in the preparation set and two each in the test set. The model was analyzed utilizing three distinctive AI techniques: Artificial Neural Network (ANN), Library for Support (SVM) Vector Machine, and Random Forest (RF).

Persuading results were acquired with basic representation strategies. Taking a gander at the case plots, we saw that healthy animals strolled a normal of around 3500m instead of the diseased animals whose normal was about 3000m. The measure of time spent lying in solid cows was generally under 750mins/day mark while that of the ailing animals was above 800mins/day. The eating minutes range for the unhealthy cows each day was 150 to 180 while that of the solid dairy cattle was 200-300 around.

The edge utilized for changing probabilities over to case recognizable proof was 0.5. A likelihood score was more noteworthy than 0.5 was marked as a positive experiment. For additional examination, the disarray lattice was analyzed which showed amazing outcomes. Every one of the three AI models impeccably separated between the sick and the solid creatures. The region under the bend (AUC) was 1 in the explicitness against the affectability bend. At the end of the day, there were no False Positive (FP) or False Negative (FN) occasions, showing that no case was misclassified.

Hence, every one of the three proficient ML calculations applied to the LP2 computational model exceptionally recognize positive and negative ailing examples. Indeed, even with a couple of preparing instances of just 12 samples, the model summed up well on the four test cows' models, and 100% exactness was accomplished. Another benefit of the model is that it depends on information gathered for four clear elements. The distance covered factor can be considered excess since it is utilized independently in the estimation of strolling distance (m) and the quantity of steps.

Use of IoT Based Sensor Unit to Monitor Rumination

The cow observing framework comprising of the Cow Device, a cloud framework, and an end-client application, presents an extraordinary IoT based domesticated animals checking framework (al., 2020). The Cow Device is appended to the collar of each creature. The information gathered from the sensors locally available the gadget is sent utilizing precise information transmission to guarantee expanded force productivity. To make up for the absence of unwavering quality of estimations, the qualities sent inside the payload of Bluetooth Low Energy (BLE) messages are refreshed every 10s. The BLE parcels are received by the Hub which are then forwarded to the server. The server speaks with the hub through the Internet and oversees information from the Cow Device. All the data is shown in an end-client application where the data of each cow can be followed.

A three organized engineering was utilized to recognize practices like rumination. Level '0' processes factual amounts to recognize practices. Level '1' predicts the current situation with the cow. Level '2' utilizes a reference model considering explicit qualities of the specific cow. The fundamental errand is to accomplish the most dependable forecast for the current cow conduct.

Image Analysis for Demeanor and Body Condition Score

Normal checking of the animals includes the perception of the creature's dietary patterns in case it is slacking its crowd or being less responsive. These strange tendencies might address indications of certain sicknesses.

Body condition scoring is a helpful method to screen the wellbeing and fat front of a cow. Notwithstanding the body size, it is a visual evaluation of the measure of fat that covers the cow's bones and isn't influenced by pregnancy. Straightforward 2D camera pictures from different points can be utilized to acknowledge slight or fat body sizes. Profound neural organizations can likewise be prepared to consequently recognize muscle versus fat without the requirement for a graph or master investigation.

The body conditioning score is specifically defined on a scale of 1-5, as shown here in Figure 1.

Figure 1. Cow body condition scoring (Howell, 2020)

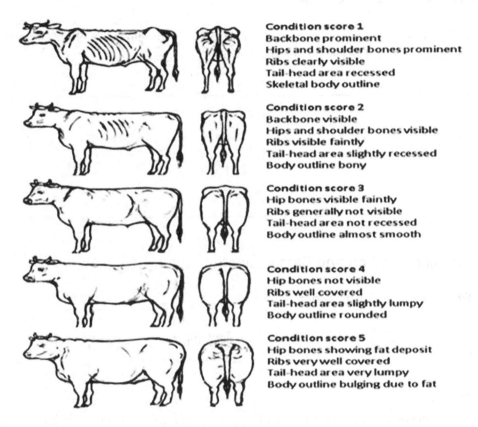

Condition score 1
Backbone prominent
Hips and shoulder bones prominent
Ribs clearly visible
Tail-head area recessed
Skeletal body outline

Condition score 2
Backbone visible
Hips and shoulder bones visible
Ribs visible faintly
Tail-head area slightly recessed
Body outline bony

Condition score 3
Hip bones visible faintly
Ribs generally not visible
Tail-head area not recessed
Body outline almost smooth

Condition score 4
Hip bones not visible
Ribs well covered
Tail-head area slightly lumpy
Body outline rounded

Condition score 5
Hip bones showing fat deposit
Ribs very well covered
Tail-head area very lumpy
Body outline bulging due to fat

Computer Vision and Deep Learning to Predict Cattle Movements and Current Positioning State

Machine vision examination has incredible potential in distinguishing the social qualities of creatures. Infra-red cameras are generally engaged with identifying temperature reactions. Pictures from basic camera video feed can be separated, prepared, and gone through a profound learning organization to identify social elements (Wurtz K, 2019). Important data like the position, number, and stance of creatures can be anticipated. Time for entering and leaving the rest region is likewise viewed as a significant element while examining creature wellbeing.

Following Figures 5 and 6 show the arrangement climate of the cow where various attributes were noticed, and its development was recorded (Yangyang Guo, 2020). Front and side view cameras recorded the cow's position and developments. Another strategy was worked to anticipate cow conduct that used foundation deduction and between outline distinction. The strategy was effectively ready to recognize the practices of entering the resting region (94.38%), leaving the resting region (92.86%), staying fixed (96.85%), pivoting (96.85), feeding (79.69%), and drinking (81.73%).

MAIN FOCUS OF THE CHAPTER

The main emphasis of the chapter is to give an overview of the development phases followed by the results and challenges faced. The key phases are shown below

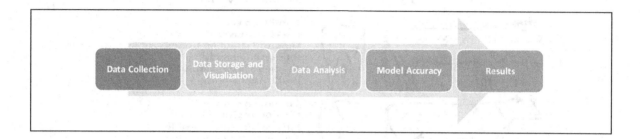

Hardware Implementation and Data Collection

Selected Physiological Parameters

The physiological parameters that are discussed here are temperature, humidity, heartbeat (pulse rate) and rumination.

Body temperature

The body's core temperature is a vital parameter that gives useful insights into the animals' health. The regular temperature of the cow is between 38-42°C. Fluctuations below the average temperature may lead to severe health deteriorations causing indigestion, milk fever, poisoning, weight shifting etc. And when it is above this range, it leads to influenza, anthrax or even cause death.

Humidity

Humidity is another crucial factor in cattle's health, especially in the determining of heat stress. The normal humidity range of animals is between 1-72, characterized as no stress. However, when the humidity is between 72-79% (mild stress) and severe stress when its value is above 80%. High humidity levels reduce the animal's ability to radiate body heat to the environment causing elevated heat stress that correlates to calcium loss/deficiencies.

Heart rate

BPM is known as Heart beats per minute and is a useful parameter for the evaluation of animal health. Average values range from 48 to 84 bpm for a healthy adult cow. Fluctuations in heart rate can be attributed to stress, aches, or multiple other diseases.

Rumination

Rumination is probably one of the most important and insightful parameters that veterinary specialists use to decipher the animal's health status. It refers to the movement of the neck and jaw muscles while chewing. Animals immediately abandon rumination when the feel uneasy because of a health issue. Adult cows can ruminate up to eight hours a day on average. Moreover, returning to normal practice of rumination after treatment is first indication of successful cure.

Hardware Components

In order to monitor vital parameters of the animals, selection of sensors is totally dependent upon the environmental condition and required accuracy. There are many sensors available in the market for monitoring said vitals with different price ranges and specifications. Keeping in view our required accuracy and harsh environment we choose the following components.

DHT11

DHT11 is a digital sensor used to measure temperature and humidity. The technology it uses is a capacitive humidity sensor for humidity measurements and a thermistor for surrounding air temperature. Comes with either a 4.7k or a 10k resistor to be used as pullup from the data pin to the VCC and has a minimum sample rate of 2 seconds. The purpose of this sensor is ambience measurements so that the relationships between them and animal activity levels can be determined and lead to the early diagnosis of heat stress.

Figure 2. DHT11 sensor

GY521 MPU6050

The MPU6050 contains both a 3-Axis Gyroscope and a 3-Axis accelerometer allowing measurements of both independently, but all based around the same axes, thus eliminating the problems of cross-axis errors when using separate devices. The use for this sensor is in detecting rumination of the animal, that is when the animal is chewing. Accelerometer and gyroscope readings are used to indicate the positional and angular accelerations, especially while chewing. Slower than average movement can be a precursor to a myriad of diseases.

Figure 3. GYU521 MPU6050 sensor

MLX90614

The Infrared Temperature Sensor is used to accurately measure animal body temperature without requiring physical contact with the animal. This is particularly useful since the kit is housed in a plastic encasing and contact would've compromised the waterproofing of the setup.

Figure 4. MLX90614 sensor

SEN 11574

The pulse sensor is used to measure the animal's heart rate. It is relatively low cost and comes at the expense of reduced accuracy. Furthermore, strict contact to the animal's neck is required for reliable readings.

Figure 5. SEN11574 pulse sensor

ATMEGA32

The microprocessor used for the animal collar is ATMega32. By executing powerful instructions in a single clock cycle, the device achieves throughputs approaching one MIPS per MHz, balancing power consumption and processing speed

Figure 6. ATMEGA32 microcontroller

ESP8266

ESP 8266 has been used as developing kit for this project. To program this we used C language and Arduino IDE. We preferred using C language because there are many open source libraries are available for the sensors we used.

Figure 7. ESP8266 NodeMCU

SX1272

The LoRaWAN module consists of the transceiver that is mounted on the animal collar and the receiver which is atop the mutlisensor board. This is particularly useful in areas where internet connectivity is not certain and a means for low cost communication between nodes is required over relatively large distances.

Figure 8. SX1272 LoRaWAN

Power Source

The power sourced used in the project is 18650mAH Lithium ion rechargeable battery. It was used particularly because of its useful portable and mounting applications. 18650mAH Lithium ion battery installed on the mobile node, powered the sensor and microcontroller assembly. After data collection, the charging module TP-4056 was used to charge the Battery through a 5V adapter.

Animal Collar (Mobile Unit)

Animal collar or the mobile node is actually the Sensor Node mounted on the cow neck. The mobile unit basically includes all the sensors on a single PCB, powered by the 18650 Lithium Ion source, and was housed in a collar-box. The collar box was wrapped around the Animal's neck with a leather animal belt and was, while making sure it fits the size but at the same does not choke the animal. The collar was rotated around the neck, randomly for getting an insight regarding the exact location at which the accurate data can be collected easily.

The images are displayed Figures 9 and 10:

Figure 9. Fabricated PCB for mobile node

Figure 10. Mobile node in plastic encasing

Data transmission

One of the challenges that was faced in the project was data collection and transmission since Wi-fi Modems are usually not available in the local dairy farms. Data transmission was supposed to be done through the Wi-fi using NodeMCU and stored and analyzed using cloud technology. However, despite the design solution the data collection on the cloud was impeded due to the indigent infrastructure of dairy farms, the data was collected locally in form of CSV file directly from the animal to local device storage (laptop) using mobile and fixed node.

The data sent from the mobile node to fix node using LoRa gateway was stored as a CSV file using Arduino COM serial port and saved concurrently in the Excel file.

Multi-Sensor Board (Fixed Unit)

The multi-sensor board is the fixed unit, which can receive the data from the multiple mobile nodes. The fixed node collects the data from the mobile node through WAN technology over a LoRa Transceiver Gateway.

The data collected from the fixed node is organized and send over to the cloud with the NodeMCU Wi-Fi Shield. In the testing farm, due to lack of resources and unavailability of Wi-Fi, the data was collected serially from fixed node and stored inside the laptop as a CSV file, later periodically uploaded inside the Power BI dashboard for visualization and analytics.

Figure 11. Fabricated PCB for fixed node

Figure 12. Fixed node in plastic encasing

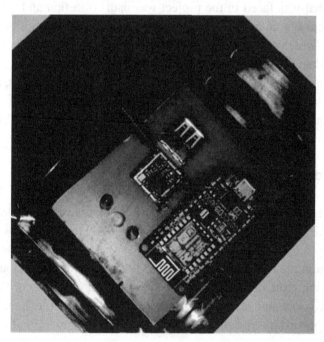

PCBs designing

Printed Circuit Boards are used for ensuring proper connectivity of electrical signals. The mobile and fixed circuits were first tested on the breadboard circuit, and then replicated on the PCBs.

PCBs design was implemented on EagleCAD and the PCBs were indigenously developed at the local lab facility. There were basically three PCBs that were developed with the following mentioned purposes.

- Fixed Node PCB for communicating with the mobile collar animal and collecting data
- Mobile Node PCB for measuring animal physiological parameters using sensors.
- The third and the final PCB was developed for charging the Mobile Node power source using the 5V power adapter.

Figure 13. Charging unit PCB

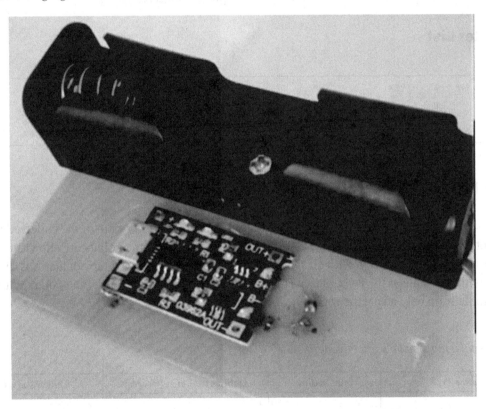

Data Storage and Visualization

The sensors were housed in the mobile node or the animal collar. The sensor data from the collar was transmitted to the fixed node through the LoRa transceiver module. Initially, we planned to use NodeMCU as the data uploading unit and Azure Cloud for the database. But, due to limited internet connectivity in the farm area, we had to resort to manual transferring the data from the fixed node to the laptop through a com port. Data was stored in the form of a relational database in a .csv file.

DATA ANALYSIS

Dataset

The collar device was tested on a buffalo in a farm near Bani Gala, Islamabad. As stated previously, the dataset is stored as a relational database in an excel file. On each day, data was collected for about 2 hours. The time interval between readings was set at 8 seconds. While such high frequency readings may seem too much for analyzing parameters like temperature and heart rate, this was vital for the storing neck movements through accelerometer and gyroscope readings. This amounts to about 900 readings for a single day. The total count of records collected was 5424, all belonging to the same animal.

The time interval for data collection for all days for at the peak of activity of the animal around mid-day. For some days, we manually labelled the activity parameter of the animal which had 4 features of walking, lying, eating, and standing. This would help in developing a supervised learning model for generalizing animal activity with the raw sensors data that we have collected.

Parameter List

Sensor Readings

Table 1 shows the accuracy of sensor readings by highlighting the outlier ratio of the collected readings and the total readings for the entire dataset. Only sensors mounted to the collars are shown. The statistics show minor sensor problem with all sensors except the pulse sensor where contact was difficult to maintain with the cattle at the right place. This required constant adjustment of the collar belt as pulse is measurable from only a specific region near the neck of the animal.

Table 1. Analyzing sensor's effectiveness

Sensor Type	Total Records	Outliers (Rejected Records)	Outlier Ratio (%)
Pulse	5724	1514	26.5
Temperature	5724	72	1.2
Accelerometer	17172	291	1.7
Gyroscope	17172	120	0.7

Analytical tools

1)   Microsoft Excel

2)   Python Libraries
3)   Power BI
4)   Azure ML Studio

Visualization

For statistical visualization, opensource python plotting libraries like Seaborn and Matplotlib were used. We also used inbuilt data visualization functionalities of Power BI. The parameter readings were plotted with respect to time, and it helped us picture how the animal functioned through a certain time period. Visualization results are discussed in detail in section the next section.

Machine learning analysis

From the above-mentioned raw features, we developed a supervised machine learning model which helped us in predicting and generalizing animal activity. The activity features used were walking, lying, standing, and eating. These labels were manually monitored and jotted down for the time period of data collection. To tackle this problem of multi-label classification from the raw features, we used different classification algorithms like logistic regression, decision tree classifiers and random forest. ML techniques were leverage through scikit-learn framework. Early results were not highly accurate due to limited data and imbalanced classes, but showed encouraging signs. With the collection of more data, the classification accuracy can be improved. These activity labels can be used as processed features to develop a lameness disease detection model as described in [4].

Model accuracy

Sensor limitations

The sensors accuracy is mentioned above where they are discussed in detail. In real-time data collection, we encountered routine problems with sensors due to which they displayed abrupt values. These were dealt with by removing the noisy readings altogether from the data. Since, data is collected every 8 seconds, discarding less than 5% of outlier data does not affect the bigger picture. The trickiest part was adjusting the belt so that the pulse sensor does not lose contact with the animal body, and therefore result in loss of heart rate data.

Machine learning models

The best classification model for predicting animal activity was developed through Decision Tree Classifiers. Data imbalance was handled by techniques like data replication. The model showed encouraging results with around 70% validation set accuracy. Thus, by leveraging more data both in quantity and diversity, we can hope to improve model validity.

Results

Visualizing parameters

The following plots for all parameters for a certain time period are shown below:

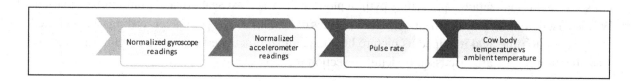

The gyroscope and accelerometer outputs are normalized as shown in Figures 14 and 15. For the gyroscope, readings are in the range of -4 to +4 while for the accelerometer they are in the range of +2 to -2. It was also observed that these readings were high at the time of peak activity like eating or walking. However, strict conclusions can only be drawn with greater amounts of data. From the above graphs, we can also conclude the sensor malfunction was very less as the readings deviated from the optimal range only a few times.

Figure 14. Gyroscope readings (Day 3)

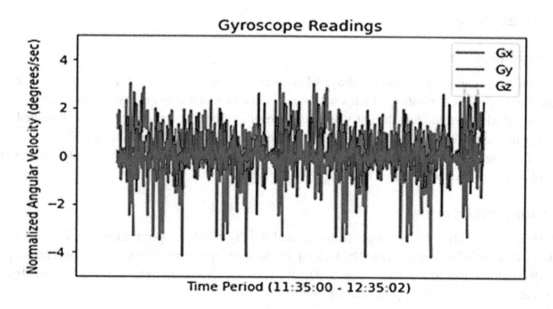

Figure 15. Accelerometer readings (Day 3)

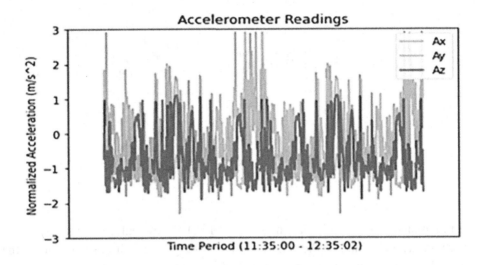

Figure 16. Cow body temperature vs environment temperature (Day 3)

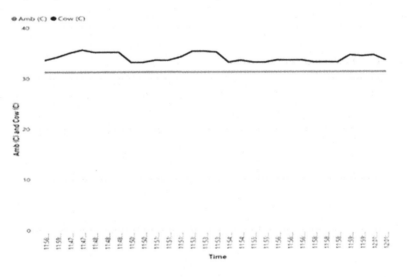

In Figure 16, we can see how the cow body temperature varied with respect to the surrounding temperature. Optimal body temperature for cows varies between (36-40) °C, depending on the type of species.

Figure 17. Pulse rate (Day 3)

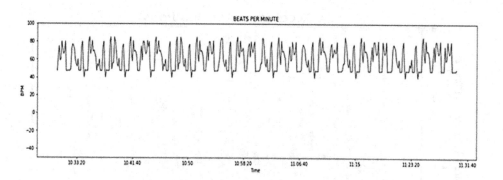

The heart rate monitoring graph is shown in Figure 17. The cow pulse rate usually varies from 48 to 84 beats per minute, depending on the activity and physiological condition. We can see the constant varying of pulse through the given time period within the optimal range. In the above graph, the outliers which occur the BPM Signal from the pulse sensor exceeds 500 were discarded.

Machine Learning Based Analysis

The labeled tabular dataset is displayed below:

Figure 18. Dataset stored as pandas data frame

	Time	BPM Signal	BPM	Amb (C)	Cow (C)	Gx	Gy	Gz	Ax	Ay	Az	DAY	Activity
1808	11:35:31	539.0	0.0	30.14	35.02	1.698125	0.190179	-0.296429	0.223326	-2.184057	-0.729682	3.0	Walking
1809	11:35:39	542.0	0.0	30.25	34.01	1.519062	0.108036	-0.056429	0.069643	-2.157857	0.012461	3.0	Walking
1810	11:35:47	545.0	0.0	30.29	35.21	1.108125	0.189464	0.066786	0.361362	-1.700759	NaN	3.0	Walking
1811	11:35:55	NaN	0.0	30.31	34.9	1.292812	0.010000	-0.057857	0.890324	-1.385664	0.952394	3.0	Walking
1812	11:36:03	539.0	0.0	30.24	35.17	1.894688	-0.057321	-0.302679	0.455446	-1.604219	-1.295725	3.0	Walking
...	...	...	...	...	...	...	...	...	...	...	...	...	...
2706	12:34:59	NaN	0.0	30.41	NaN	-1.361429	0.011607	0.299286	2.252946	-1.472137	-1.503633	3.0	Lying
2707	12:35:07	529.0	49.0	30.33	36.17	-0.061071	-0.346071	0.064464	2.908616	-1.579068	-1.175809	3.0	Lying
2708	12:35:15	528.0	72.0	30.12	55.21	0.036429	0.894286	-0.122500	2.225011	-1.886390	-0.494280	3.0	Lying
2709	12:35:23	527.0	77.0	NaN	76.21	0.316429	1.207321	0.517857	1.385045	-1.956099	-0.844749	3.0	Lying
2710	12:35:31	526.0	0.0	NaN	33.61	0.388929	2.284107	1.097500	1.056339	-1.493108	-1.694347	3.0	Lying

The .csv file was read into a Pandas data frame and set for preprocessing before machine learning techniques can be applied. From the above diagram, we can see a few missing values of BPM as well as accelerometer readings.

Predicting animal activity

Features

The raw sensors data formed the features which included the variables highlighted in Figure 18. Their correspondences and description is provided below:

BPM Signal: Amplitude value from the pulse sensor

BPM: Heart rate in beats per minute

Amb (C): Surrounding temperature in C

Cow (C): Cow Body temperature in C

Ax: Normalized Acceleration in the x-direction

Ay: Normalized Acceleration in the y-direction

Ay: Normalized Acceleration in the y-direction

Gx: Normalized Angular Velocity in the x-direction

Gy: Normalized Angular Velocity in the y-direction

Gz: Normalized Angular Velocity in the z-direction

Labels

The labels constituted the animal activity which was monitored manually and tabulated with the sensor collected data. They were divided into 4 categories: walking, eating (and drinking), standing, lying. The data was collected around midday around which all activities were performed by the animal, from walking to resting in a shed to avoid the scorching heat. There was a class imbalance as greater time was spent lying. To avoid this from effecting our classifier, we used repetitive data and weights correction techniques.

Results

The best results so far were achieved using the random forest classifier algorithm. The collected data was cleaned by neglecting the missing values and outliers. In the given instance, we had the liberty to discard the whole row as the reading record is taken and updated every 8 seconds. The data cleaning was followed by a Label Binarize which assigned binary codes to the activity labels i.e. walking, standing, lying, and eating (if needed). Different classifiers were used to develop and train the best possible model. All algorithms were tested with multiple train-test split thresholds and the best results was selected considering the bias and variance evaluation metrics.

Decision trees

The results shows a simple model without high variance when it comes to developing a classification criterion. Results can be further improved by more data which can be collected using multiple collars. With different animals, there is a better chance to generalize and develop a forecast model that achieves higher accuracies. The training and test accuracies are shown in Table 2.

Random forest

The results show less overfitting on the test set as compared to the train set. However, both the test and train accuracy are closer to 80% as shown in Table 2. Further improvements can be made with more data collection as highlighted in the above analysis as well.

Naïve Bayes

The algorithm showed comparable results with the random forest classifier. Both the random forest and naïve Bayes suit best for the purpose of this classification problem and should be used with further samples for better accuracy and desirable results. The training and test accuracies are shown in Table 2.

Table 2. Test and training accuracy of different classifiers

Classifier	Train Set Accuracy	Test Set Accuracy
Decision Tree Classifier	0.6716%	0.6548%
Random Forest Classifier	0.7890%	0.7864%
Naïve Bayes Classifier	0.7850%	0.7850%

Accuracy metrics

The confusion matrix graphs for the best model predictions are displayed, which show the true positive, false positives, true negatives, and false negatives for the predictions. The precision, recall, and F1 score for each activity is separately calculated and shown Figure 19.

Figure 19. RF classifier confusion matrix plots

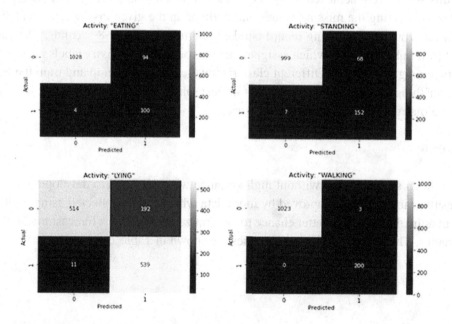

Table 3. Evaluation metric

Activity	Precision	Recall	F1
EATING	0.52	0.96	0.68
LYING	0.73	0.98	0.87
STANDING	0.70	0.96	0.82
WALKING	0.98	1.00	0.99

We can clearly see that the activity of walking was easier to predict, partly because of the high bpm rate and the increase in amplitude of neck movements, captured by the gyroscope and accelerometer. Eating was the most difficult activity to predict because this activity had the least significant records in the dataset. The evaluation metrics are tabulated in Table 3.

Comments

With our ML model to predict and classify activity, we can calculate time spent in a certain position for all the days. This is where we can further our research by contacting domain experts for gathering lameness data and developing a supervised computational model (as shown in Figure 20) for detecting lameness in cattle. The LP2 computational model as described in [4] is based on the features shown in the diagram below. The model achieves accuracy closer to a 100% when it comes to detecting lameness.

Figure 20. LP2 computational model for detecting lameness in cattle

Computational Model LP2			
Feature 1	Steps per day		
Feature 2	Walking per day (m)		
Feature 3	Lying per day (min)		
Feature 4	Eating per day (min)		

State of cattle	Steps per day (dimensionless)	Walking per day (m)	Lying per day (min)	Eating per day (min)
Healthy	2900	3700	660	178
With problem in hooves	600	2350	830	168

LIMITATIONS AND CHALLENGES

There were certain limitations that hindered the flow of the project. First and foremost being the lockdown imposed due to the COVID-19 pandemic that restricted access to labs at key phases of hardware development and stalled progress.

The next problem encountered was the search of a private farm that would allow the team to the test the hardware kit on a live animal. After repeated rejections, a local farm in Bani Gala, Islamabad, agreed to provide an animal for product testing. Upon the insistence of the farm workers to not take responsibility of the hardware kit of extended durations of time, the team was forced to schedule multiple visits for hours at a stretch to the venue for data collection and take the kit back with them.

There was also an issue encountered from the pulse sensor that required firm contact with the animal's neck to be able to provide accurate data, so the original design of the collar had to be changed to accommodate for this obstacle. Furthermore, the obtrusion of the pulse sensor out of the plastic encasing of the animal collar compromised the water proofing of the hardware kit and ensured that in any case the kit was not to be left behind to care of any potential damages caused by rainstorms and erratic movements of the animal that could damage the kit.

The original hardware design incorporated a front and back IP camera to be used for visual analysis of the animal's movement patterns and behaviors to detect for a condition called lameness that correlates reduced activity levels with a lower milk yield. However, the farm the team was provided access to was had no firm vantage point on which the camera could be installed. Also, the animal had to be taken inside routinely away from the sweltering heat, so sight of the animal would be lost.

FUTURE RECOMMENDATIONS

The future work for the projects entails the usage of better Machine/Deep Learning algorithms and collecting larger amounts of data both on the same animal as well as animals of varying health statuses for better accuracy, redesigning the mobile and fixed nodes to include the accurate detection of more health parameters and making the sensors more cost effective. Furthermore, after sufficient data is collected AI at the Edge can be deployed to significantly reduce cloud overhead costs. It can be particularly useful for dairy organizations seeking to maximize yield per animal and government organizations seeking extensive data on the livestock sector for better policy making and management.

Another significant aspect of future development is the incorporation of a front and back IP camera which can be installed at a point high enough at the farm and could visually monitor the animals' movement activity and behavior to detect for lameness and accurately predict animal yield. Similar technology may be used for other entirely different purposes such as a pet monitoring application or a crop/land monitoring application.

CONCLUSION

With the development of the proposed system and the data that will be generated from it will reshape the livestock monitoring by providing an indigenous and intelligent solution that will support the farmers and the agricultural industry to enhance animal productivity in the future. The proposed system will provide useful information regarding the health and physiology of an animal and will contribute to alleviating the harmful effects of infectious outbreaks by providing cost-effective diagnostic methods. With our solution, we can generalize animal activity based on raw data from sensors with high accuracies. This will help us in detection of diseases like lameness in cattle. In the long run, the collected data can also be correlated with the milk productivity of the animal and the reasons for declining yield can be answered.

REFERENCES

Aswini, D., Santhya, S., Nandheni, T. S., & Sukirthini, N. (2017). Cattle health and environment monitoring system. *International Research Journal of Engineering and Technology*, *4*(3), 1899–1903.

Costa, D., Turco, S., Ramos, R., Silva, F., & Freire, M. (2018). Electronic monitoring system for measuring heart rate and skin temperature in small ruminants. *Engenharia Agrícola*, *38*(2), 166–172. doi:10.1590/1809-4430-eng.agric.v38n2p166-172/2018

Guo, Y., He, D., & Chai, L. (2020). A Machine Vision-Based Method for Monitoring Scene-Interactive Behaviors of Dairy Calf. *Animals (Basel)*, *10*(2), 190. doi:10.3390/ani10020190 PMID:31978962

Howell, A. (2011). *Snail-borne diseases in bovids at high and low altitude in Eastern Uganda: Integrated parasitological and malacological mapping*. Liverpool School of Tropical Medicine.

Liakos, K., Moustakidis, S. P., Tsiotra, G., Bartzanas, T., Bochtis, D., & Parisses, C. (2017). Machine Learning Based Computational Analysis Method for Cattle Lameness Prediction. *HAICTA*, *128*, 139.

Majeed, A. (2017, Jan. 20). *Multi-sensor board*. https://www.electronicproducts.com/multi-sensor-board-speeds-up-wearable-iot-designs/#

Unold, O., Nikodem, M., Piasecki, M., Szyc, K., Maciejewski, H., Bawiec, M., Dobrowolski, P., & Zdunek, M. (2020). IoT-Based Cow Health Monitoring System. *Computational Science – ICCS 2020: 20th International Conference, Amsterdam, The Netherlands, June 3–5, 2020. Proceedings*, *5*(12141), 344–356. doi:10.1007/978-3-030-50426-7_26

Wurtz, K., Camerlink, I., D'Eath, R. B., Fernández, A. P., Norton, T., Steibel, J., & Siegford, J. (2019). Recording behaviour of indoor-housed farm animals automatically using machine vision technology: A systematic review. *PLoS One*, *14*(12), e0226669. doi:10.1371/journal.pone.0226669 PMID:31869364

Chapter 4
How Artificial Intelligence Can Enhance Predictive Maintenance in Smart Factories

María A. Pérez-Juárez
University of Valladolid, Spain

Miguel Alonso-Felipe
University of Valladolid, Spain

Javier M. Aguiar-Pérez
University of Valladolid, Spain

Saúl Rozada-Raneros
University of Valladolid, Spain

Javier Del-Pozo-Velázquez
University of Valladolid, Spain

Mikel Barrio-Conde
University of Valladolid, Spain

ABSTRACT

The Fourth Industrial Revolution, under the name of Industry 4.0, focuses on obtaining and using data to facilitate decision-making and thus achieve a competitive advantage. Industry 4.0 is about smart factories. For this, a series of technologies have emerged that communicate the physical and the virtual world, including Internet of Things, Big Data, and Artificial Intelligence. These technologies can be applied in many areas of the industry such as production, manufacturing, quality, logistics, maintenance, or security to improve the optimization of the production capacity or the control and monitoring of the production process. An important area of application is maintenance. Predictive maintenance is focused on monitoring the performance and condition of equipment during normal operation to reduce the likelihood of failures with the help of data-driven techniques. This chapter aims to explore the possibilities of using artificial intelligence to optimize the maintenance of the machinery and equipment components so that product costs are reduced.

DOI: 10.4018/978-1-7998-9201-4.ch004

INTRODUCTION

The Fourth Industrial Revolution, under the name of Industry 4.0, focuses on obtaining and using data to facilitate decision-making and thus achieve a competitive advantage. Industry 4.0 is about building connected and autonomous factories, that is, Smart Factories. Smart Factories and Industry 4.0 have attracted the interest of many researchers and practitioners in recent years (Antunes et al, 2018; Jardim-Goncalves et al, 2016; Jerman & Dominici, 2018; Liao et al, 2017; Osterrieder et al, 2020; Sony & Naik, 2019; Strozzi et al, 2017).

To implement Industry 4.0 and Smart Factories, a series of technologies have emerged that communicate the physical and the virtual world. There are a series of technologies that are being key in the digital transformation that the industry is experiencing. These technologies include the so-called Internet of Things (IoT), Big Data and Artificial Intelligence, Robotics or Augmented and Virtual Reality.

These technologies can be applied in many areas of the industry such as production, manufacturing, quality, logistics, maintenance or security to improve the optimization of the production capacity or the control and monitoring of the production process. An important area of application is maintenance. Predictive Maintenance is focused on monitoring the performance and condition of equipment during normal operation to reduce the likelihood of failures with the help of data-driven techniques. Lee et al. (2019) highlight that the emergence of Industry 4.0 and smart systems is leading to increasing attention to Predictive Maintenance strategies that can decrease the cost of downtime and increase the availability (utilization rate) of manufacturing equipment. Predictive Maintenance also has the potential to foster sustainable practices in manufacturing by maximizing the useful lives of components.

In this chapter the authors aim to explore the possibilities of using Artificial Intelligence to optimize the maintenance of the machinery and equipment components so that this maintenance is increasingly predictive and consequently operation and maintenance costs are reduced.

SMART FACTORIES ARE THE FACTORIES OF THE FUTURE

The so-called Industry 4.0 is here to stay. More and more organizations are deciding to turn their manufacturing centres into Smart Factories. The so-called Smart Factory presents multiple advantages that are essential to promote the survival and growth of the industry in turbulent times, marked by rapid social, economic and technological transformation.

In a Smart Factory, both the facilities and the processes of the factory have been digitized, connected and automated, through various technologies that include the Internet of Things, Robotics, Cloud Computing, Big Data or Artificial Intelligence, among others.

Through the use of these technologies, it is possible to integrate smart machines capable of executing repetitive actions that were previously carried out by humans within the factory. These machines are also integrated into a network system that connects all departments and processes. Thus, the Smart Factory is also characterized by a constant flow of information updated in real time, allowing greater control and optimized decision-making.

The objective of Smart Factories is to increase productivity, reduce physical effort, improve quality and monitor processes through the collection of data in real time.

As can be expected, the advantages of the Smart Factory are numerous, among which the following stand out:

- Quality: Through intelligent processes, the risk of human error is reduced. Machines execute repetitive tasks with precision and accuracy, regardless of how many hours they have been running.
- Efficiency: By automating processes, plants can increase their production rates.
- Flexibility: Smart Factories allow modular and flexible production to be carried out quickly and with satisfactory results.
- Control: In Smart Factories all departments are connected, which results in efficient and real-time information flows. Processes are monitored at all times and deviations can be detected and corrected effectively and efficiently.
- Savings: Smart Factories allow reducing costs in different areas such as production and exploitation.
- Profitability: By increasing the quality of production and reducing production costs, Smart Factories allows companies' operating margins to increase.

Another important aspect is that Smart Factories are committed to renewable energy. If in the 18th and 19th centuries the factories were located near rivers or near areas for obtaining fossil fuels, Smart Factories for their part seek to locate in areas where there are renewable energy sources. In addition, it seems to be in the minds of technologists and industry visionaries to go further and implement totally new concepts in the infrastructural field, so that in the future we will be able to see different ways of understanding a factory far removed from what we have today in our minds.

The development of Smart Factories is a point of no return for the industrial sector, since, although it initially involves a great effort and investment, its implementation brings succulent benefits. Production becomes more efficient, with a significant reduction in costs and greater flexibility, better integrating all links in the chain, from suppliers to customers.

TECHNOLOGY IS THE PILLAR OF SMART FACTORIES

To implement Industry 4.0 and Smart Factories, there are a series of technologies that are turning out to be key. These technologies include the so-called Internet of Things which refers to a network of interconnected objects and devices that can receive and transmit data through the Internet. Ali et al. (2015) highlight that Internet of Things seeks to implement the connectivity concept of anything from anywhere at any time. All the data generated is usually stored in a central server hosted in the cloud that the operators can consult at any time. Internet of Things is an essential driver for data-driven optimization and automation in all industrial sectors and is therefore considered an essential technology for the implementation of Industry 4.0.

Along with the Internet of Things, it is worth mentioning Big Data, which is the set of data whose size (volume), complexity (variability) and speed of growth (speed) make it difficult to capture, manage, process or analyse, using conventional technologies and tools, within the time necessary for them to be useful. Data is the foundation that Artificial Intelligence needs in order to reach its full potential. Currently, the volume of such data is of such magnitude that it is necessary to have analysis tools that can separate the most useful data, in each context, and at each moment. The unstoppable flow of data or Big Data is today a crucial asset for any organization, and in the specific case of the industry, the data and its subsequent analysis are essential to optimize products and their production processes.

Big Data and Artificial Intelligence are together giving Industry 4.0 a big boost. Artificial Intelligence is the ability of a computer to present the same capabilities as a human at the level of information processing, learning and decision making. Its objective is to tackle complex problems by imitating human logic and reasoning through algorithms. For its part, Machine Learning refers to the study of computer algorithms that improve automatically through experience. Machine Learning is seen as a subfield of Artificial Intelligence. And finally, Deep Learning is a subfield of Machine Learning where neural networks that function much like the biological neural connections in our brains are used.

An important role is also played by industrial Robotics, which implies the presence of sensorized robots, capable of analysing the environment and adapting to it to execute programmed tasks automatically.

Another important actor is Computer Vision, which focuses on the methods necessary to acquire, process, analyse and understand images from the real world in order to produce numerical or symbolic information that can be processed by a computer. For its part, Augmented Reality allows to improve experiences by adding virtual components such as digital images, graphics or sensations as a new layer of interaction with the real world. In contrast, Virtual Reality creates its own world, which is completely computer-generated and driven. Finally, Mixed Reality, also known as hybrid reality, combines, in real time, virtual reality, augmented reality and physical reality, thus multiplying its possibilities. In this way, new spaces can be created where both real and virtual objects / people interact.

To end up with, it is worth mentioning the so-called Cloud Computing, which is the technology that has made the software independent of the hardware, and which is also being of great help. Kumar Paul and Ghose (2012) define Cloud Computing as a model for enabling convenient, limitless, on demand network access to a shared pool of computing resource. Through this technology it is possible to handle certain software, hosted on a server, from different devices and with total security.

THE IMPACT OF TECHNOLOGY IN FACTORIES

The result of jointly using these technologies opens up numerous possibilities in the industrial field and allows Smart Factories to be a reality. Artificial Intelligence can be applied in many areas of the industry such as production, manufacturing, quality, logistics, maintenance or security. Some concrete examples would be, for example, the optimization of production capacity, the control and monitoring of the production process, the optimization of consumption and resources, the improvement of the quality of both the product and the customer service, supply chain optimization, inventory and pricing, or enhanced maintenance.

In the field of production, manufacturing and quality, it is worth highlighting the complexity of production systems, and the highly increasing demand for personalized products that make it necessary to introduce flexible solutions that can face problems in a changing environment led by uncertainty. In the field of logistics, the supply chain generates a huge amount of structured and unstructured data every day that can only be analysed and managed thanks to data-driven techniques. Logistics is based on physical and digital networks that cannot be optimized by humans due to their high complexity. Therefore, at this specific point, the objective of data-driven techniques is to transform reactive behaviours into proactive ones, manual ones into automatic ones, and standardized ones into personalized ones.

In fact, according to Coleman et al. (2017), the benefits of digital transformation in factories far outweigh the risks. These benefits include, among others, the following:

- Material cost savings (5-10 percent).
- Reduced inventory carrying costs.

- Less time spent on brute-force information extraction and validation.
- More time spent on data-driven problem solving.
- More confidence in data and information leading to ownership of decisions.
- Increased equipment uptime and availability (10-20 percent).
- Reduced maintenance planning time (20-50 percent).
- Reduced overall maintenance costs (5-10 percent).

IMPROVING INDUSTRIAL MAINTENANCE: FROM PREVENTIVE AND CORRECTIVE MAINTENANCE TO PREDICTIVE MAINTENANCE

Industrial maintenance encompasses all those actions aimed at extending the useful life and proper functioning of the machinery and equipment involved in the production process. Consequently, industrial maintenance generates significant benefits for organizations, including the following:

- Prevents and avoids work accidents thus increasing safety for the operators involved in the production process.
- Avoids or reduces losses due to production stoppages.
- Allows to have a documentation and follow-up of the necessary maintenance for each equipment.
- Prevents irreparable damage to industrial facilities.
- Increases the useful life of the equipment.
- Keeps equipment and machinery in good condition.
- Reduces costs.
- Improves the quality of industrial activity.

There are different approaches to carry out maintenance activities to extend the useful life of equipment and machinery including the following:

- Preventive maintenance: It is maintenance focused on the prevention of failures in equipment and facilities with the aim of reducing risks. It tries to reduce errors or breakdowns with a constant and planned review according to the needs of each industry.
- Corrective maintenance: It is that maintenance whose purpose is to correct any defect that appears in the equipment or machinery. It corrects actions only when the fault has been detected.
- Predictive maintenance: Like preventive maintenance, Predictive Maintenance seeks to get ahead of failure. The difference is that it is based on the application of tools or techniques for detecting different variables that are an indication of the state of an equipment and that anticipate a future failure such as vibration, pressure or temperature.

Dilmegani (2021) explains that in the case of preventive maintenance (planned or scheduled maintenance) the maintenance schedule is fixed, and maintenance is completed at regular intervals. Complete control and maintenance of all machine components is conducted over time, but not all machine components may be checked with the same frequency though.

According to Sipos et al. (2014), scheduled maintenance is labour-intensive and ineffective in identifying problems that develop between technician's visits.

Lee et al. (2019) highlight that scheduled maintenance replaces components frequently to avoid unexpected equipment stoppages, which increases the time associated with machine non-operation and maintenance cost.

Predictive Maintenance is a better approach because it allows the organization to prevent problems without incurring the cost of unnecessary frequent maintenance. So, over the lifetime of a manufacturing facility, some components may never be checked if they are not predicted to cause problems.

However, some authors like Mobley (2001) holds that Predictive Maintenance is not a substitute for the more traditional maintenance management methods, but that it is, however, a valuable addition to a comprehensive maintenance program. Where traditional maintenance management programs rely on routine servicing of all machinery and fast response to unexpected failures, a Predictive Maintenance program schedules specific maintenance tasks, as they are actually required by plant equipment.

In fact, companies should try to optimize the maintenance function in order to achieve the highest levels of availability and reliability at the lowest possible cost by combining corrective, preventive and predictive strategies. The correct distribution of maintenance tasks (predictive, preventive and corrective) is essential, although the current trend is a progressive migration towards Predictive Maintenance.

For Industry 4.0 to become a reality, proper industrial maintenance is essential. In fact, according to Chiu et al. (2017), Predictive Maintenance has been featured as a key theme of Industry 4.0. Predictive Maintenance is based on the early detection of failures by identifying failure patterns. The elimination of unforeseen failures is pursued so that the availability and reliability of assets can be increased. Its philosophy is to intervene in machines only when necessary. In this way, dismantling of the machine that does not provide greater reliability is avoided, since a good part of the failures can occur at any time, with which preventive maintenance at a fixed interval begins to be quite questionable.

Predictive Maintenance provides important advantages, including the following:

- Equipment availability improvement.
- Fewer losses of raw materials caused by emergency shutdowns when some equipment fails.
- Reduction of the number of system repair interventions.
- Reduction of the cost of buying spare parts.
- Accidents are reduced and plant safety increases.
- Greater global reliability.

The experience performed by He et al. (2017) showed that Predictive Maintenance can achieve approximately 26.02% cost improvement over periodic preventive maintenance.

The results of an IBM study showed that companies using Predictive Maintenance tools have a tenfold return on investment of 20-25% reduced maintenance costs, 70-75% failure reduction, 35-45% downtime reduction and increased production by 20-25% compared to companies with traditional approach to maintenance (Negandhi et al, 2015).

There are different indicators that can be used to diagnose the operating status of the equipment and machinery and that are very useful in Predictive Maintenance. Some of these indicators are the following:

- Visual inspection: The machines and equipment are observed paying special attention to the search for cracks, fissures or noises.
- Vibration analysis: This technique allows to analyse problems of unbalance and eccentricity, damage to impellers and pumps or misalignment of shafts and pulleys, among others.

- Ultrasound: This technique allows to analyse evacuation systems in equipment, humidity, environments and their isolation, among others.
- Thermography: They are made using thermal imaging cameras, the price of which has decreased significantly in recent years and, thanks to them, it is possible to detect hot spots.
- Temperature measurement: Alarms are implemented that warn the operators of an inappropriate excess of temperature in certain equipment.
- Pressure control: The pressure is evaluated to detect hydraulic failures due to leaks, obstructions in filters or closures, among others.
- Shock impulses: This technique is used to carry out the control of the state of the bearings. The collision speed between the rolling elements and the raceways is measured.
- Penetration liquids: This technique is used to search for cracks.
- Magnetic elements: This technique also seeks to detect cracks. The material to be analysed is magnetized.
- X-rays: They are used to detect defects in the interior part of the equipment. Special attention is paid to the part in which the pieces and their welds are joined.

To be able to apply any of these Predictive Maintenance techniques, it is necessary to have the equipment and machinery monitored and to have a data history that allows analysing the operation.

TECHNOLOGY CAN ENHANCE PREDICTIVE MAINTENANCE

Predictive Maintenance of all types of machinery and equipment components is increasingly important, since it allows optimizing the use of said machinery and equipment, as well as minimizing the negative impact it would have on business activity, if any machinery or equipment would be rendered useless due to not having received maintenance suitable. The application of Big Data and Artificial Intelligence techniques to implement Predictive Maintenance has attracted the interest of many researchers and practitioners (Amruthnath & Gupta, 2018; Bousdekis et al, 2017; Bousdekis et al, 2019; Poór & Basl, 2019; Selcuk, 2017). In fact, many experts believe that Big Data and Artificial Intelligence will be essential for industrial organizations to compete in today's market. Among other reasons, because Big Data and Artificial Intelligence allow fast and accurate decision making, even in the field of maintenance.

Companies usually have a large database with the history of maintenance operations carried out on different components. However, for the analysis of such a broad set of data to be effective, a comprehensive process that includes the use of data-driven techniques is required. The objective is to find patterns related to maintenance work already carried out, with the aim of predicting the maintenance work that will need to be carried out in the future.

Moreover, in recent years, the cheaper sensors and their smaller size have facilitated the obtaining of valuable information on the state of the machines. Specifically, by measuring different points and characteristics of a machine, it is possible to have an almost real-time view of its status. And thanks to the use of Big Data and Artificial Intelligence techniques it is possible to create models from this data in order to detect possible anomalies before they occur. This Predictive Maintenance is one of the most reliable ways to prevent equipment and machinery from failing and damaging the production process.

It is true that the concept of Predictive Maintenance is not new, but, without any doubt, Big Data and Artificial Intelligence have contributed to creating a new perspective on it. The possibility of using data from numerous sources processed with increasingly complex algorithms is allowing to significantly reduce maintenance costs.

Lee et al. (2019) highlight that often, manufacturing equipment is utilized without a planned maintenance approach and that such a strategy frequently results in unplanned downtime, owing to unexpected failures, and propose the use of Artificial Intelligence based algorithms to implement Predictive Maintenance.

Ayvaz & Alpay (2021) performed a study in which a data-driven Predictive Maintenance system was developed for production lines in manufacturing. By utilizing the data generated from Internet of Things sensors in real-time, the system was able to detect signals for potential failures before they occur with the help of Machine Learning methods. Consequently, operators were notified earlier, so that preventive actions can be taken prior to a production stop. The evaluation results indicated that the Predictive Maintenance system was successful in identifying the indicators of potential failures and that it could help prevent some production stops from happening.

Eren (2017) focused on bearing faults that are the biggest single source of motor failures. This author explains that Artificial Neural Networks (ANNs) and other decision support systems are widely used for early detection of bearing faults. The typical decision support systems require feature extraction and classification as two distinct phases. Extracting fixed features each time may require a significant computational cost preventing their use in real-time applications. Furthermore, the selected features for the classification phase may not represent the most optimal choice. Instead, this author proposed the use of 1D Convolutional Neural Networks (CNNs) for a fast and accurate bearing fault detection system.

Janssens et al. (2017) focused on vibration analysis which is a well-established technique for condition monitoring of rotating machines as the vibration patterns differ depending on the fault or machine condition. Different manually-engineered features, such as the ball pass frequencies of the raceway, kurtosis an crest, are used for automatic fault detection. Unfortunately, engineering and interpreting such features requires a significant level of human expertise. To enable non-experts in vibration analysis to perform condition monitoring, the overhead of feature engineering for specific faults needs to be reduced as much as possible. Therefore, these authors proposed a feature learning model for condition monitoring based on Convolutional Neural Networks. The goal of this approach was to autonomously learn useful features for bearing fault detection from the data itself. Furthermore, these authors compared the feature-learning based approach to a feature-engineering based approach using the same data to objectively quantify their performance. The results indicated that the feature-learning system, based on Convolutional Neural Networks, significantly outperformed the classical feature-engineering based approach which used manually engineered features and a random forest classifier. The former achieved an accuracy of 93.61 percent and the latter an accuracy of 87.25 percent.

Susto et al. (2015) proposed a multiple classifier Machine Learning methodology for Predictive Maintenance. The proposed Predictive Maintenance methodology allowed dynamical decision rules to be adopted for maintenance management, and could be used with high-dimensional and censored data problems. This was achieved by training multiple classification modules with different prediction horizons to provide different performance tradeoffs in terms of frequency of unexpected breaks and unexploited lifetime, and then employing this information in an operating cost-based maintenance decision system to minimize expected costs. The effectiveness of the methodology was demonstrated using a simulated example and a benchmark semiconductor manufacturing maintenance problem.

Stodola and Stodola (2020) presented a mathematical model in Predictive Maintenance of equipment based on differentiated machine care. Their model included automatic classification of machines by labour intensity, determination of labour intensity standards, and drawing up monthly and yearly maintenance plans for manufacturing lines and technical equipment in an engineering company. Their model reduced human error, clarified accounting and operational records of machines, evaluated the actual maintenance labour intensity, eliminated routine administrative work, enabled the use of cloud storages, and included automatic reporting of problems in the case of on-board diagnostic systems.

Other experts, like Dillon et al. (2020), have focused on the Big Data and Artificial Intelligence techniques based ecosystems and architectures needed for the implementation of fault detection and diagnosis in Predictive Maintenance that try to overcome multiple challenges, including big data ingestion, integration, transformation, storage, analytics, and visualization in a real-time environment using various technologies.

Zheng et al. (2020) proposed a next generation Artificial Intelligence and Internet of Things based maintenance framework composed of different elements, including 1) Machine Learning algorithms - including probabilistic reliability modelling with Deep Learning -, 2) real-time data collection, transfer, and storage through wireless smart sensors, 3) Big Data technologies, 4) continuously integration and deployment of Machine Learning models, 5) mobile device and Augmented Reality/Virtual Reality applications for fast and better decision-making in the field. Particularly, these authors proposed a novel probabilistic Deep Learning reliability modelling approach and demonstrated it in the Turbofan Engine Degradation Dataset.

THE IMPORTANCE OF THE HUMAN FACTOR

It is important to emphasize that the roles of employees will change as Smart Factories establish themselves in the industrial fabric. Mechanical, tedious and repetitive tasks will increasingly be the responsibility of machines, while the personnel will take on actions of greater responsibility.

It should also be noted that, although Artificial Intelligence manages to analyse data and learn extremely effectively, it is important to highlight the role of humans when developing and improving maintenance-related tools and processes. Both, the experience of the personnel and the historical data of the machines are basic information for the systems to work. In addition, feedback from an experienced person is needed to adjust the algorithms and validate the results they show. After all, decisions are made by humans based on predictions made by Artificial Intelligence. However, the use of Big Data and Artificial Intelligence techniques seeks to minimize the dependence of maintenance operation on the experience and memory of the operators, and above all, to take advantage of the possibilities that Big Data and Artificial Intelligence techniques offer.

Finally, it is important for organizations to create a culture of trustworthiness. Industry 4.0 is a big change, which involves many aspects. However, the implementation of new technologies begins with people. In addition to exceptional work in training and organizing maintenance personnel, they must be prepared for the changes that will come with new systems and processes.

FINAL DISCUSSION AND FUTURE RESEARCH DIRECTION

Industrial maintenance is essential in any industry. A breakdown can be catastrophic because it not only involves a large outlay, but it will also stop the industrial production process, with all the inconveniences and consequences that this can lead to. Therefore, a detailed industrial maintenance plan is essential to ensure the survival of the business.

With the development of technology, the way companies produce has evolved and manufacturers are now trying to make the most of their equipment and human capital. Advanced systems and technologies, combined with greater access to data, make Industry 4.0 possible.

Predictive Maintenance is considered the most advanced of the types of industrial maintenance due to its predictive nature. No need to guess when faults will appear, by using certain objective parameters, it is possible to measure variations in the way equipment and machinery works and detect irregularities before they become bigger problems. With a Predictive Maintenance program, a failure will never appear unexpectedly causing multiple problems.

Industry 4.0 is based on the fact that machinery and equipment is connected to the Internet. This means that each part of the industrial production process interacts with the rest to adapt to the circumstances and achieve optimal performance at all times. This implies that it is not necessary to carry out interventions in the machinery to be able to take measurements of these physical parameters, since the machines already communicate with each other, assuming this a great advantage.

To carry out Predictive Maintenance with guaranteed success, it is essential to collect good quality data. Data is the cornerstone of Industry 4.0. Advanced technology cannot do its job without detailed and accurate information. Having high-quality data makes it easy to use Industry 4.0 systems to their full potential. Of course, another key issue is ensuring data integrity at all times.

Predictive Maintenance techniques are primarily designed to help manufacturers determine the health of their equipment and accurately forecast when maintenance will need to be carried out. This approach can reduce costly downtime due to unforeseen machine failures, as well as help save money on maintenance costs. Its benefits are so many that it is not surprising that emerging technologies and new business models are being applied to strengthen the sector.

However, the data collected by the sensors is not only used to minimize losses, but also to create significant value for companies by improving productivity. For example, Predictive Maintenance enables manufacturers to extend the life of their equipment or reduce the health and safety hazards posed by machine failures.

According to Çınar et al. (2020), with the appearance of Industry 4.0, the concept of Prognostics and Health Management (PHM) has become unavoidable tendency in the framework of industrial Big Data and smart manufacturing; plus, at the same time, it offers a reliable solution for handling the industrial equipment health status. In fact, one important challenge is generating the so-called health factors, or quantitative indicators, of the status of a system associated with a given maintenance issue, and determining their relationship to operating costs and failure risk.

Ballinger (2020) identifies three transformative trends in Predictive Maintenance that are plug and play technology, remote monitoring and Predictive Maintenance as a service. According to this practitioner, plug and play devices are becoming increasingly popular for Predictive Maintenance applications. Most manufacturers rely on legacy equipment to run critical applications in their factories, and these machines are not typically equipped with connectivity capabilities to communicate data in real time. However, plug and play devices allow manufacturers to connect legacy machines without going through a cost prohibitive factory overhaul. Additionally, plug and play technology doesn't necessarily need configuration or testing, minimising downtime. On the other side, remote maintenance and management systems were originally developed to monitor applications in isolated or hazardous locations, such as oil and gas platforms located offshore or in polar regions. But today, remote monitoring is routinely used to assess the conditions of manufacturing machines and reduce the cost of unnecessary or premature maintenance. Finally, Predictive Maintenance as a service, is especially important for Original Equipment Manufacturers (OEMs): since industrial assets can be monitored remotely, manufactures can collect performance data from their customer base and have access to much more comprehensive data than is available to individual users. This gives them a distinct advantage when it comes to predictive analytics.

In Machine Learning, algorithms do not depend on explicit programming, but improve their performance based on internal data analysis. This means that the Machine Learning application actively observes what is happening and what is the result. From this data, the app has the power to form predictions and continually learn and improve based on tracking and analysis of success and failure.

When exploring how this can be applied to maintenance processes, Machine Learning offers the ability to take advantage of predictive algorithms to optimize maintenance processes, reduce downtime, and increase production. In other words, Machine Learning offers the organization the ability to gain a deeper understanding of the ins and outs of its physical assets. This enables companies to adapt and respond to changing dynamics in real time by detecting patterns and deriving alternative courses of action in the future to avoid similar problems. Machine Learning helps monitor assets, predict problems, and change routines to help maintenance departments optimize their maintenance processes.

The increase in Machine Learning capabilities for maintenance has changed the way it is perceived. Çınar et al. (2020) made a comprehensive review of the recent advancements of Machine Learning techniques widely applied to Predictive Maintenance for smart manufacturing in Industry 4.0 by classifying the research according to the Machine Learning algorithms, Machine Learning category, machinery, and equipment used, device used in data acquisition, classification of data, size and type, and highlighted the key contributions of the researchers. These authors reached interesting conclusions that should be taken into account by authors and practitioners for further research:

- Extraction of real time data using intelligent data acquisition system can help to automate Predictive Maintenance.
- Combination of more than one Machine Learning models can provide better prediction compared to use of individual model.
- Machine Learning model implementations based on cloud should be further studied.
- Classification and anomaly detection algorithms can be combined to maintain precision of classification models without losing anomaly detection advantages. Thus, Predictive Maintenance can be applied to equipment or machinery which does not have large dataset.

CONCLUSION

Taking into consideration the discussion presented in this chapter, the main conclusions are the following:

- Smart Factories are the factories of the future.
- Industrial facilities and processes must be digitized, connected and automated.
- A number of emergent technologies, including Internet of Things, Big Data, Artificial Intelligence, Augmented Reality, Virtual Reality, Mixed Reality, Robotics or Cloud Computing, among others, are becoming essential for industries.

- The emergent technologies that are at the basis of the Fourth Industrial Revolution are disintegrating the boundaries between the physical world and the digital world.

- The use of technology opens up numerous possibilities in the industrial field and can be applied in many areas including production, manufacturing, quality, logistics, maintenance or security.

- Industrial maintenance has evolved in recent years, in which it has gone from being seen as a cost centre to being seen as a profit centre whose activities add value by avoiding the appearance of other costs linked to the malfunction of the production equipment or losses caused by the unavailability of such equipment.

- Industrial maintenance is essential in any industry. Companies should try to optimize the maintenance function in order to achieve the highest levels of availability and reliability at the lowest possible cost by combining corrective, preventive and predictive strategies.

- The correct distribution of maintenance tasks (predictive, preventive and corrective) is essential, but the current trend must be a progressive migration towards Predictive Maintenance.

- The objective of Predictive Maintenance is to be able to predict the failure of a component of a machine, in such a way that said component can be replaced, based on a well-determined plan, just before it fails. In this way, it will be possible to minimize the dead time of the equipment, while maximizing the lifetime of the component.

- Artificial Intelligence and Big Data have contributed to creating a new perspective on Predictive Maintenance.

- To be able to apply any Predictive Maintenance technique, it is necessary to have the machines monitored and to collect good quality data. Advanced Artificial Intelligence algorithms cannot do its job without detailed and accurate information. Of course, another key issue is ensuring data integrity at all times.

- Although Artificial Intelligence manages to analyse data and learn extremely effectively, it is important to highlight the role of humans when developing and improving maintenance-related tools and processes.

- The possibilities of the application of Machine Learning on Predictive Maintenance have to be further researched including, for example, the combination of more than one Machine Learning model to provide better predictions.

REFERENCES

Ali, Z., Ali, H., Badawy, M., Alam, S., Chowdhury, M., Noll, J., Kalmar, A., Vida, R., Maliosz, M., Gubbi, J., Buyya, R., Marusic, S., Palaniswami, M., Li, D., Chen, Y., Ning, H., Liu, H., Wang, N., Wu, W., ... García, C. G. (2015). Internet of Things (IoT): Definitions, challenges and recent research directions. *International Journal of Computers and Applications*, *128*(1), 37–47. doi:10.5120/ijca2015906430

Amruthnath, N., & Gupta, T. (2018). A research study on unsupervised machine learning algorithms for early fault detection in Predictive Maintenance. In *2018 5th International Conference on Industrial Engineering and Applications (ICIEA)* (pp. 355-361). IEEE. 10.1109/IEA.2018.8387124

Antunes, J. G., Pinto, A., Nogueira Reis, P., & Henriques, C. (2018). Industry 4.0: A challenge of competition. *Millenium*, (6), 89–97. Advance online publication. doi:10.29352/mill0206.08.00159

Ayvaz, S., & Alpay, K. (2021). Predictive Maintenance system for production lines in manufacturing: A machine learning approach using IoT data in real-time. *Expert Systems with Applications, 173*. doi:10.1016/j.eswa.2021.114598

Ballinger, N. (2020). *Three transformative trends in predictive maintenance.* https://www.electronic-specifier.com/news/three-transformative-trends-in-predictive-maintenance

Bousdekis, A., Lepenioti, K., Apostolou, D., & Mentzas, G. (2019). Decision making in Predictive Maintenance: Literature review and research agenda for industry 4.0. *IFAC-PapersOnLine, 52*(13), 607-612. doi:10.1016/j.ifacol.2019.11.226

Bousdekis, A., Papageorgiou, N., Magoutas, B., Apostolou, D., & Mentzas, G. (2017). A proactive event-driven decision model for joint equipment Predictive Maintenance and spare parts inventory optimization. *Procedia CIRP, 59*, 184–189. doi:10.1016/j.procir.2016.09.015

Chiu, Y. C., Gheng, F. T., & Huang, H. C. (2017). Developing a factory-wide intelligent Predictive Maintenance system based on Industry 4.0. *Zhongguo Gongcheng Xuekan, 40*(7), 562–571. doi:10.1080/02533839.2017.1362357

Çınar, Z. M., Abdussalam Nuhu, A., Zeeshan, Q., Korhan, O., Asmael, M., & Safaei, B. (2020). Machine Learning in Predictive Maintenance towards sustainable smart manufacturing in industry 4.0. *Sustainability, 12*(8211). Advance online publication. doi:10.3390u121982

Coleman, C., Damofaran, S., & Deuel, E. (2017). *Predictive Maintenance and the Smart Factory.* https://www2.deloitte.com/content/dam/Deloitte/us/Documents/process-and-operations/us-cons-predictive-maintenance.pdf

Dilmegani, C. (2021). *Predictive Maintenance (PdM): Why it Matters & How it Works.* https://research.aimultiple.com/predictive-maintenance/

Eren, L. (2017). Bearing fault detection by one-dimensional convolutional neural networks. *Mathematical Problems in Engineering, 2017*, 1–9. Advance online publication. doi:10.1155/2017/8617315

He, Y., Gu, C., Chen, Z., & Han, X. (2017). Integrated Predictive Maintenance strategy for manufacturing systems by combining quality control and mission reliability analysis. *International Journal of Production Research, 55*(19), 5841–5862. doi:10.1080/00207543.2017.1346843

Janssens, O., Slavkovikj, V., Vervisch, B., Stockman, K., Loccufier, M., Verstockt, S., Van de Walle, R., & Van Hoecke, S. (2016). Convolutional neural network based fault detection for rotating machinery. *Journal of Sound and Vibration, 377*, 331-345. doi:10.1016/j.jsv.2016.05.027

Jardim-Goncalves, R., Romero, D., & Grilo, A. (2017). Factories of the future: Challenges and leading innovations in intelligent manufacturing. *International Journal of Computer Integrated Manufacturing, 30*(1), 4–14. doi:10.1080/0951192X.2016.1258120

Jerman, A., & Dominici, G. (2018). Smart factories from business, management and accounting perspective: A systemic analysis of current research. *Management, 13*(4), 355–365. doi:10.26493/1854-4231.13.355-365

Kumar Paul, P., & Ghose, M. K. (2012). Cloud Computing: Possibilities, challenges and opportunities with special reference to its emerging need in the academic and working area of information science. *Procedia Engineering, 38*, 2222-2227. doi:10.1016/j.proeng.2012.06.267

Lee, W. J., Wu, H., Yun, H., Kim, H., Jun, M. B. G., & Sutherland, J. W. (2019). Predictive maintenance of machine tool systems using artificial intelligence techniques applied to machine condition data, *Procedia CIRP, 80*, 506-511. doi:10.1016/j.procir.2018.12.019

Liao, Y., Deschamps, F., Freitas Rocha Loures, E., & Pierin Ramos, L. F. (2017). Past, present and future of Industry 4.0 - a systematic literature review and research agenda proposal. *International Journal of Production Research, 55*(12), 3609–3629. doi:10.1080/00207543.2017.1308576

Mobley, R. K. (2001). *Plant Engineer's Handbook*. Butterworth-Heinemann.

Negandhi, V., Sreenivasan, L., Giffen, R., Sewak, M., & Rajasekharan, A. (2015). *IBM Predictive Maintenance and Quality 2.0 Technical Overview*. IBM Redbooks.

Osterrieder, P., Budde, L., & Friedli, T. (2020). The smart factory as a key construct of industry 4.0: A systematic literature review. *International Journal of Production Economics, 221*. doi:10.1016/j.ijpe.2019.08.011

Poór, P., & Basl, J. (2019). Predictive maintenance as an intelligent service in Industry 4.0. *Journal of Systems Integration, 10*, 3–10. doi:10.20470/jsi.v10i1.364

Selcuk, S. (2017). Predictive maintenance, its implementation and latest trends. *Proceedings of the Institution of Mechanical Engineers. Part B, Journal of Engineering Manufacture, 231*(9), 1670–1679. doi:10.1177/0954405415601640

Sipos, R., Fradkin, D., Moerchen, F., & Wang, Z. (2014). Log-based predictive maintenance. In *Proceedings of the 20th ACM SIGKDD international conference on Knowledge discovery and data mining* (pp. 1867-1876). 10.1145/2623330.2623340

Sony, M., & Naik, S. S. (2019). Key ingredients for evaluating Industry 4.0 readiness for organizations: A literature review. *Benchmarking, 27*(7), 2213–2232. doi:10.1108/BIJ-09-2018-0284

Stodola, P., & Stodola, J. (2020). Model of Predictive Maintenance of Machines and Equipment. *Applied Sciences (Basel, Switzerland), 10*(1), 213. doi:10.3390/app10010213

Strozzi, F., Colicchia, C., Creazza, A., & Noé, C. (2017). Literature review on the 'Smart Factory' concept using bibliometric tools. *International Journal of Production Research, 55*(22), 6572–6591. doi:10.1080/00207543.2017.1326643

Susto, G. A., Schirru, A., Pampuri, S., McLoone, S., & Beghi, A. (2015). Machine Learning for Predictive Maintenance: A multiple classifier approach. *IEEE Transactions on Industrial Informatics, 11*(3), 812–820. doi:10.1109/TII.2014.2349359

Yu, W., Dillon, T., Mostafa, F., Rahayu, W., & Liu, Y. (2020). A global manufacturing big data ecosystem for fault detection in predictive maintenance. *IEEE Transactions on Industrial Informatics, 16*(1), 183–192. doi:10.1109/TII.2019.2915846

Zheng, H., Paiva, A. R., & Gurciullo, C. (2020). *Advancing from Predictive Maintenance to Intelligent Maintenance with AI and IIoT.* ArXiv, abs/2009.00351.

KEY TERMS AND DEFINITIONS

Artificial Intelligence: A wide-ranging branch of computer science concerned with building smart machines capable of performing tasks that typically require human intelligence.

Big Data: It refers to the possibility of analyzing and systematically extracting information from, or otherwise deal with data sets that are too large or complex to be dealt with by traditional data-processing application software.

Deep Learning: It refers to networks capable of learning unsupervised from data that is unstructured or unlabeled. It is also known as deep neural learning. It is seen as a subset of Machine Learning in Artificial Intelligence.

Industry 4.0: It refers to the digitization of the industry and all the services related to it. The goal is to achieve effective automation and smarter factories. Terms such as cyber industry, smart industry or fourth Industrial Revolution are used synonymously with Industry 4.0.

Internet of Things: It refers to the digital interconnection of all kinds of objects or devices with the Internet, such as sensors and mechanical devices, but also everyday objects such as household appliances, footwear or clothing. The goal is to reach Machine to Machine (M2M) interaction.

Machine Learning: It refers to the study of computer algorithms that improve automatically through experience. It is seen as a subset of Artificial Intelligence.

Predictive Maintenance: It refers to a type of maintenance that is able to predict the failure of a component of a machine, in such a way that said component can be replaced, based on a well-determined plan, just before it fails. It allows to minimize the dead time of the equipment, while maximizing the lifetime of the component.

Smart Factory: It refers to a highly digitized and connected production plant using different technologies such as Big Data, Artificial Intelligence, Internet of Things, Robotics or Augmented and Virtual Reality in order to function with minimal human intervention, learn and be able to adapt to changes in real time.

Chapter 5
A Review of Artificial Intelligence Models in Prognosticating Abdominal Aorta Aneurysms

Shier Khee Saw

Department of Surgery, School of Medical Sciences, Universiti Sains Malaysia Health Campus, Kubang Kerian, Malaysia

Syaiful Azzam Sopandi

Department of Surgery, Hospital Raja Perempuan Zainab II, Kota Bharu, Malaysia

Rosnelifaizur bin Ramely

Department of Surgery, School of Medical Sciences, Universiti Sains Malaysia Health Campus, Kubang Kerian, Malaysia

Chow Khuen Chan

Department of Biomedical Engineering, Universiti Malaya, Kuala Lumpur, Malaysia

Michael Pak Kai Wong

Department of Surgery, School of Medical Sciences, Universiti Sains Malaysia Health Campus, Kubang Kerian, Malaysia

Shier Nee Saw

Department of Artificial Intelligence, Faculty of Computer Science and Information Technology, Universiti Malaya, Kuala Lumpur, Malaysia

ABSTRACT

Abdominal aorta aneurysm (AAA) is defined as an abnormal dilatation of the aorta at least 50% more than the adjacent normal vessel diameter. AAA is usually asymptomatic until complications occur such as aorta dissection and ruptured AAA, which has a direct relationship with the size of the aneurysm. Early detection with early intervention of AAA reduces the mortality rate related to rupture. In the era of digitalization, medical data such as electronic medical record, ultrasound images, and physical measurements are available for analysis. Furthermore, with the advancement of artificial intelligence (AI) technologies, numerous AI models have been proposed and shown to improve AAA diagnosis and prognostication. AI technologies, with no doubt, possess an infinite potential to improve the services of healthcare providers. Hence, this chapter targets the audience from all professions: clinicians, radiologists, and computer scientists. This chapter aims to close the gap between the medical profession and computer scientists and thus to design an AI model that can be clinically used.

DOI: 10.4018/978-1-7998-9201-4.ch005

INTRODUCTION

Abdominal aorta aneurysm (AAA) is the thirteenth leading cause of death in USA and causes 1.3% of all deaths among men age between 65-85 years in developed countries (Sakalihasan et al., 2005). In developing country such as Malaysia, it has a prevalence of 15,000 cases among the total population. The annual mortality rate in Malaysia is 5.2 per 100,000, which makes Malaysia one of the three countries within Asia with highest mortality rate (Soon et al., 2019). AAA is usually asymptomatic until catastrophic event of rupture, which often requires surgical emergency. Study showed that the mortality of elective AAA repair is 5%, which is far better than the emergency AAA repair of 50% (Cota et al., 2005). Early detection of AAA is thus imperative to enable proper clinical management to reduce morbidity and mortality.

It has been a major challenge for physicians to diagnose AAA clinically due to indirect clinical presentation of AAA. Asymptomatic AAA has been detected via physical examination and radiological images from ultrasound, computed tomography (CT) and magnetic resonance imaging (MRI) machines. AAA patients may present with pulsatile mass at the epigastric region during physical examination. However, the sensitivity and specificity of physical examination to diagnose AAA are low, at 68% and 75%, respectively (Wijeyaratne, 2011). Abdominal ultrasound is a good modality to diagnose AAA due to its high sensitivity (97.7-100%) and specificity (94.1-100%) (Lema et al., 2017). CT and MRI are preferred during operative planning as these imaging modalities offer better visualization of the entire aorta, which allows the detailed aorta morphological examination.

With the technology advancement, artificial intelligence (AI) models have shown promising results in various fields, including medicine. The number of publications in medicine has increased exponentially within this two decades (Rong et al., 2020). AI research in cardiovascular medicine is also gaining interest with an aim to improve patient's healthcare quality. However, to develop a clinically relevant AI model which can be used in clinics, basic understanding, and communication of both worlds (medicine and computer science) is the key to success. This paper aims to close the gap between medical profession and computer scientists. We anticipate that this paper could provide a necessary information to physicians, radiologists, and computer scientists and thus able to design an optimal and useful AI models that drive real clinical benefits to patients.

ABDOMINAL AORTA ANEURYSM RISK FACTORS

Risk factors such as age, gender, hypertension, dyslipidaemia, coronary arterial disease (CAD), smoking and family history of AAA are associated with AAA. The following paragraph discussed the findings related to each risk factor.

According to the report from the Hospital Kuala Lumpur, which is the capital city of Malaysia, the median age group of AAA is 69 years old, which 87.9% of the cohort is 60 years old and above (Zainal & Yusha, 1998). In the state of Sarawak, which is located at the east of Malaysia, the median age group of AAA is slightly higher, 70 years old but the incident rate of AAA rises significantly if patient is 70 years old and above (Yii, 2003). It was also reported that progressive increase in abdominal aorta diameter occurs with advancing age (Dixon et al., 1984). Guidelines have been in placed to manage AAA patients. For example, NICE guidelines recommended that above 66 years old with certain risk factors should be subjected for AAA screening (Powell & Wanhainen, 2020) and Society Vascular Surgery recommends that age between 65 and 75 years with a history of smoking or healthy patients older than 75 years old should undergo AAA screening (Chaikof et al., 2018).

Gender, in fact, plays a role in affecting the AAA risk. Epidemiologic studies show that male, in fact, suffers 4-6 times more frequent in developing AAA (Mladenovic et al., 2012). Similar findings are also found in Zainal et al. study, where 84.7% of their study cohorts re male (Zainal & Yusha, 1998) and Ming Kon Yii's study, where male to female ratio of the AAA incidence is 3.5 to 1 in Sarawak (Yii, 2003).

In a population-based study, hypertension patients are found to have higher risks of AAA (Cornuz et al., 2004), and hypertension increases the risk of developing AAA by 66% (Kobeissi et al., 2019). For every 20mmHg rise in systolic blood pressure, AAA risk increased by 14% while for every 10mmHg rise in diastolic blood pressure, AAA risk rose by 28% (Kobeissi et al., 2019). The reason of substantial increased risk of AAA caused by hypertension is due to the change of hemodynamic flow in the blood vessel (Sheidaei et al., 2011). Abnormal blood flow leads to abnormal wall shear stress that cause shifts in the elastin and collagen fibres within vessel walls and smooth muscle cells (Wagenseil & Mecham, 2012), and such alteration could lead to an increased risk of AAA. This leads to the fourth risk factors – elevated lipoprotein concentration, which is known as cholesterol level. High cholesterol level increases AAA risk because cholesterol deposited in the vascular intima layer that will damage vascular endothelium and cause inflammation, which leads to vessel degeneration. Study reports that there is a correlation between elevated lipoprotein concentration and AAA risk in a prospective population-based cohort study (Kubota et al., 2018). Moreover, there is a significant association between low-density lipoproteins (LDL) and small aneurysm, suggesting that the initiation process of AAA is due to LDL (Hobbs et al., 2003).

Smoking is one of the most common risk factors of AAA. Smokers are associated with 7.6 times increased in the prevalence of AAA than non-smokers and the relative risk of AAA increased 4% with every increase of the number of year in smoking (Wilmink et al., 1999). Beside, Teun et al. study shows that the amount of cigarette smoked did not reached statistically significant after adjustment is made for the duration of smoking (Wilmink et al., 1999). Juraj et al. study shows that smoking is an independent risk factor for AAA (Madaric et al., 2005).

Following that, a meta-analysis concludes that CAD is a strong predictor of AAA occurrence (Hernesniemi et al., 2015). Patients with CAD have 2.4 times increased risk of AAA compared to those patients without CAD (Li et al., 2017). Lastly, patients with family history of AAA possess double the risk in developing AAA especially the first-degree relatives (Larsson et al., 2009). There is a 4.33-fold increased risk of AAA in the individual with family history of AAA (Salo et al., 1999).

BASIC UNDERSTANDING OF ARTIFICIAL INTELLIGENCE MODEL

AI uses computational approaches to learn natural patterns in data that generate insight to make better decisions and predictions. The use of AI has been applied in predicting cardiovascular risk using various data sources such as patient demographics and physiological data. AI techniques include machine learning and deep learning. The most used supervised machine learning models comprises the Support Vector Machine (SVM), Artificial Neural Network (ANN), Logistic Regression and Random Forest. Deep learning includes Convolutional Neural Network (CNN),in which image data is often utilized instead of tabular physiological data (*data in excel*). In supervised model, a function is designed to map between the input and output from data. For example, if we want to develop an AI model to predict AAA risk based on physiological data. The inputs to the model would be risk factors such as age, gender and CAD, while output would be AAA risk (low, medium, or high risk of AAA). This is a classification problem, whereby the AI model will be trained based on the data such that it learns the data patterns. Once it has beens trained with high accuracy, this AI model can be used in the future.

Figure 1. An overview of different type of AI models. Logistic Regression uses logistic function to map between input (independent) and output (dependent) variables. SVM determines the best hyperplane that separates two outcomes with maximum gap. Random Forest uses multiple decision trees to obtain the final prediction via majority voting policy. ANN consists of nodes that are connected to each other and optimizes the weights in network to determine the best function for classification. Deep learning extracts useful features from image via convolution technique. The last second layer of deep learning is similar to ANN network which consists of multiple nodes that connect to the output nodes, which yield the outcome probability.

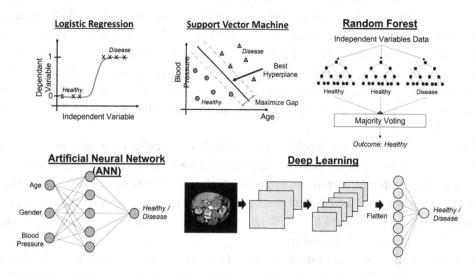

Logistic regression is one type of supervised machine learning model that uses logistic function to describe the relationship between the independent (input) and dependent (output) variables (Menard, 2002). It is commonly used for classification in which the output is categorical, instead of continuous data. The input values (independent variables, x) are paired with coefficients (b_0, b_1, b_2, ..., b_n) to predict the output value (dependent variable, y). The following equation represents logistic regression equation. Input and output variables are obtained from the datasets while coefficients will be learned based on the underlying data patterns. Different datasets will have different coefficients as the data patterns are different. The value of y will be squashed within a value between 0 and 1 using the logistic function as shown in Figure 1.

$$y = \frac{e^{(b_0 + b_1 x_1 + b_2 x_2 + b_3 x_3)}}{1 + e^{(b_0 + b_1 x_1 + b_2 x_2 + b_3 x_3)}} \qquad (1)$$

where y is the output variable, x_1, x_2, x_3 are the input variables, b_1, b_2, b_3 are the coefficients with respect to independent variables and b_0 is the bias.

The idea of SVM lies in finding the optimal hyperplane that best separates between two classes (i.e healthy vs disease) judging from the gap distance (Figure 1). The hyperplane can also be illustrated as decision boundaries that aid in classify the data points into correct classes. The dimension of the hyperplane depends on the number of input variables. When there are two input variables, the hyperplane is a line as shown in Figure 1. When there are three input variables, the hyperplane is a 2-dimensional plane. SVM model makes use of different type of kernel functions (linear, polynomial, radial basis functions) to map between the input and output variables. In another words, SVM enables non-linear separation which makes it a good classifier in classification problem (Pisner & Schnyer, 2020).

Random Forest algorithm is an ensemble (averaging) method that constructs multiple decision trees and identifies the predicted classes (healthy/disease) with the highest votes across the decision trees (Zhang & Ma, 2012). Decision tree is made up of a set of rules that split the datasets into smaller subsets using different rules at every layer. The rules are constructed using the Classification and Regression Tree (CART) algorithm by computing the GINI Impurity from the datasets. The Gini Impurity for each variable will be computed. The variable with the lowest Gini Impurity will be chosen to split the data. Then, Gini Impurity is calculated again using the remaining variables and the variable with the lowest Gini Impurity will be chosen again to split the data. This process is repeated until all variables are used, or a pure node is reached.

ANN consists of multiple layers with different number of nodes. Each node consists of weight and each layer consists of bias. Weight and bias are the hyperparameters that will be optimized during model training. In each layer, an activation function will be applied. This activation function will transform the data, which thus able to capture non-linear relationships and aids in convert the data into more useful output (Noriega, 2005; Suzuki, 2011). It is noteworthy to realize that the number of nodes and number of hidden layers can be modified. The deeper the network (more hidden layers), the ability of the network to capture complex data is higher. However, deeper network would require more training data to prevent overfitting issue. Deep learning uses the same concept as ANN where the nodes in each layer in deep learning can be in 2-dimension (Figure 1) or 3-dimension. Deep learning uses convolution technique to extract useful features from image, these features will be flattened and used for classification purpose.

ARTIFICIAL INTELLIGENCE STUDIES RELATED TO AAA

To improve AAA patient's outcome, multiple machine learning models have been proposed to predict in hospital mortality in AAA patients (Hadjianastassiou et al., 2006; Monsalve-Torra et al., 2016; Wise et al., 2015). In Hadjianastassiou et al study, a total of 1751 patients were included and physiological variables such as age, acute physiology score, emergency operation and chronic health evaluation were used as input variables for model development (Hadjianastassiou et al., 2006). They showed that area under receiver operating curve (AUROC) of ANN and logistic regression model were higher than clinicians' diagnosis (AUROC of ANN: 0.870; Logistic Regression: 0.869; Clinicians: 0.816). However, further analysis showed that ANN tend to overestimate and underestimate in-hospital mortality risk in the low and high-risk patients' cohort, respectively. Logistic regression was shown to be the best model, achieving the lowest difference between prediction and observed mortality rate (Hadjianastassiou et al., 2006). Another study from Wise et al adopted ANN model to predict in-hospital mortality (death) from ruptured AAA patients. Their sample size was lower, recruiting a total of 107 patients and used four input variables (age at rupture, loss of consciousness, shock, and cardiac arrest) in model developments. They showed that ANN model achieved higher accuracy as compared to the Glasgow Aneurysm Score and the logistic regression model (Accuracy of ANN: 85%; Glasgow Aneurysm Score: 74%; Logistic Regression Model: 78%) (Wise et al., 2015). Lastly, Torra et al. combined three machine learning models – ANN, Radial Basis Function and Bayesian network to predict in-hospital mortality in patients undergoing open repair AAA. They selected nine important features – American Society of Anesthesiologists (ASA) Score, elective, vascular surgery, cod bleeding, complication in surgery, renal, pulmonary, vascular, other and showed that 87% sensitivity was achieved (Monsalve-Torra et al., 2016).

There are 20-30% of patients who undergo endovascular aortic repair (EVAR) requires re-intervention (Columbo et al., 2019). Post EVAR patients need to be followed up with lifelong surveillance. Frequent imaging surveillance of post-EVAR not only increases hospital cost and healthcare workers' burden but it also exposes extra radiation towards the patients. An AI re-intervention risk prediction system could be used as a risk-stratified tool to aid in patient's selection for surveillance after surgery. Studies used machine learning for survival analysis in clinical trials, particularly in predicting the survival risk for AAA patients who undergo endovascular aortic aneurysm surgery (Attallah et al., 2017; Attallah & Ma, 2014). Attallah et al. studies showed that the Bayesian model achieved an average AUROC of 0.75 in predicting the risk of re-intervention after surgery in two centres using morphological features measured from CT scan (Attallah & Ma, 2014). Another study by Attallah et al showed that a hybrid classifier outperformed a single classifier (Sensitivity: Hybrid model - 80.8%; SVM: 73.08%; ANN: 73.08%) in predicting risk of endovascular aortic aneurysm repair re-intervention using variables selection from Cox's model (Attallah et al., 2017). On top of that, ANN model was proposed to predict endograft complications after surgery (Karthikesalingam et al., 2015). A total of 19 pre-operative variables measured from CT scan, such as angulation, length, areas, diameter, volume, tortuosity of aneurysm in neck / sac / iliac segments were used as inputs to model with a total of 761 patients' data. (Karthikesalingam et al., 2015).

The expansion rate of AAA is an important factor of AAA rupture risk, and the average annual expansion rate increases with the size of the AAA. Decision for surgical repair of AAA is based on diameter, AAA expansion rate and patient's fitness (Keisler & Carter, 2015). AAA expansion rate is affected by many factors and thus it is difficult to predict patient-specific AAA expansion (Cronenwett et al., 1990) using traditional statistical models that describe features' relationship as linear functions. AI, on the other hand, can model features' relationship in a non-linear manner, which could be benefit in developing a personalized AAA expansion risk prediction model. One study used SVM model to predict AAA growth in 12- and 24-months after, using flow mediated dilatation and AAA diameter as inputs to model. SVM model prediction achieved errors within 2mm for 85% and 71% of patients at 12- and 24-months, respectively (Lee et al., 2018). Another study, with smaller sample size of 50 subjects, showed that machine learning model (XGBoost, an ensemble model) achieved an AUROC of 0.86 in predicting rapid expansion of AAA. Furthermore, XGBoost showed that the major axis of AAA was the most important variable to reflect AAA expansion (Hirata et al., 2020). Kerut et al study recruited a total of 10,329 patients who underwent ultrasound scan in their study. In their study, they adopted ANN to predict aneurysm that was larger than 3cm in diameter using 18 variables such as medical problems (heart problem, hypertension, high cholesterol, diabetes mellitus, family history of AAA or brain aneurysm, BMI, age, gender, ethnicity, and smoking status). However, this study did not report any accuracy or AUROC but instead conducted a sensitivity analysis to determine the most important feature that have impact to the machine learning model. Surprisingly, results showed that the most important factor was ethnicity - Caucasian, but authors acknowledged that this could be due to the high prevalence of AAA in Caucasian population (Kerut et al., 2019). Another AI study was conducted on Asian population (312 patients) where they showed that Naïve Bayesian model had the best AUROC (0.974) in classifying three classes of aneurysm (unruptured abdominal aortic aneurysm, ruptured abdominal aortic aneurysm, and normal aorta without aneurysm) using clinical and geometrical features measured from CT scans (Canchi et al., 2018).

Another aspect of AI studies related to AAA is to develop automated segmentation or quantification of AAA morphological measurements from radiography images. To date, measurements of AAA are done by radiologists manually. More often, these measurements such as AAA diameter is used to determine AAA severity. It has been shown that deep learning can automate the measurements process, which will hence reduce radiologist's burden and inter-observer variability (Caradu et al., 2021; Graffy et al., 2019; Hong & Sheikh, 2016; Jiang et al., 2020; López-Linares et al., 2018; López-Linares et al., 2019; Lu et al., 2019). Such models would be beneficial for automatically determine AAA diameter and open opportunities for more complex aneurysm analysis.

Caradu et al. developed a fully automated software (PRAEVAorta) that allows fast and robust detection of aortic lumen and infrarenal AAA. The measurements computed from the software correlated well with measurements done by experts. The segmentation and measurement procedure were significantly faster than manual labelled by human and was suggested to have strong clinical applications in practice and research (Caradu et al., 2021). Furthermore, 3D U-Net with ellipse fitting was proposed to perform aorta segmentation and AAA prediction (using 3cm as threshold to define AAA). A total of 321 abdominal-pelvic CT images were used as training data and 57 images were used as testing data. The quality of segmentation is evaluated using Dice score. A Dice score of 1 indicates a perfect segmentation. Segmentation results achieved a Dice score of 0.90±0.05 and their proposed AI model, the Deep AAA model, achieved 91% sensitivity and 95% specificity in predicting AAA. Their model was reported to have higher sensitivity than radiologists (Lu et al., 2019). Moreover, 3D convolutional neural network for segmenting of both preoperative and postoperative aneurysm was proposed. The performance was validated via diameter's, volume's measurements using Dice score. It was reported the 3D convolutional neural network generated a mean diameter error of 3.3mm and a Dice score of 0.87 (López-Linares et al., 2019). Furthermore, a fully automatic using Deep Convolutional Neural Network (DCNN) was proposed to segment post-operative abdominal aortic thrombus. The model had a Dice score of 0.82. In their study, their approach of localizing region of interest first then proceed for thrombus segmentation had lower memory used and shorter computation time (López-Linares et al., 2018). A convolutional neural network (CNN) was developed to automatically segment aortic calcification. Aortic calcium volume, mass and Agatston score were assessed using automatic CNN model and semi-automated method. Results showed that the R^2 agreement value in Agatston score between automated CNN method and semi-automated method was 0.84. Such tool will facilitate the assessment of aortic calcification and aid in cardiovascular risk (Graffy et al., 2019). Another study by Jiang et al, they used deep learning to predict AAA expansion using longitudinal data. However, it is well known that large datasets are needed to train a deep learning model. To solve this problem, they leveraged computational model to generate *in silico* data. Combining both *in silico* and patient data, the Deep Belief Network were able to predict AAA enlargement with an error of 3.1%. (Jiang et al., 2020).

Endovascular aortic repair (EVAR) is a minimally invasive procedure by placing a stent graft into the aneurysm to treat AAA. Complication rate post-EVAR is up to 30% which includes endoleak (*General Complications of EVAR*). Thus, post-EVAR surveillance imaging is important to reduce the morbidity and mortality. For example, in Sage et al study, they developed a deep learning model to detect endoleak and measurement of aneurysm's diameter, area, and volume from CT angiography. 191 patients were included in the study and their model achieved a ROC for binary endoleak detection of 0.94±0.03 and an accuracy of 0.89±0.03. The pipeline consists of localization network to detect AAA locations on all axial slices of CT scan, followed by endoleak detection modules and multiclass AAA segmentation modules (Hahn et al., 2020). One interesting study used machine learning approach to estimate wall shear stress distribution in AAA (Jordanski et al., 2018).

IMPACT OF AI IN AAA MANAGEMENT

Despite AI is widely used in patients with AAA to improve radiological imaging, to predict the risk of AAA rupture, to determine the prognosis of patients with AAA (Raffort et al., 2020), none is deployed in clinics to predict the occurrence of AAA in guiding physicians on early management of AAA. If there is an effective AI model to predict AAA accurately, it gives clues to primary centre physicians to pay more attention to those patients and they could carry out appropriate managements at early stage to improve outcomes. Besides that, such model can optimize hospital resources by reducing unnecessary ultrasound scanning or tertiary referral.

In developing countries, there are many false negatives whereby patients are frequently misdiagnosed as healthy individuals in rural district hospitals or healthcare centres. One reason is due to the inaccessibility to imaging devices such as ultrasound, CT, and MRI machines. Physicians in district hospitals or healthcare centres diagnose AAA through history taking or physical examination, which is reported to have low sensitivity (Wijeyaratne, 2011). As such, patients did not receive proper clinical AAA treatment and therefore resulted in enlarged aneurysm. Some progressed to the stage of near rupture as well. Failure to detect AAA at the early stage and delayed in applying early prevention measures may increase the risk of complicated AAA. AI models have shown capability in AAA prediction using only physiological data, but more prospective clinical trials must be conducted to validate model's performance. It is undoubtedly that AI will revolutionize cardiovascular management practice in the near future.

REFERENCES

Attallah, O., Karthikesalingam, A., Holt, P. J., Thompson, M. M., Sayers, R., Bown, M. J., Choke, E. C., & Ma, X. (2017). Using multiple classifiers for predicting the risk of endovascular aortic aneurysm repair re-intervention through hybrid feature selection. *Proceedings of the Institution of Mechanical Engineers. Part H, Journal of Engineering in Medicine, 231*(11), 1048–1063. doi:10.1177/0954411917731592 PMID:28925817

Attallah, O., & Ma, X. (2014). Bayesian neural network approach for determining the risk of re-intervention after endovascular aortic aneurysm repair. *Proceedings of the Institution of Mechanical Engineers. Part H, Journal of Engineering in Medicine, 228*(9), 857–866. doi:10.1177/0954411914549980 PMID:25212212

Canchi, T., Ng, E. Y. K., Narayanan, S., & Finol, E. A. (2018). On the assessment of abdominal aortic aneurysm rupture risk in the Asian population based on geometric attributes. *Proceedings of the Institution of Mechanical Engineers. Part H, Journal of Engineering in Medicine*, *232*(9), 922–929. doi:10.1177/0954411918794724 PMID:30122103

Caradu, C., Spampinato, B., Vrancianu, A. M., Bérard, X., & Ducasse, E. (2021). Fully automatic volume segmentation of infrarenal abdominal aortic aneurysm computed tomography images with deep learning approaches versus physician controlled manual segmentation. *Journal of Vascular Surgery*, *74*(1), 246–256.e246. doi:10.1016/j.jvs.2020.11.036 PMID:33309556

Chaikof, E. L., Dalman, R. L., Eskandari, M. K., Jackson, B. M., Lee, W. A., Mansour, M. A., Mastracci, T. M., Mell, M., Murad, M. H., & Nguyen, L. L. (2018). The Society for Vascular Surgery practice guidelines on the care of patients with an abdominal aortic aneurysm. *Journal of Vascular Surgery, 67*(1), 2-77.

Columbo, J. A., Kang, R., Hoel, A. W., Kang, J., Leinweber, K. A., Tauber, K. S., Hila, R., Ramkumar, N., Sedrakyan, A., & Goodney, P. P. (2019). A comparison of reintervention rates after endovascular aneurysm repair between the Vascular Quality Initiative registry, Medicare claims, and chart review. *Journal of Vascular Surgery*, *69*(1), 74–79.e76. doi:10.1016/j.jvs.2018.03.423 PMID:29914838

Cornuz, J., Sidoti Pinto, C., Tevaearai, H., & Egger, M. (2004). Risk factors for asymptomatic abdominal aortic aneurysm: Systematic review and meta-analysis of population-based screening studies. *European Journal of Public Health*, *14*(4), 343–349. doi:10.1093/eurpub/14.4.343 PMID:15542867

Cota, A., Omer, A., Jaipersad, A., & Wilson, N. (2005). Elective versus ruptured abdominal aortic aneurysm repair: A 1-year cost-effectiveness analysis. *Annals of Vascular Surgery*, *19*(6), 858–861. doi:10.100710016-005-7457-5 PMID:16177868

Cronenwett, J. L., Sargent, S. K., Wall, M. H., Hawkes, M. L., Freeman, D. H., Dain, B. J., Curé, J. K., Walsh, D. B., Zwolak, R. M., McDaniel, M. D., & Schneider, J. R. (1990). Variables that affect the expansion rate and outcome of small abdominal aortic aneurysms. *Journal of Vascular Surgery*, *11*(2), 260–269. doi:10.1016/0741-5214(90)90269-G PMID:2405198

Dixon, A., Lawrence, J., & Mitchell, J. (1984). Age-related changes in the abdominal aorta shown by computed tomography. *Clinical Radiology*, *35*(1), 33–37. doi:10.1016/S0009-9260(84)80228-7 PMID:6690178

General Complications of EVAR. (n.d.). *Thoracic Key*. https://thoracickey.com/general-complications-of-evar/

Graffy, P. M., Liu, J., O'Connor, S., Summers, R. M., & Pickhardt, P. J. (2019). Automated segmentation and quantification of aortic calcification at abdominal CT: Application of a deep learning-based algorithm to a longitudinal screening cohort. *Abdominal Radiology*, *44*(8), 2921–2928. doi:10.100700261-019-02014-2 PMID:30976827

Hadjianastassiou, V. G., Franco, L., Jerez, J. M., Evangelou, I. E., Goldhill, D. R., Tekkis, P. P., & Hands, L. J. (2006). Informed prognosis after abdominal aortic aneurysm repair using predictive modeling techniques. *Journal of Vascular Surgery*, *43*(3), 467–473. doi:10.1016/j.jvs.2005.11.022 PMID:16520157

Hahn, S., Perry, M., Morris, C. S., Wshah, S., & Bertges, D. J. (2020). Machine deep learning accurately detects endoleak after endovascular abdominal aortic aneurysm repair. *JVS: Vascular Science, 1,* 5-12.

Hernesniemi, J. A., Vänni, V., & Hakala, T. (2015). The prevalence of abdominal aortic aneurysm is consistently high among patients with coronary artery disease. *Journal of Vascular Surgery, 62*(1), 232-240.

Hirata, K., Nakaura, T., Nakagawa, M., Kidoh, M., Oda, S., Utsunomiya, D., & Yamashita, Y. (2020). Machine Learning to Predict the Rapid Growth of Small Abdominal Aortic Aneurysm. *Journal of Computer Assisted Tomography, 44*(1), 37–42. doi:10.1097/RCT.0000000000000958 PMID:31939880

Hobbs, S., Claridge, M., Quick, C., Day, N., Bradbury, A., & Wilmink, A. (2003). LDL cholesterol is associated with small abdominal aortic aneurysms. *European Journal of Vascular and Endovascular Surgery, 26*(6), 618–622. doi:10.1016/S1078-5884(03)00412-X PMID:14603421

Hong, H. A., & Sheikh, U. U. (2016, March 4-6). Automatic detection, segmentation and classification of abdominal aortic aneurysm using deep learning. *2016 IEEE 12th International Colloquium on Signal Processing & Its Applications (CSPA).*

Jiang, Z., Do, H. N., Choi, J., Lee, W., & Baek, S. (2020). A deep learning approach to predict abdominal aortic aneurysm expansion using longitudinal data. *Frontiers in Physics, 7,* 235.

Jordanski, M., Radovic, M., Milosevic, Z., Filipovic, N., & Obradovic, Z. (2018). Machine Learning Approach for Predicting Wall Shear Distribution for Abdominal Aortic Aneurysm and Carotid Bifurcation Models. *IEEE Journal of Biomedical and Health Informatics, 22*(2), 537–544. doi:10.1109/JBHI.2016.2639818 PMID:28113333

Karthikesalingam, A., Attallah, O., Ma, X., Bahia, S. S., Thompson, L., Vidal-Diez, A., Choke, E. C., Bown, M. J., Sayers, R. D., Thompson, M. M., & Holt, P. J. (2015). An Artificial Neural Network Stratifies the Risks of Reintervention and Mortality after Endovascular Aneurysm Repair; a Retrospective Observational study. *PLoS One, 10*(7), e0129024. doi:10.1371/journal.pone.0129024 PMID:26176943

Keisler, B., & Carter, C. (2015). Abdominal aortic aneurysm. *American Family Physician, 91*(8), 538–543. PMID:25884861

Kerut, E. K., To, F., Summers, K. L., Sheahan, C., & Sheahan, M. (2019). Statistical and machine learning methodology for abdominal aortic aneurysm prediction from ultrasound screenings. *Echocardiography (Mount Kisco, N.Y.), 36*(11), 1989–1996. doi:10.1111/echo.14519 PMID:31682022

Kobeissi, E., Hibino, M., Pan, H., & Aune, D. (2019). Blood pressure, hypertension and the risk of abdominal aortic aneurysms: A systematic review and meta-analysis of cohort studies. *European Journal of Epidemiology, 34*(6), 547–555. doi:10.100710654-019-00510-9 PMID:30903463

Kubota, Y., Folsom, A. R., Ballantyne, C. M., & Tang, W. (2018). Lipoprotein (a) and abdominal aortic aneurysm risk: The Atherosclerosis Risk in Communities study. *Atherosclerosis, 268,* 63–67. doi:10.1016/j.atherosclerosis.2017.10.017 PMID:29182987

Larsson, E., Granath, F., Swedenborg, J., & Hultgren, R. (2009). A population-based case-control study of the familial risk of abdominal aortic aneurysm. *Journal of Vascular Surgery, 49*(1), 47–51. doi:10.1016/j.jvs.2008.08.012 PMID:19028058

Lee, R., Jarchi, D., Perera, R., Jones, A., Cassimjee, I., Handa, A., Clifton, D. A., Bellamkonda, K., Woodgate, F., Killough, N., Maistry, N., Chandrashekar, A., Darby, C. R., Halliday, A., Hands, L. J., Lintott, P., Magee, T. R., Northeast, A., Perkins, J., & Sideso, E. (2018). Applied Machine Learning for the Prediction of Growth of Abdominal Aortic Aneurysm in Humans. *EJVES Short Reports*, *39*, 24–28. doi:10.1016/j.ejvssr.2018.03.004 PMID:29988820

Lema, P. C., Kim, J. H., & St James, E. (2017). Overview of common errors and pitfalls to avoid in the acquisition and interpretation of ultrasound imaging of the abdominal aorta. *Journal of Vascular Diagnostics and Interventions*, *5*, 41–46. doi:10.2147/JVD.S124327

Li, W., Luo, S., Luo, J., Liu, Y., Ning, B., Huang, W., Xue, L., & Chen, J. (2017). Predictors associated with increased prevalence of abdominal aortic aneurysm in Chinese patients with atherosclerotic risk factors. *European Journal of Vascular and Endovascular Surgery*, *54*(1), 43–49. doi:10.1016/j.ejvs.2017.04.004 PMID:28527818

López-Linares, K., Aranjuelo, N., Kabongo, L., Maclair, G., Lete, N., Ceresa, M., García-Familiar, A., Macía, I., & González Ballester, M. A. (2018). Fully automatic detection and segmentation of abdominal aortic thrombus in post-operative CTA images using Deep Convolutional Neural Networks. *Medical Image Analysis*, *46*, 202–214. doi:10.1016/j.media.2018.03.010 PMID:29609054

López-Linares, K., García, I., García-Familiar, A., Macía, I., & Ballester, M. A. G. (2019). *3D convolutional neural network for abdominal aortic aneurysm segmentation*. arXiv preprint arXiv:1903.00879.

Lu, J.-T., Brooks, R., Hahn, S., Chen, J., Buch, V., Kotecha, G., Andriole, K. P., Ghoshhajra, B., Pinto, J., & Vozila, P. (2019). DeepAAA: clinically applicable and generalizable detection of abdominal aortic aneurysm using deep learning. *International Conference on Medical Image Computing and Computer-Assisted Intervention*.

Madaric, J., Vulev, I., Bartunek, J., Mistrik, A., Verhamme, K., De Bruyne, B., & Riecansky, I. (2005). Frequency of abdominal aortic aneurysm in patients> 60 years of age with coronary artery disease. *The American Journal of Cardiology*, *96*(9), 1214–1216. doi:10.1007/978-3-030-32245-8_80

Menard, S. (2002). *Applied logistic regression analysis* (Vol. 106). Sage. doi:10.4135/9781412983433

Mladenovic, A., Markovic, Z., Grujicic-Sipetic, S., & Hyodoh, H. (2012). Abdominal Aortic Aneurysm in Different Races Epidemiologic Features and Morphologic-Clinical Implications Evaluated by CT Aortography. *Aneurysm*, 109.

Monsalve-Torra, A., Ruiz-Fernandez, D., Marin-Alonso, O., Soriano-Payá, A., Camacho-Mackenzie, J., & Carreño-Jaimes, M. (2016). Using machine learning methods for predicting inhospital mortality in patients undergoing open repair of abdominal aortic aneurysm. *Journal of Biomedical Informatics*, *62*, 195–201. doi:10.1016/j.jbi.2016.07.007 PMID:27395372

Noriega, L. (2005). *Multilayer perceptron tutorial. School of Computing*. Staffordshire University.

Pisner, D. A., & Schnyer, D. M. (2020). Support vector machine. In A. Mechelli & S. Vieira (Eds.), *Machine Learning* (pp. 101–121). Academic Press. doi:10.1016/B978-0-12-815739-8.00006-7

Powell, J. T., & Wanhainen, A. (2020). Analysis of the differences between the ESVS 2019 and NICE 2020 guidelines for abdominal aortic aneurysm. *European Journal of Vascular and Endovascular Surgery, 60*(1), 7–15. doi:10.1016/j.ejvs.2020.04.038 PMID:32439141

Raffort, J., Adam, C., Carrier, M., Ballaith, A., Coscas, R., Jean-Baptiste, E., Hassen-Khodja, R., Chakfé, N., & Lareyre, F. (2020). Artificial intelligence in abdominal aortic aneurysm. *Journal of Vascular Surgery, 72*(1), 321-333.

Rong, G., Mendez, A., Bou Assi, E., Zhao, B., & Sawan, M. (2020). Artificial Intelligence in Healthcare: Review and Prediction Case Studies. *Engineering, 6*(3), 291–301. doi:10.1016/j.eng.2019.08.015

Sakalihasan, N., Limet, R., & Defawe, O. D. (2005). Abdominal aortic aneurysm. *Lancet, 365*(9470), 1577–1589. doi:10.1016/S0140-6736(05)66459-8 PMID:15866312

Salo, J. A., Soisalon-Soininen, S., Bondestam, S., & Mattila, P. S. (1999). Familial occurrence of abdominal aortic aneurysm. *Annals of Internal Medicine, 130*(8), 637–642. doi:10.7326/0003-4819-130-8-199904200-00003 PMID:10215559

Sheidaei, A., Hunley, S. C., Zeinali-Davarani, S., Raguin, L. G., & Baek, S. (2011). Simulation of abdominal aortic aneurysm growth with updating hemodynamic loads using a realistic geometry. *Medical Engineering & Physics, 33*(1), 80–88. doi:10.1016/j.medengphy.2010.09.012 PMID:20961796

Soon, G. T. J., Zhi, P. K. L., Krishnan, S. M., & Meng, C. K. (2019). A review of aortic disease research in Malaysia. *The Medical Journal of Malaysia, 74*(1), 67. PMID:30846666

Suzuki, K. (2011). *Artificial neural networks: Methodological advances and biomedical applications.* BoD–Books on Demand. doi:10.5772/644

Wagenseil, J. E., & Mecham, R. P. (2012). Elastin in Large Artery Stiffness and Hypertension. *Journal of Cardiovascular Translational Research, 5*(3), 264–273. doi:10.100712265-012-9349-8 PMID:22290157

Wijeyaratne, S. M. (2011). Diagnosis of Aortic Aneurysm. *Diagnosis, Screening and Treatment of Abdominal, Thoracoabdominal and Thoracic Aortic Aneurysms,* 69.

Wilmink, T. B., Quick, C. R., & Day, N. E. (1999). The association between cigarette smoking and abdominal aortic aneurysms. *Journal of Vascular Surgery, 30*(6), 1099–1105. doi:10.1016/S0741-5214(99)70049-2 PMID:10587395

Wise, E. S., Hocking, K. M., & Brophy, C. M. (2015). Prediction of in-hospital mortality after ruptured abdominal aortic aneurysm repair using an artificial neural network. *Journal of Vascular Surgery, 62*(1), 8–15. doi:10.1016/j.jvs.2015.02.038 PMID:25953014

Yii, M. K. (2003). Epidemiology of abdominal aortic aneurysm in an Asian population. *ANZ Journal of Surgery, 73*(6), 393–395. doi:10.1046/j.1445-2197.2003.t01-1-02657.x PMID:12801335

Zainal, A., & Yusha, A. (1998). Profile of patients with abdominal aortic aneurysm referred to the Vascular Unit, Hospital Kuala Lumpur. *The Medical Journal of Malaysia, 53*(4), 423–427. PMID:10971988

Zhang, C., & Ma, Y. (2012). *Ensemble machine learning: methods and applications.* Springer. doi:10.1007/978-1-4419-9326-7

Chapter 6
Enhanced Water Quality Monitoring and Estimation Using a Multi-Modal Approach

Aamir Farooq Khan
National University of Sciences and Technology, Pakistan

Rafia Mumtaz
National University of Sciences and Technology, Pakistan

Muhammad Usama
National University of Sciences and Technology, Pakistan

Taimoor Khan Mahsud
National University of Sciences and Technology, Pakistan

ABSTRACT

Remote sensing through satellites and internet of things (IoT) technology are two widespread techniques to assess inland water quality. However, both these techniques have their limitations. IoT provides point data, which is insufficient to represent entire water body, especially if the water body has complex terrain and hydrology. Through remote sensing, we can sample data of a large area, but data acquisition is constrained by satellite. Revisit time and quality of estimates can be affected by image resolution. Moreover, non-optical properties that might affect water quality cannot be sensed through satellites. To complement this, GIS data from labs can be useful for providing higher resolution and accurate data and can be used as ground truth. Thus, in this chapter, the authors aim to integrate both these data collection techniques followed by estimation and prediction through machine learning models. The accumulated datasets are used to train machine learning (ML) models deployed at a server. The selected ML model is an artificial neural network with train accuracy of 97% and test accuracy of 95%.

DOI: 10.4018/978-1-7998-9201-4.ch006

INTRODUCTION

Water quality plays a vital role in a heathy ecosystem. Environmentalist and researchers have been actively working to improve methodologies of water quality analysis. Although, remote sensing has been used since 1970s to examine wide range of water quality constituents. However, with the advent of new technologies such as the Internet of Things (IoT), the water quality can be measured and reported in real-time. The limitation of this approach is that it provides point data which is insufficient to represent the entire water body quality unless the water samples are collected from many points and for long durations. The collection of data from multiple points in a water body is not feasible particularly if the data collection sites are difficult to access due to complex terrain or other resources. In contrast to this approach, the data collected from remotely sensed satellite imagery provide ample details about the entire water body, however, the data acquisition is constrained by satellite revisit time and the quality of estimates is affected by image resolution. In addition to this, the GIS data collected from labs can be useful for providing higher resolution and accurate data and can be used as a ground truth. However, the lab measurements are done manually which consume a substantial amount of time and creates a delay for data synthesis and assimilation. In view of the above, an enhanced water quality monitoring and estimation system is proposed that integrates the above approaches to overcome the limitations of these existing methods. This chapter gives a broad overview of the domain of water quality monitoring and estimation by providing insights about water quality factors, different models used for water quality estimation, role of spectroscopy, and spectral bands in remote sensing. A comparison of two satellites- Landsat-8 and Sentinel-2 is also included as a part of this chapter.

BACKGROUND

Water is an essential resource, despite its pivotal role in the sustenance of life on planet earth, water quality is constantly compromised and degraded by certain human activities. Poor water quality is a problem of great concern worldwide. Particularly in countries like Pakistan, the available water for drinking is impure and polluted by human, agricultural and industrial waste. Poor governance and management in the water monitoring sector further adds to the miseries.

Unfortunately, Pakistan ranks 80th among 122 nations regarding water quality (M.K. Daud., Muhammad Nafees. & Shafqat Ali. (2017) *Drinking Water Quality Status and Contamination in Pakistan*). The two primary water sources, surface and ground water are heavily contaminated by toxic metals, pesticides, and coliforms. Besides, human factors such as improper disposal of waste, frequent use of agrochemicals, and urbanization have badly affected the water quality in Pakistan, leading to 40% deaths and 50% diseases. Given the lack of water treatment facilities, the water quality problem could get worse in the future if Pakistan's government continues to overlook this issue. In the current situations, researchers and environmentalists would require real-time and frequent monitoring of quality of a water body, which is often hampered by conventional lab testing. The current lab testing for water quality is a slow, costly and overall inefficient method for a research setup. One of the novel solutions is detecting water quality in real-time through IoT sensors. IoT sensors quickly monitor and report quality parameters saving a great amount of time spent in lab to get the same results. However, IoT provides point data only and could be insufficient to reflect the water quality status of a large water body. Additionally, accessing a complex terrain or harsh areas could be challenging for data collection.

Alternatively, sensing water quality remotely through satellites imagery is a viable technique to gather sufficient data from a large water body. However, remote sensing techniques have their own drawbacks. Firstly, those water parameters that are not optically active cannot be monitored through satellites bands or imagery. Secondly, the data acquisition is bounded by satellite revisit time. Lastly, satellite image resolution, and environmental & climatic effects like cloud covering, back-scattering, top-of the atmosphere reflectance, could significantly influence the results derived from the satellite imagery.

In addition to this, the GIS data collected from labs can be useful for providing higher resolution and accurate data and can be used as a ground truth. Keeping in view the limitations of both of these techniques, this research work combines both IoT and remote sensing data through satellite and overcomes the drawbacks of both methods. Predictive and statistical modeling allow integrating both these data modalities and deriving reliable results that could allow environmentalists to remotely monitor a water body and have access to real-time water quality data.

The study area of this research work is Rawal Lake, located in Islamabad, the capital of Pakistan. It is one of the three major sources of drinking water to Islamabad. Unfortunately, the water of the lake is highly polluted, mainly by anionic pollutants (Ahmad, I., Ali, S., Tariq, M., & Ikram, M. (2001). Water pollution in Rawal lake Islamabad (part-1). Pakistan Journal of Analytical Chemistry, 2(1), 66-69.). Anionic pollutants can cause dangerous health issues including diarrhea, hepatitis, and cancer. The parameters analyzed in this work are temperature, turbidity, pH, total dissolved solids and dissolved oxygen. These parameters are optically active which implies that they could be detected through remote sensing, and contributes significantly towards water quality.

The project started with downloading satellite data from Sentinel and Landsat repositories. This data is basically Sentinel's and Landsat's band image data and required different levels of processing before they could be used further. The IoT-sensor based data from 2016-2020 was provided to us in the form of excel sheets. After processing satellite data, it was modelled with IoT-data using co-relation and regression models, providing parameter mapping relations between the two. These mathematical relations were in terms of band ratios and subsequently used in further processing in QGIS, an open-source GIS software. Following this phase, the processed image pixel values were averaged and reported in an excel file.

After all these phases, the water quality parameters were mapped to a water quality index or WQI based on Canadian Water Quality Index. This index reflects the quality status of water. Finally, after mapping each record to its quality label, the data was used to train machine learning models including Random Forest, Artificial Neural Networks, and SVM. The artificial neural network outperformed the other models in terms of several performance parameters, and it was finally deployed in the backend to support water quality classification through the web portal.

The web portal is powered with basic and advanced data analytics for improved data visualizations including a heat map which uses color codes to differentiate between points of varying water quality index. The water quality parameters extracted from multiple data sources are also archived on the web portal.

The project has several phases, starting from research phase to data collection, processing. After these phases when the dataset is ready, the project moves to model training, optimization, and testing, and finally development and deployment of a web portal.

LITERATURE REVIEW

Researchers around the globe have made efforts in the field of water quality monitoring. Since, this is a vast field, and some research papers are highly specialized in one aspect. Therefore, in the research phase a thorough literature review was conducted that provided the knowledge beneficial for the subsequent phases.

Water Quality Monitoring

Quality Indicators Based on Remote Sensing

Researchers classify the water quality indicators into two types, based on the remote sensing techniques. The first type is indicators with active optical properties, for example, Chl-a, TSS, CDOM, which are sensitive to electromagnetic waves and absorbs different segments of the spectrum.

The second type is indicators without optical properties like, Total Nitrogen, DO, Total Phosphorus. To assess these indicators, researchers often use statistical relationships with other indicators. (Topp, S. N., Pavelsky, T. M., Jensen, D., Simard, M., & Ross, M. R. V. (2020). Research Trends in the Use of Remote Sensing for Inland Water Quality Science: Moving Towards Multidisciplinary Applications. Water, 12(1), 169.).Satellite observations of reflectance must be corrected for atmospheric effects to derive water quality.

Some major techniques for remote sensing of water bodies are following:

1. Simple image interpretation to derive qualitative information about water quality
2. Use of various types of algorithms combining atmospherically corrected satellite images and in situ measurements to derive quantitative information about water quality

Modeling Approaches

Number of different types of models are used for water quality monitoring and estimation. The models which relate different optical properties, and its concentration of optically active indicators are known as bio-optical algorithms. For inland waters they are classified as empirical, semi analytical and machine learning models All these models have their own pros and cons. (Topp, S. N., Pavelsky, T. M., Jensen, D., Simard, M., & Ross, M. R. V. (2020). Research Trends in the Use of Remote Sensing for Inland Water Quality Science: Moving Towards Multidisciplinary Applications. Water, 12(1), 169.).

Empirical Methods

Empirical and semi-empirical methods involve statistical analyses of spectral information. Here the inversion algorithm is obtained between selected bands and water quality parameters through statistical analyses. Here the correlation is calculated between different parameters and corresponding reflectance values. Empirical methods include linear regression, single band method artificial neural network etc. One limitation of this model is that it is non-generalizable across large temporal and spatial scales. This type of model is restricted to confident predictions within the input data range and limits its application across spatiotemporal domains. These shortcomings usually take priority over the simplicity of model and its minimal computational requirements. Semi-empirical approaches also play important role in water quality monitoring and estimation. They use multi band values along with some basis in the physical properties of the constituents.

Semi-empirical Methods

Semi-empirical approach is designed to improve the spectral properties of the constituents which reduce noise from extraneous optical parameters. They focus on the measurement of water clarity, chlorophyll-a, cyanobacteria and total suspended solids (TSS). Notable semi-empirical indexes include the normalized difference chlorophyll index, the Floating Algal Index, and the normalized difference suspended sediment index etc. Semi-empirical models are more generalizable due to their reliance on physical properties. Application of this type of approach includes determining the presence of harmful cyanobacteria concentrations associated with eutrophication, robust algal bloom detection, and modelling sediment concentrations in rivers and deltas.

Semi-analytical Models

Semi-analytical methods require theoretical analyses of spectral information. As compared to semi-empirical and empirical models, these models make appropriate assumptions about light physics and are generalizable outside the scope of a given study.

In semi-analytical models, these properties of light are modelled through the backscattering and absorption coefficients of all the optically active elements found within the target area. One advantage of this inverse modelling procedure is that it can simultaneously estimate many parameters of water quality. But the development of model is still complex and requires information about atmospheric composition, bottom reflectance etc.

Machine Learning Models

Empirical and machine learning approaches are differentiated by their ability to operate through non-linear mathematic relationships in a multi-dimensional space. Within inland water remote sensing, different machine learning models are Naïve Bayes, random forest, SVM, artificial neural networks, genetic algorithms etc. have shown good results estimating parameters of water quality. Like empirical models, machine learning models are only applicable within the scope of the dataset.

The machine learning models tends to over fit with large number of input features and to avoid overfitting, distinct training and testing datasets that contain balanced samples of the parameters are required. These algorithms can generalize well and capture complicated non-linear relationships. Many researchers found decrease in accuracy and increase in the classification accuracy while using different machine learning models as compared to other empirical models.

Satellite Imagery and Spectral Bands

Electromagnetic wave is one of the wonderful phenomena of the universe. It is a mean of transfer of energy through a combination electric and magnetic waves. Electromagnetic wave is the reason human can see. It can travel through vacuum at the speed of light. Electromagnetic waves enable the transmission of signals and thus it is used in telecommunication and satellites. The whole technological setup involves use of electromagnetic waves directly or indirectly. Electromagnetic wave is divided into spectrums based on wavelengths/frequencies. The region which humans can perceive with our eyes is called the "visible region". The figure below illustrates different spectrums of electromagnetic waves.

Although humans can only see the visible region of the spectrum, all the other regions in the spectrum are invisible to the naked eye but it doesn't mean they are useless. In fact, each spectral band is useful in its own ways. Our eyes cannot perceive the invisible, but there are sensors that can pick up the other spectrums. For instance, near-infrared(NIR) are detectable through sensors, but it is invisible to human eye. Monitoring vegetation in an area is based on data from these sensors. Spectroscopy is the study of interaction of matter and electromagnetic waves. Although, the document will not discuss spectroscopy, but the main concept of spectroscopy called spectral signature gives insight about why study EM and why study the invisible spectrum. (GISGeography, *Spectral Signature Cheatsheet*, https://gisgeography. com/spectral-signature/)

A Survival Guide to Landsat Processing

This paper thoroughly explains the data preprocessing of Landsat products. Landsat provides products in different levels. The higher-level products are already pre-processed and could be ready-for-use data or it may require a little pre-processing. The main products used in this project was level-1 and level-2.

A pre-processing pipeline for Landsat product is explained in detail in this paper. Landsat 7 and 8 has a dedicated thermal band for temperature sensing, other bands like short wave infrared and the visible spectrum are also a significant part of satellite band imagery.

The research paper covers different concepts pertaining to satellite pre-processing like different conversion units, corrections, calculations of indices, and spectral analysis. (Young, N.E., Anderson, R.S., Chignell, S.M., Vorster, A.G., Lawrence, R. and Evangelista, P.H. (2017), *A survival guide to Landsat preprocessing*. Ecology, 98: 920-932. https://doi.org/10.1002/ecy.1730)

Machine Learning Approach for Predicting the Quality of Water

In this research paper five different models were used for water quality prediction. They are Naïve Bayes method, decision trees, support vector machine (SVM), K-Nearest Neighbor (KNN), artificial neural networks, and Multi-layer perceptron (MLP). Table 1 shows performance of the models on the top eight chosen attributes.

Table 1. Faiyaz Ahmad, M. A. D. R. R. S. M. M. (2020).

Technique	Accuracy	Precision	F1-Score	Recall
SVM	93.6%	0.93	0.92	0.92
KNN	92.2%	0.92	0.91	0.90
NB	88.6%	0.87	0.86	0.86
DT	94.4%	0.94	0.94	0.94
MLP	96.1%	0.95	0.95	0.95

From table 1, MLP performed best on the given attributes with an accuracy of 96.1%.

Classification Model for Water Quality using Machine Learning Techniques

Many classifiers were used namely using Naive Bayes algorithm,, Trees model using different algorithms, and Meta model using Bagging algorithms. Using different attributes produced different results in some cases Naïve Bayes was good, in some cases Bagging algorithm performed better than others but the classifier which performed well more than one time was K Star algorithm. The most significant features among the attributes were these six features namely Biochemical Oxygen Dissolve (BOD), Dissolved Oxygen (DO), Suspended solid (SS), Chemical Oxygen Dissolve (COD), Ammoniacal Nitrogen (NH3-N), and pH value.

Table 2. Muhammad, Salisu, et. al (2015).

Classifiers	Naïve Bayes	Conjunctive Rule	J48	K Star	Bagging
Accuracy	85.19%	74.81%	81.48%	86.67%	71.85%
Corrected Classified Instances	115	101	110	119	97
In Corrected Classified Instances	20	34	25	16	38

From the table 2, K Star algorithm outperformed the others in these selected features.

Comparison of Machine Learning Algorithms

Four machine learning techniques were used including support cubist regression trees (CB), random forest (RF), vector regression (SVR), artificial neural network (ANN), and, based on their accuracies in estimating WQI concentration. 10- fold cross-validation and Leave-one-out cross-validation (LOOCV) and was used to validate in situ data and satellite derived dataset. Three water quality indicators were examined namely Chl-a, SS, and turbidity.

Table 3. Hafeez, Sidrah, et.al (2019)

WQI	ANN (RMSE)	SVR (RMSE)	CUBIST (RMSE)	RF (RMSE)
In Situ Reflectance Data				
Chl-a (0.5-5.0 µg/L)	0.27	0.66	0.68	0.72
SS (0.7-8.0 mg/L)	0.7	1.18	1.18	1.29
Turbidity (1.3-12.0 NTU)	0.94	0.97	0.94	1.6
Satellite Derived Data				
Chl-a (0.3-28 µg/L)	0.27	0.66	0.68	0.72
SS (0.8-33.0 mg/L)	0.7	1.18	1.18	1.29
Turbidity (0.8-31.3 NTU)	0.94	0.97	0.94	1.6

After comparing all these models with better performing model parameters, table 3 suggests ANN outperformed all other models in almost every water quality indicator (WQI) in both in situ data and satellite derived data.

A Model for Water Quality Analysis Using Decision Tree

WEKA is the data analysis tool used for this study and five different decision tree classifiers were tested which are J48, LMT, Random forest, Hoeffding tree and Decision Stump. Their accuracies were compared using four parameters which are PH level, Alkalinity, conductivity and color.

Table 4. (Consolata, G., & Jeniffer, J. (2019)

Decision Tree	Accuracy
J48	93.6%
LMT	89.9%
Random Forest	91.7%
Hoeffding Tree	80.7%
Decision Stump	83.4%

From the table 4, J48 decision tree has outperformed all other decision tree classifiers in this case with an accuracy of approximately 94%.

MAIN FOCUS OF THE CHAPTER

The main emphasis of the chapter is to give an overview of the project phases followed by the issues and problems associated with them. The project had phases, namely

1. Project Phases

    a. Data collection
    b. Data Processing
    c. Modelling and optimization
    d. Development of the web portal

Data Collection

In the data collection phase, the researchers downloaded Landsat-8 data from UGS earth explorer, a satellite data repository. The timeline varied from 2006- 2020 for our data. This five-years data is sufficient to model the study for quality classification A shape file was used to extract target location's data. It is made in GIS software and is a polygon reflecting the shape of the target location.

Figure 1. Shape file of Rawal Lake

16.00002

10.03222

Sentinel-2 data was provided by a group of researchers who were working on a similar project. Moreover, GIS was provided to us in the form of excel sheets. The GIS had data for many different parameters.

1. Water Quality Parameters

    a. Temperature
    b. pH
    c. Turbidity
    d. Dissolved Oxygen
    e. Total Dissolved Solids

Table 5. Data Collection Summary

S.No	Data	Source
1	Data for 2016-2020	Landsat 7 & 8
2	Data for 2016-2020	Sentinel-2
3	Data for 2016-2020	GIS data from Rawal Lake filtration plants
4	IoT-sensor based data (2019)	IoT sensors at different testing points at Rawal Lake

It is important to consider temporal resolution of a satellite, which is 16 days for Landsat and 5 days for Sentinel. Obviously, Sentinel would provide more yearly data records because of its shorter temporal resolution.

Data Processing

This phase was the most challenging phase in this project and caused different types of problems during the project. This phase had different sub-processes. The chapter enlist them in the right sequence and briefly describe them. The figure 2 is a flow diagram that shows the processing pipeline for this project. The chapter will discuss each of these sub-processes in this section.

Figure 2. Phases of data processing

QGIS is the main software that has been used for processing satellite imagery. It is an open-source software with useful features relevant to our processing requirements like masking and raster processing. Moreover, for statistical calculations, researchers have used Python.

Masking

Masking is the process of selecting the target area only from a satellite image. Satellite images are in the form of .TIF file and it could include surrounding or unwanted areas besides the target location. The shape file subsets the target location only and passes it on to the subsequent stages in the processing pipeline.

Raster Processing for Different Band Ratios

Satellite image files are also known as raster files in GIS jargon. In this phase, researchers derive different values based on different band ratios. These band ratios are not taken arbitrarily but based on a target feature, for example, green band for vegetation index or blue for water bodies.

Different band ratios provide different information as described in the literature review section. Once this information is available, researchers can move towards co-relation stage, which binds the satellite data with the IoT- sensor based data.

Correlation among Satellite, GIS and IoT-sensor Data

It is a crucial step because the project's core idea is to fuse the IoT-sensor based data, GIS and satellite data and form an accumulated data set, to overcome the shortcomings of these datasets when treated/ processed individually. In this regard, a statistical correlation is derived from amalgamated dataset of satellite, GIS, and IoT-data. The data is unified based on the matching dates and days.

The GIS data is obtained from lab and reflects the ground realities. Similarly, IoT-data is a point data which is very precise. Therefore, correlating and regressing them with satellite data would provide us reliable relationships to model our case study.

After correlation, the regression is performed to obtain a set of relations based on band ratios for each parameter. These relations are the ones with minimum RMSE. The relations are shown in Table 6.

Table 6. Parameters equations

Parameter	Equation
Temperature	1201.1442/ln((480.8883/B11) + 1)
pH	11.82+411280000*(B3)10-1255600000*(B4)10-0.6244*(1/B5)+0.059250*(1/B6)+ 0.016603*(1/B5)2
Turbidity	log(10.22*(B4+ B5) - 0.18359)
DO	-6.16178*(B2-B2/B4-B6) – 0.12025
TDS	-222.44*(B4/B5) + 146.81

Raster Processing Based on Relations

The relations obtained because of co-relation and regression are made input to the raster processing equation manipulation feature. As a result, the pixel values of each .TIF file is transformed. These values reflect the accumulated mode of data. Finally, all these values are averaged and reported in an excel sheet.

WQI Calculations

Water Quality Index is a number assigned to a data record based on what features are under consideration. This number summarizes of what is there in the data record and maps it to a water quality label, as defined by a selected standard.

There are different WQI standard, and every standard defined its own set of rules for obtaining water quality index. Similarly, different WQI standards provide different ranges and respective labels for mapping water quality.

This project uses Canadian Water Quality Index (CWQI) defined Canadian Council of Ministers of the Environment (CCME). The CWQI maps ranges to different quality labels, the table 7 shows it.

Table 7. Canadian water quality index

CCME WQI Categories	Water Quality Status
Excellent (CCME WQI Value 95-100)	Water quality is protected with a virtual absence of threat or impairment; conditions very close to natural or pristine
Good (CCME WQI Value 80-94)	Water quality is protected with only a minor degree of threat or impairment; conditions rarely depart from natural or desirable levels
Fair (CCME WQI Value 65-79)	Water quality is usually protected but occasionally threatened or impaired; conditions sometimes depart from natural or desirable levels
Marginal (CCME WQI Value 45-64)	Water quality is frequently threatened or impaired; conditions often depart from natural or desirable levels
Poor (CCME WQI Value 0-44)	Water quality is almost always threatened or impaired; conditions usually depart from natural or desirable levels

The reason for choosing Canadian Index is tackling the issue of label imbalance problem in the subsequent model training and optimization phase. CWQI defined broad ranges and multiple labels, keeping in mind the water quality of different water bodies.

Most of the records fall in the fair, marginal and poor category and this is obvious because the water quality status of Rawal lake is not on par with what has been defined for excellent and good labels in the figure.

Once, the WQI is applied on the dataset, labels are derived for each data record. The final dataset is in the form of feature and labels, implying that it is ready for supervised machine learning models.

Modeling and Optimization

In this phase, different machine learning models were trained based on our datasets. The features included {Temperature, Turbidity, pH, Dissolved Oxygen, Total Suspended Solids} and labels were {Good, Fair, Poor}.

1. Trained Machine Learning Models

    a. Support vector machine
    b. Random forest
    c. Artificial neural network

After optimization, the parameters and hyperparameters of these models, the final statistics obtained are shown in table 8.

Table 8. Model accuracies

Model Name	Testing Accuracy	Training Accuracy
Support Vector Machine	95%	82%
Random Forest	90%	88%
Artificial Neural Network	97%	95%

Support Vector Machine

Support vector machine (SVM) separates the data points linearly and is suitable for data points that are linearly separable. Different kernels and C parameters have been used. However, the model accuracy reflects that the dataset cannot be linearly separated very well. Therefore, a complex algorithm was required to model it.

Random Forest

Random Forest is quite a robust algorithm with potential for parallel processing. It has a built-in feature selection, bagging and boosting properties and combines accumulated response by training many different decision trees. However, even after applying these features, it stands second to the ANN.

Artificial Neural Network

Artificial Neural Network (ANN) comes in different architecture. A deep-learning architecture with six hidden layers, a learning rate of 0.001, and Adam's optimizer has been used. L1-regularization is done to solve the bias-variance problem. After all this setup, ANN outperforms others, and this is the model deployed for water quality classification.

Development of Web Portal

The development of the web portal was initiated in parallel with the modeling phase. Its UI design is based on standard modern dashboard and spans many features required to support data analytics and visualization.

1. Web Portal Functionalities

    a. Dashboard
    b. Parametric analysis
    c. Data records
    d. Result analysis

Dashboard

Figure 3. Dashboard screenshot

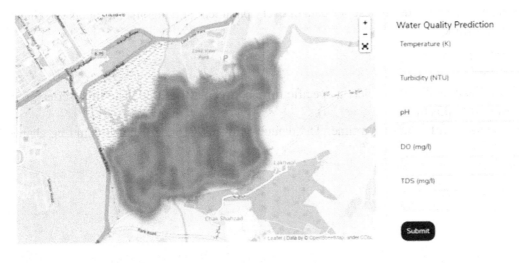

Feature Description

The **form** placed at the right side of the screen allows a user to input parameters data. On clicking the submit button, the parameters are passed to the trained model in the backend. The model predicts and returns data back, and finally displays it in the same area.

A **heatmap** on the left of the form displays Rawal Lake and plots the data points according to latitudes and longitudes. The color codes show quality status. Green is for good water quality; yellow is for fair, and red is for poor.

Parametric Analysis

Figure 4. Parametric analysis

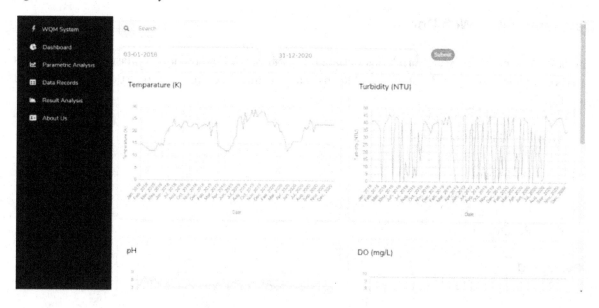

The parametric analysis provides timeseries data of every parameter, giving insights about the changes that took place overtime.

Feature Description

The **date picker** feature allows to select a specific time between 2016-2020. The time-series graphs change accordingly to this time.

The **time-series graphs** data for a time. The default is from 2016-2020 but it could be changed through the date picker.

Data Records

Figure 5. Data records dashboard

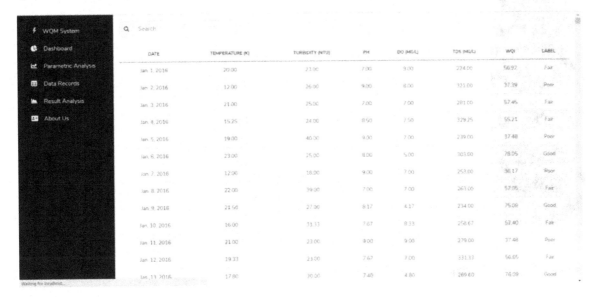

The data records section shows the excel sheet structure along with the data in a table format.

Result Analysis

The result analysis section displays two types of graphs. Both these graphs are based on the label and water quality index reading.

The **water quality count** is a bar graph which shows the frequency of occurrence of each quality label and the **water quality index** is a time-series graph based on water quality index and shows the trends in water quality overtime.

Figure 6. Parametric analysis

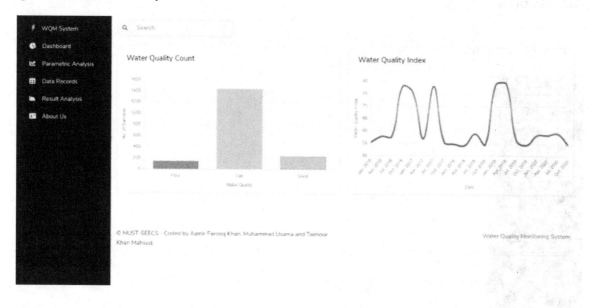

Issues, Controversies, and Problems

The data processing and modelling phase had major challenges and problems. Satellite data products comes at different level of editing. The more abstract a data product is, the less efforts are required to process it. Moreover, the inconsistencies within the data products invalidated the idea of a single data processing pipeline and researchers had to build unique pipelines for a data product.

The modelling phase had issues that are quite frequent in the machine learning domain. However, the main reason behind the occurrence of these problems was the dataset itself. After analyzing the dataset, some crucial facts were discovered. In the subsequent text, the chapter discusses these problems in detail.

Standardized Data Processing Pipeline

Satellite data is available at varying degrees of abstraction, which means that each data product is pre-processed by the vendor at different levels. Level-3 products are ready-to-use products and does not require any processing. However, level-2 and 1 requires certain processing steps.

Level-2 is more abstract than level-1 and didn't require solar and atmospheric correction while level-1 required both these corrections along with topographic, and radiometric corrections. Therefore, data pipeline build for level-2 data was invalid for level-1 data.

Researchers realized this later when they found out discrepancies in processed level-1 data. After two to three weeks of research it was realized that the processing pipeline was working correctly but the level-1 data required additional processing.

Radiometric Corrections

Radiometric correction involves aligning the image pixels with the underlying shape line. A python script was performing this correction. However, the pixels didn't align well. After debugging the script, the issue was detected.

The script used xml metadata file for each data product in to extract band constants. However, it was performing the computations in a wrong way. The band constant was shuffled in the script, and it ended up shuffling the fractions. This created a discrepancy in the script which was resolved successfully.

Scaling Issues

Satellite data product includes a metadata XML file which includes band constants. The scaling issue occurred while computing the temperatures. The temperature unit was Kelvin. However, after computing temperature, the values were out of range.

In processing script, the temperature band constant and co-efficient were incorporated but zenith angle was missing. This was realized after thorough research and finally incorporated into the script. The result was correct temperature values.

Label Imbalance Issue

After applying WQI, the researchers observed that some categories had very few labels. Likewise, the top two categories had no representation in the dataset. This was an underlying problem for creating bias in model training.

This problem is known as label imbalance, and it was quite evident in this case. Techniques like random sampling, K-fold cross validation were tried but they had little effect. The researchers realized that the problem is inherent to the target area because the water quality can never be excellent or good. The implied that some labels would never occur. Given the fact, the labels were restricted to the categories which had occurred frequently in the dataset and the supervised model had target variables representing the selected water quality labels.

SOLUTIONS AND RECOMMENDATIONS

The project requires GIS and remote sensing domain knowledge. It is recommended to take courses which would set the journey for the project because for absolute beginners it could be challenging. Secondly, a recommended way to avoid problems and inconsistencies is to read the product specification for the data products.

Band constants and co-efficient play a great role in computation therefore a good practice is to cross check equations and relations before processing the file. Finally, documenting the problems along the way always help in understanding crucial concepts and causes related to the GIS.

FUTURE RESEARCH DIRECTIONS

A viable future step to enhance this project would be redirecting IoT-sensor readings from IoT-devices to the web portal. This would allow for real time visualization and more integration between the web portal and the IoT-sensors.

Moreover, a novel WQI could be devised for the water quality in Pakistan. Many WQIs are specific and doesn't work perfectly when generalized. Canadian water quality assumes pristine and excellent conditions. However, this is not the case in Pakistan where the water bodies are more polluted in general.

CONCLUSION

Pakistan is a developing country with insufficient measures for water treatment and quality measurement. The measurement techniques mostly used for water quality are outdated, costly and time-consuming. Conventional lab techniques require data acquisition from the site and performing complex procedures to derive the status of water quality.

These days IoT sensors are widely used for water quality measurement. However, IoT-devices collect point data from a water body. Therefore, a large water body would require data collection from many points which could be time-consuming and inefficient. Moreover, accessing a complex terrain is another issue for data acquisition.

Alternatively, satellite can remotely monitor a water body, but its specifications could limit its usefulness. Therefore, an ideal solution would be combining these two techniques. The accumulated datasets are used to train machine learning models which was deployed to a server Finally, the project completed with the development of web portal. The web portal provides data analytics and visualization elements and communicates with the trained model in the backend through an interface.

REFERENCES

Ahmad, I., Ali, S., Tariq, M., & Ikram, M. (2001). Water pollution in Rawal lake Islamabad (part-1). *Pakistan Journal of Analytical Chemistry, 2*(1), 66–69.

Consolata, G., & Jeniffer, J. (2019) *A classification model for water quality analysis using decision tree.* Academic Press.

Daud, M. K. (2017). Drinking Water Quality Status and Contamination in Pakistan. Academic Press.

Faiyaz Ahmad, M. (2020). Machine Learning Approach for Predicting the Quality of Water. *International Journal of Advanced Science and Technology, 29*(5s), 275–282.

Hafeez, S., Wong, M. S., Ho, H. C., Nazeer, M., Nichol, J., Abbas, S., Tang, D., Lee, K. H., & Pun, L. (2019). Comparison of machine learning algorithms for retrieval of water quality indicators in case-II waters: A case study of Hong Kong. *Remote Sensing, 11*(6), 617. doi:10.3390/rs11060617

Muhammad, S., Makhtar, M., Rozaimee, A., Aziz, A., & Jamal, A. A. (2015). Classification Model for Water Quality using Machine Learning Techniques. *International Journal of Software Engineering and Its Applications*, *9*(6), 45–52. doi:10.14257/ijseia.2015.9.6.05

Topp, S. N., Pavelsky, T. M., Jensen, D., Simard, M., & Ross, M. R. V. (2020). Research Trends in the Use of Remote Sensing for Inland Water Quality Science: Moving Towards Multidisciplinary Applications. *Water (Basel)*, *12*(1), 169. doi:10.3390/w12010169

Young, N. E., Anderson, R. S., Chignell, S. M., Vorster, A. G., Lawrence, R., & Evangelista, P. H. (2017). A survival guide to Landsat preprocessing. *Ecology*, *98*(4), 920–932. doi:10.1002/ecy.1730 PMID:28072449

KEY TERMS AND DEFINITIONS

Adam's Optimizer: Adam's optimizer is a supplement for stochastic gradient decent algorithm which is commonly used for optimization purposes.

Atmospheric Corrections: It is the process of removing atmospheric effects on the reflectance values of a satellite image.

GIS: Geographic Information System or GIS is a system for accumulating, managing, and analyzing geo-spatial data.

Masking: It is the process of cropping a satellite image based on a shape file.

MLP: Multi-layer perceptron or MLP refers to a feedforward artificial neural network with many hidden layers.

WQI: A water quality index or WQI is a standard which derives the status of water quality based on given parameters.

XML: Extensible markup language or XML is a markup language for encoding documents in a human and machine-readable format.

Chapter 7
Machine Intelligence in Customer Relationship Management in Small and Large Companies

Ainul Balqis Sawal
Universiti Malaya, Malaysia

Muneer Ahmad
Universiti Malaya, Malaysia

Mitra Anusri Muralitharan
Universiti Malaya, Malaysia

Vinoshini Loganathan
Universiti Malaya, Malaysia

N. Z. Jhanjhi
iD https://orcid.org/0000-0001-8116-4733
Taylor's University, Malaysia

ABSTRACT

The current problem with CRM is weak marketing, as there is no individualization strategy, low productivity, and blurred marketing objectives. Next, companies are failing to completely capitalize on and extract useful information from a vast collection of databases. Organizations were also unable to adequately evaluate consumer behaviors and customer expectations that contribute to poor vendor-customer relationships and decrease customer loyalty. Most companies are seeking innovative ways to develop their CRM as it helps to challenge new ways of marketing and growing income, as customer loyalty and sales rely on one another. Whereas some businesses are incorporating data mining techniques in the management of CRM, there are several disadvantages in the market basket analysis, and one of the main drawbacks is that it is difficult to distinguish interesting patterns, as the number of rules obtained is very high. However, we might assume that it is computationally efficient as a minimum support value of 60% with a minimum confidence value of 80%.

DOI: 10.4018/978-1-7998-9201-4.ch007

INTRODUCTION

Recently, there has been an emergence of the modern business community. Customer relationships in economics are changing in a fundamental way, and businesses face the need to integrate new techniques and business strategies to address these shifts. Furthermore, the rapid growth of the Internet and its associated technology has significantly increased marketing opportunities and has changed the way businesses managed with their customers. The idea of large-scale manufacturing and marketing campaigns is now being transformed by the current ideas wherein the customer relationships are the central business concerns. Today, businesses are worried with the rising customer significance through customer lifecycle analysis. The methods and innovations of data mining and data warehousing in customer relationship management (CRM) offer the latest approaches for business sectors to adapt to the idea of marketing strategy. Ergo, to search useful information in the vast amount of data and incorporate the information in the practical operations becomes the issues that need to be tackled promptly, hence contributing to the emerging use of data mining. In meantime, economic development and the huge changes of e-commerce are shifting the actual rules of rivalry among the companies with unparalleled complexity and scope thus, more companies are starting to focus exclusively on customer relationship management (CRM), the latest philosophy of marketing management. The variety of customer information as well as business data available becomes the core customer relationship management (CRM) issues that need to be addressed urgently for modern businesses to obtain reliable information and optimize the value of the customer. It may appear that customer relationship management (CRM) may tend to be pertinent mostly in handling the relationship between businesses and customers. A further analysis reveals that for business customers it is even more vital (Rygielski, Wang, & Yen, 2002). Customers are mindful of all being delivered, and they demand better. In order to conform with the scenario, businesses must distinguish the goods to prevent them from being trifling commodities as an unwanted outcome. A vast amount of information is shared daily in a business-to-business (B2B) environment. For instance, there are increasingly myriad transactions, more varied customized agreements, and more intricate pricing structures. Collecting demographics and behavior data from customers makes accurate targeting feasible. There are a variety of diverse ways in which researchers can conduct a systematic study of datasets from various viewpoints, such as correlation and regression analysis, time series analysis, classification and prediction analysis, and cluster analysis. The application of data mining tools in customer relationship management (CRM) using association rules is also known as "market-basket analysis". Market basket analysis is a process that searches for association between individuals and objects that occur most often together. For customer-centric economy, market basket analysis evaluates product types to find commonalities that are crucial in various contexts (Kotu & Deshpande, 2019). This process analyses customer buying behaviors by identifying associations between the different customers placed in their shopping carts. The discovery of association rules along with other algorithms in customer relationship management (CRM) can help the business create marketing strategies to obtain and preserve prospective customers and optimize the value of the customer. In the global company, there is an evolving trend with the use of data mining techniques in customer relationship management (CRM). Many businesses have obtained and retained large quantities of customer data, however, the difficulty of finding important information concealed in data averts the businesses from converting the information into meaningful and practical aspects. The right use of data mining tools using association rules is among the greatest supporting tools for creating multiple CRM decisions as it's fantastic for retrieving and recognizing valuable knowledge and information from huge customer databases. As such in a customer-centric economy, it is worthwhile to pursue the use of data mining methods using association rules in customer relationship management (CRM) (Ngai, Xiu, & Chau, 2009).

LITERATURE REVIEW

Citation	Problem(s) identified	Suggested Solution(s)	Significance (e.g. evaluated as performance)
Run-Qing Liu, & Hong-Lei Mu. (2018). Customer Classification and Market Basket Analysis Using K-Means Clustering and Association Rules. Evidence from Distribution Big Data of Korean Retailing Company. 10.15813/kmr.2018.19.4.004	Client information and information mining examination have steadily ruled the cycle of Customer Relationship Management (CRM). This wonder shows that client information alongside the utilization of data strategies (IT) have become the reason for building an effective CRM methodology. Notwithstanding, a few organizations cannot find important data through a lot of client information, which prompts the disappointment of making suitable business techniques. Without appropriate procedures, the organizations may lose the upper hand or most likely fail. The motivation behind this examination is to propose CRM methodologies by sectioning clients into VIPs and Non-VIPs and distinguishing buy designs utilizing the VIPs' exchange information and information mining procedures (K-implies bunching and affiliation rules) of internet shopping centers in Korea. Un-fulfillment or clients' stir will prompt unforeseen misfortune, including both monetary misfortune and non-monetary misfortune.	K-implies grouping is to standard perceptions into K bunches and every one of the perceptions has a place with the group which has the closest mean. K-implies bunching calculation has been used broadly, including information mining, factual information examination, and other business settings. Subsequent to dividing clients into VIPs utilizing RFM esteems and K-implies bunching, our examination would utilize the market bin investigation to distinguish VIPs qualities and buy examples to assist SMEs with creating methodologies. The main use of the market bin examination is the affiliation rule. Affiliation is one of the significant procedures to distinguish and separate helpful information from enormous scope exchange information. It can address a few inquiries like what sort of items will in general be bought together by clients. Rules with high confidence and solid help can be alluded to as solid principles.	The use of information mining techniques in CRM is an arising pattern in the indus-attempt. It has pulled in the consideration of specialists and scholastics. In the advancement cycle of the huge information period, information mining applications can additionally improve the exactness of rules and examples, so that make it simpler for endeavors to improve customers' fulfillment and devotion and lessen client agitating aims. Second, this investigation chiefly analyzes clients' practices from the point of view of SMEs, subsequently called the hole in the SMEs setting. Since the majority of the past examinations have utilized CRM under the huge organizations' current circumstance, while a couple of studies zeroed in on SMEs. The consequence of this investigation is useful for SMEs that have little blemish ket share. It can likewise assist SMEs with distinguishing VIP client gatherings to hold important customers and give a significant technique to assist undertakings with improving a piece of the overall industry, crm market position and keep a positive advancement measure.
Gurudath, S. (2020, August). Market Basket Analysis & Recommendation System Using Association Rules. https://www.researchgate.net/publication/343558578_Market_Basket_Analysis_Recommendation_System_Using_Association_Rules#pf8	Information base innovation advancement has adequately developed to keep these data stacks strong, in any case, it is huge not to just keep that data, yet to survey the data to expand the estimation of the association. In the present client focused business sectors, business needs to build up sufficient and low publicizing methods that can respond to changes in client discernments and requests for items. It may likewise help business to perceive a totally different market technique that can adequately target.	Information Mining is the way toward refining significant information from colossal Databases which incorporates a gigantic combination of factual and computational techniques, neural organization investigation, grouping, characterization and summarizing data. The calculation of affiliation rules, which is one of the Data Mining strategies actualized by Market Basket Research. The examination is completed on the Grocery stores' exchange information for the clients. The exploration objective is to consider the classification of item that is probably going to be advertised related by executing Apriori and FP-Growth calculations	Apriori is the most mainstream and compelling calculation. The center standard of Apriori is to make different goes through informational collections or data sets that store exchanges or information. The report likewise recognized a property clarifying that the entirety of the super sets neglect to finish the assessment if the framework can't breeze through the base help assessment. The Breadth First Search (BFS) is utilized for the Apriori calculation. It additionally utilizes downstream bolting property (any super assortment of a bizarrely broken thing is abnormal). The exchange information base is generally on a level plane spread out. The recurrence of the thing set is estimated in each exchange. FP Growth calculation utilizes divided and vanquishing methods taking everything into account, and the structure of FP information is utilized to accomplish a streamlined value-based data set portrayal. No customary thing sets are needed for applicants.

Continued on following page

Table Continued

Citation	Problem(s) identified	Suggested Solution(s)	Significance (e.g. evaluated as performance)
Cavique, L. (n.d.). Next-Item Discovery in the Market Basket Analysis. http://pwp.net.ipl.pt/escs/lcavique/	Current information base limits related with scanner tag innovation and development of the Internet has prompted an immense assortment of client exchange information. The strategically pitching advertising system and the information on the following thing is basic. Nonetheless, the sheer number of affiliation rules may make the translation of the outcomes troublesome. Thus, it is hard for the advertiser to foresee the following thing that every client will purchase. To get continuous market bushels in diminished computational occasions the Similis calculation can be utilized, since this calculation lessens the quantity of ignores the information base. This paper tends to the issue of finding the following thing for every client in a huge information base promoting and can be viewed as an expansion of the market bushel investigation.	Consecutive example mining is a significant information mining issue with a wide scope of utilizations. Arrangement investigation is utilized to decide information designs all through a succession of worldly states. a grouping examination is generally applied to discover clickstream information in the Web destinations. In this work we will apply it to the market bushel grouping. A few creators have been attempting to discover clients' buy examples of merchandise in the retail and in the monetary areas. For the market bushel investigation, in the information base, every exchange speaks to a buy, which happened in a particular time and place and the exchange needs in any event two credits: client and thing. For this new issue every exchange needs three credits: client, thing and time. The yield of the calculation will be the most continuous succession of things.	The point of this work is to computerize the strategically pitching procedure. We might want to disentangle crafted by the advertiser, maintaining a strategic distance from the examination of thousands of rules in partner clients with their next thing. To locate the successive examples, we utilize a state change calculation that profits the most likely thing grouping. With the given arrangement in the information base advertising, it is conceivable to find the following thing for every client. The versatility of S1 and S2 calculations performs well, permitting their consideration in a business information base showcasing.
M, M., & T, R. (2014). Customer Relationship Management System A Case Study on Small- and Medium-sized Companies in North Germany. In Information Systems for Small and Medium-sized Enterprises., 169-198.	Client relationship the board is a typical instrument in enormous endeavors; yet it appears to be not to be effective with little and medium-sized undertakings (SME). further perception during our examination uncovered that even with CRM-framework accessibility, most endeavors don't exploit the maximum capacity of the frameworks as they either do not have a full reconciliation in different frameworks or specialty units or are not prepared to utilize the usefulness effectively or by any stretch of the imagination, individually. All project supervisors utilizing CRM approaches notice the presence of still unsolved issues, missing framework wide mix of CRM, representatives' acknowledgment, comprehension of CRM and client center, missing of highlights, squandered potential, and no agreeable measurement to assess accomplishment of targets. SMEs with particular CRM-frameworks, clients are altogether more fulfilled than with offices.	Led a study (blended methodology with a mysterious online poll and adjusted master meeting to confirm and increase the review results) to dissect client relationships, the executive's frameworks and the level of invasion in SME. The review covers 253 SMEs from North Germany and assesses the accessible frameworks, wanted and really conveyed usefulness, joining in the Information Technology scene, customer fulfillment, and contentions about advantages and disadvantages of current CRM-frameworks.	Notwithstanding the chosen target gathering of SMEs from a modern zone not known to have a solid fondness for CRM-frameworks, we could notice a moderately high conveyance of particular and incorporated CRM-frameworks. The outcomes additionally confirm our desire that conventional office application usefulness is frequently coordinated with the CRM prerequisites. The huge number of SMEs not having and not anticipating a CRM-framework contend that the low dissemination is brought about by the various requirements contrasted with huge ventures as the quantity of relations is either too low or not needing a serious connection with the executives. CRM-frameworks would surpass the need, subsequently a support for interest in programming and preparing isn't given.
K, P., & Siakas. (2017). Social Customer Relationship Management. A Case Study. International Journal of Entrepreneurial Knowledge, 5(1), 20-34. https://doi.org/10.1515/ijek-2017-0002	The disadvantage was discovered to be the time escalated presence required. Someone must be mindful to screen each organization, post important item data, answer questions and react to client remarks. There is an enormous rivalry in the area in the field. Hence, we endeavor to turn out to be better and to show it through Social Networks. Concentrate on the posts, yet it is currently simpler and more moderate gratitude to cell phones.	Executed the InCISIV Framework through a Social CRM technique. Three privately-run companies were acquainted with a reasonable Social CRM technique made. The assignment was to make a Social CRM Strategy in a joint exertion with an association. The points were likewise to help the organizations actualize the system reasonable to the organization points and to make proportions of the results.	The two organizations are extremely happy with the results and furthermore plan to broaden their quality via web-based media later on. Social CRM, another example that requires dynamic responsibility by customers and various accomplices, was broken down. The results consider the inclinations for an enterprise of the looks by means of electronic media are cost-related because of the free thought of their use and viral nature of online media that extended the customer base.

Continued on following page

Table Continued

Citation	Problem(s) identified	Suggested Solution(s)	Significance (e.g. evaluated as performance)
Awan, A. G., & Azhar, M. S. (2014, November). Customer Relationship Management System: A Case Study of Floor Mills in Bahawalpur District. *British Journal of Marketing Studies*, 2(7), 1-13. Retrieved from European-American Journals: https://www.eajournals.org/journals/british-journal-of-marketing-studies-bjms/vol-2issue7-november-2014/customer-relationship-management-system-case-study-floor-mills-bahawalpur-district-3/	In the business world, customer relationship management is a huge issue to discuss because every company's business and success relies upon this. The goal of this research paper is to examine the Customer Relationship Management of Bahawalpur Flour Mills and how these mills strategically manage their customer relationship. A major issue in CRM implementation is that it is viewed by employees as a technical innovation or application and not as a policy.	CRM strategy development is a systematic phase involving several strategic-level steps. Similarly, designing the business model is the first step. The analytical CRM system can generate the value of information by evaluating the data previously collected and by segmenting and targeting the analytical CRM customers. Exploratory analysis has been applied to the deductive method. For data collection, structured questionnaire approaches have been used. In order to record their responses to the CRM in SMEs, the respondents are directly contacted to fill out the questionnaire.	For the long term, CRM is well strategized, as it is systematically formulated, applied and exercised. The study also showed that the employee's doubts could be due to the fact that very little attempts have been made to test CRM yet. There is no system for monitoring and regulation, which often contributes to chaos in customer relationship management. The lack of proper assessment framework often contributes to multi-dimensional strategies of undirected CRM that ultimately contribute to consumer dissatisfaction. The management of flour mills is determined by the existence of the foregoing discrepancies and makes every reasonable opportunity to practice CRM in all its functional areas.
Ariffin, N. H. M. (2013). The Development of a Strategic CRM-i Framework: Case Study in Public Institutions of Higher Learning. *Procedia - Social and Behavioral Sciences*, 75, 36–43. https://doi.org/10.1016/j.sbspro.2013.04.005	The existing educational institutions emphasized on the management of customers. The vital arrangement for National Higher Education in Research and Development's objective is to assemble HR to pick up a culture of value exploration and global acknowledgment. Client Relationship Management (CRM) is hence observed as another viewpoint to assist instructive establishments with advancing client asset centralization. Examination has indicated that there is no reasonable structure accessible in Malaysia for the implementation of human-based methods using CRM in public higher learning institutions (IPTA).	Establish a conceptual structure for the management of customer relationship. The research technique is qualitative, using the approach of iterative triangulation. The architecture is called CRM-i and incorporates three separate approaches which are CRM, Strategic Information System Planning and Humanity Management. Furthermore, in the CRM approach, this research adds a new dimension which is human dimension. The creation of components and stages for the system is discussed in this article. As a suitable customer management framework, the CRM-i framework has been recognized.	CRM and Strategic Information System Approach shows that these two principles are chosen to emphasize the model in the CRM approach as a collection of value chain that adds value to an enterprise and is appropriate for the use in the IPTA in the context of this analysis. This implies that three approaches that have been identified and explained in this study were used to establish the system development known as CRM-i framework development. In the use of CRM systems and applications, there is still more room for development and the future for CRM appears to be encouraging on the grounds that its executions have acquired consideration among the top leadership. In addition, the importance of CRM implementation in their organization was highly perceived by all the organizations.
Seyedeh Pardis Bagherighadikolaei, &, Abdorrahman Haeria. (2020, February). A Case Study in an Automated Teller Machine (ATM) Manufacturing Company. A Data Mining Approach for Forecasting Failure Root Causes, 13(2), 101-121. 10.22094/JOIE.2020.1863364.1630	Based on the results of the Massachusetts Institute of Technology, information from organizations doubles every five years. The rate of data consumption, however, is 0.3. The method of information extraction from a welter of data has now been greatly accelerated by data mining techniques. A hybrid model using data obtained from an ATM manufacturing business is introduced in this paper. The same Machines' Code and the same customers in various provinces are grappling with the Code of Issues, which may be due to environmental conditions, the culture of using ATMs, and so on. In addition, the same Machines' Code and the same Code of Issues, as well as variations in the skills of technicians, appear to be some causes of substantially different repair times. This may be due to the extent of their competence and so on in the training history of technicians. Finally, in terms of its strategic decision-making, the organization will benefit from the results of this model.	Data Mining is one of the methods, through which the effective patterns of data with the least interference of users are discovered. In addition, the outputs of using data mining methods enable managers to make decisively crucial decisions. Decision Tree plays the most essential piece in C5.0 algorithm and is a common and powerful tool to predict and classify data. Moreover, Decision Tree produces rules within its structure, which explain the obtained results. algorithms of Clustering, Association Rules, and Classification can be used to compare the obtained results with the current research's result. Moreover, Fuzzy Clustering can be deployed to cluster different pieces.	Other clustering, association rules, and classification algorithms could be considered, and their results could be compared with algorithms used in this work. Regarding the results of clustering, it is suggested to use fuzzy clustering to obtain results that are more accurate.

Continued on following page

Table Continued

Citation	Problem(s) identified	Suggested Solution(s)	Significance (e.g. evaluated as performance)
D, T. (2018). Customer relationship management in business to business marketing. (Vol. 22). https://doi.org/10.30892/gtg.22204-291	The wide scope of administrations and merchandise available makes it hard for purchasers to settle on a choice and powers them to question their loyalty. To have the option to endure and keep up their situation on the lookout, organizations are very much aware that they need to adjust to the necessities of the cutting-edge period. In this vein, both to draw in buyers and to guarantee their dedication, they make significant client related strides. The board of client connections has become a methodology that advances absolutely because of these elements. Represents the usage of client relationship the executive's frameworks in the travel industry organizations working together to-business advertising. The investigation endeavors to explain the standards of business-to-business advertising and the executives of client relations. To have any kind of effect in a globalizing world where rivalry conditions are progressively developing, the presence and usage of client relationships and the board frameworks, especially in business-situated associations, is fundamental.	Assessed how the management systems for customer relationships are developed, how they should be applied, and what benefits they offer by performing interviews with employees in business-to-business promoting organizations in the travel industry area and business zones.	The representatives collectively expressed that they are aware of the upsides of CRM tasks as far as execution, time the board and benefits, just as stressed the estimation of CRM frameworks to the organization, adding that they should be actualized. Therefore, it is clear how significant CRM is to organizations with organizations as their center market. In order to retain their marketplace, to accomplish customer satisfaction and long-term customer loyalty, it is critical to provide an up-to-date strategy for customer management for tourism companies engaged in business-to-business marketing.
Gil-Gomez, H., Guerola-Navarro, V., Oltra-Badenes, R., & Lozano-Quilis, J. A. (2020). Customer relationship management: digital transformation and sustainable business model innovation. *Economic Research-Ekonomska Istraživanja, 33*(1), 2733–2750. https://doi.org/10.1080/1331677x.2019.1676283	Existing examinations on maintainability, which give a practical exploration model to survey and approve the possible impact of each CRM part (deals, advertising, and administrations) on the three elements of supportability (financial, natural, and social). Endless supply of our speculations, the ensuing approval of a particular model ought to bring a superior comprehension of the manner by which CRM-related advantages may build the positive effect of its segments on each component of maintainability. CRM can thus be viewed as such a Green IT, arranged toward computerized change and reasonable plan of action development.	CRM arose during the 1970s as another device for overseeing and enhancing deals power computerization inside organizations. From that point onward, it has gotten quite possibly the most famous devices for big business data the executives, for deals and showcasing purposes, yet additionally for more successful Customer Interaction and client information the board, just as for the comprehension of authoritative conduct, zeroing in especially on client maintenance and relationship the board, CRM is the latest integrational approach accessible for relationship the board. CRM a proficient mechanical answer for help organizations in the current abuse of their assets, just as to investigate and improve in all territories prompting supportable monetary and monetary development.	Most existing examinations have effectively portrayed the advantages of CRM execution on firm execution. By the by, a merged way to deal with the previously mentioned, misuse investigation duality of present and future CRM benefits remains undertheorized. With respect to manageability, CRM could be viewed as a key instrument and a beneficial answer for more feasible plans of action. CRM seems, by all accounts, to be a vital answer for decreasing the natural effect of the executive's choices given its conclusiveness on, for example, paper-saving cycles. In social terms, the comprehension of CRM as an administration arrangement, permitting the centralization of client information on a solitary information base with bound together access, could likewise be a central issue for regular great administration hypotheses, because of the ensuing accomplishment of more proficient between organization measures with regards to client merchant relations. Natural manageability suggests the usage of successful arrangements conveying a positive effect on a business' biological measurement.

Continued on following page

Table Continued

Citation	Problem(s) identified	Suggested Solution(s)	Significance (e.g. evaluated as performance)
Nugroho, A., Suharmanto, A., & Masugino. (2018). Customer relationship management implementation in the small and medium enterprise. *AIP Conference Proceedings, 1941*(1), 1–440. https://doi.org/10.1063/1.5028076	CRM isn't an action restricted just to the promoting office. Or maybe, it includes manageable coordinated improvement in the association's way of life and cycles. As matter of reality that these days advertising turns out to be all the more firmly adjusted to overseeing vital connections – inside and remotely – it would be fascinating for researchers to look at the conditions under which these organizations are powerful and those under which they are bound to perform inadequately (from at probably some organization members' point of view). It would likewise be clever to assess the connections among inside and outer organizations – both formal and casual – to all the more likely comprehend the connection between network linkages and business execution in various circumstances. Altering CRM can be dangerous and interface instruments of CRM sellers can't give wanted degrees of combination. CRM framework usage ought but rather take customary programming bundles, yet with enormous and complex variations it might require even a year. Little and medium undertaking will execute a dependable data innovation application to help their client information base, creation and deals just as showcasing the executives. to build up great correspondence with their client.	This investigation found that CRM usage at little and medium endeavors can help them in reengineering their information base administration incorporates the leads, account, contacts, bargain, task and meeting scheduler, report and examination. This chain cycle is planned in a CRM stage through Zoho CRM. This is executed because of its usefulness and the costumed model. The usefulness of the CRM stage that is set up by the supplier isn't ready to straightforwardly use by the client, it must be reengineered and upgraded through arrangement to meet the prerequisite and the requirements of the client. In this manner, a specific measure of arrangement is expected to guarantee the CRM would be handily executed to pick up the depicted advantages.	CRM execution in the SMEs gives numerous advantages to build their business, promoting technique, profitability, improve data sharing inside the selling organization and set up solid close connections with the client that lead to the client devotion. CRM innovation execution can improve their information base administration framework, locate the required record and contact effectively, discover the business information effectively on fingers activity without finding as much accounting as they did in the past time. CRM innovation execution killed their pressure and stress of losing their significant business information since the information has been logged and overseen in a cloud base client application. Adjacent to it improves their business efficiency both in promoting, deals and business investigation. One of the conceivable data innovation applications for little and medium endeavor is Customer Relation Management known as CRM. A decent CRM (client relationship the executives) program that helps organizations in fulfilling the client, the exploration study would investigate various strategies and procedures for setting up powerful CRM to fulfill the clients. Client relationship the board characterized as a business technique, it is viewed as emphatically identified with the standards of relationship showcasing and it depends on a strong direction of the entire association on customers.
M.S., & M.A.-S. (2019). *The effect of customer relationship management practices on airline customer loyalty.* (5th ed., Vol. 2). Journal of Tourism, Heritage & Services Marketing, Tourlab, the InternationalHellenic University. https://doi.org/10.5281/zenodo.3601669	Relationship advertising is one angle fortifying the business nonexclusive system to pick up supportable serious advantage. Henceforth it is fundamentally critical for business to hold clients and to keep them steadfast, particularly in a help industry, for example, carriers. Nonetheless, it is getting progressively hard to make separation in this serious carrier administration market. Data that is anything but difficult to acquire and utilize, an enormous field of alternatives, with qualities of immaterial administrations replace the trade. Value, publicizing, with advancement techniques are generally used to pull in clients and keep up associations with them, despite the fact that they are exceptionally simple to impersonate and supplant.	Carriers need to think in CRM as a drawn-out speculation, with the genuine advantages acquired through productive long lasting client connections. CRM contains the securing and arrangement of information about clients to empower the aircraft to sell a greater amount of their administrations all the more productively. CRM is a basic segment of the corporate system of aircrafts to separate themselves from contenders. CRM is a technique for transporters to individualize their organization, improve correspondence channels with customers and assure satisfaction of customers as a base of dependability. The reasonability of the technique grasped maintains the affiliation's capacity to win customers from competitors which is the base for business continuation. Different advantages of CRM to aircrafts incorporate capacity to target beneficial clients, tending to individualize advertising messages, and redid items and administrations.	CRM is shared quality. Shared qualities allude to the convictions of travelers towards the administrations given by EgyptAir and how fitting they are. The underlying model uncovered critical connection between shared qualities and traveler fulfillment. Travelers see the administrations given by EgyptAir to merit the worth paid. EgyptAir includes clients in their client related choices and offers motivating forces to purchase the organization administrations. These practices are positively influencing traveler fulfillment and move them to be faithful travelers through time. The more noteworthy the business is focused on its objective and mission, the more prominent the clients are fulfilled and steadfast. The respondents of travelers in this examination imagine that EgyptAir is focused on a great administration with satisfied responsibilities for their clients and is resolved to keep a good and high standard value. The responsibility of EgyptAir prompts its consumer loyalty as an outcome.

Continued on following page

Table Continued

Citation	Problem(s) identified	Suggested Solution(s)	Significance (e.g. evaluated as performance)
L.X., & A.N.D. (2019). Implementation of CRM Strategies to Increase Customer Loyalty, Case of Kazakhstan Companies. *The Influence of Customer Relationship Management on Customer Retention in the Insurance Sector, Malaysia, 2019*(42), 33–57. http://intijournal. newinti.edu.my	Clients become additionally requesting for predominant assistance and are being drawn closer by numerous contenders with equivalent or far and away superior help. Thus, the test isn't to make fulfilled clients; yet to build up pleased and faithful clients to expand the client maintenance. Malaysia's protection industry has moved to an accentuation towards client steadfastness and maintenance to get productive and practical improvement throughout some undefined time frame. Thus, Malaysia's protection industry needs to reposition itself towards the client arranged teaching to achieve benefit. The protection associations are re-planning to meet the cravings of the current clients through strong CRM strategies and projects to stay in business, increment the benefit, and upgrade the client maintenance. In 2018, Insurance infiltration of Malaysia has been drowsy. In the course of recent years, entrance in the business remained low at 4.8%. Under 40% of Malaysian residents own extra security. It is a genuine test to get precise measurements on the Malaysia protection market. Consequently, embraced learns about the training and adequacy of CRM in the protection business is essential to improve the usage and practice.	Client relationship the executives (CRM) is the procedure that encourages the protection associations to adjust to be serious in the protection business. CRM encourages the protection associations to hold the clients through the CRM practice to focus on the necessities of individual clients in the drawn-out period and contribute a ton to consumer loyalty, reliability, and maintenance.	CRM innovation expands the protection association's benefit and acquires significant assets for the workers to redo the client support. Personalisation administration empowers us to fulfill the requirements of clients. When the protection associations have finished client, data set and the arrangement of data stockpiling, CRM innovation enables the protection associations to have a productive promoting procedure for holding the focused-on clients, create solid associations with them, and gain benefit to the associations by utilizing the portable application to offer quality administrations and comfort to the clients. CRM innovation offers important client data to the protection associations for upgrading the advertising limits. Henceforth, CRM innovation empowers the protection associations to achieve the particular advertising objectives effectively in the protection business.
Singh, P., Yadav, S. K., & Pall, L. (2018). Sales Trend Analysis of Products Using Customer Relationship Management Tool. *International Journal of Scientific Research in Computer Science, Engineering and Information Technology*, 289–294. https://doi. org/10.32628/cseit183888	CRM is a modern strategy in the business where the dealer is seeking to establish a long partnership with their customers. CRM is trying to get its customer to climb the step ladder of dependability. Company first attempted to create the set-up with the public, that have a strong attention to the products and are happy to pay for it. Companies are seeking to adjust consumers' lookouts so that they could retain their buyers for a lifetime. The corporate sector is trying to connect people who demonstrate their clear value to the product and are always willing to pay for any product. The organization is constantly trying to adjust the look of its customers so that they can tailor their customers to repeat customers for the first time. The business then aims to adapt these consumers to everyday users. Then these consumers purchase only from the business in the related product categories. The transformation from manual to automatic processes is typically one of the most common. Significant problems that they may face while implementing CRM is in the execution of CRM software systems where the team might well have issues about the possibility of being hacked.	In the latest business developments, we will boost products for payment and the quality of products, promote the product and make it easier for consumers. The procedure for establishing a supportive and combined bond between the client and the provider is called customer relationship management, which is soon to be called CRM. As a result, we require CRM to keep our business well-organized and grow our company's profits. We need CRM to be an eminent operation. We would like a CRM as a function of which we prefer to take a square measure seeking to find an active advantage in an incredibly multi-channel atmosphere. We want to consider the needs of buyers that the square measure is increasingly tough and square, neglecting the previous ways in which we prefer to accustomate market and sell to them.	CRM is highly beneficial when it comes to sales and stock holding in line with industry dynamics and research. Builds longer connections with customers, through small yet important tasks, such as giving them birthday needs. Assure however and wherever they want to make changes to the company.

Continued on following page

Table Continued

Citation	Problem(s) identified	Suggested Solution(s)	Significance (e.g. evaluated as performance)
D.L.P.R., & T.S. (2017). Sumedha Journal of Management. A Study on Problems Faced by the Customer in Relation to Customer Relationship Management Practices, 6(3), 24–38. https://www.indianjournals.com	Customer service is of considerable importance to the banking field. Financial systems are the leading provider of financial services and also acts as an essential platform for the distribution of financial services. Although coverage has been widening on a regular basis, the content and quality of the dispensation of customer service has been experiencing immense pressure, largely due to the inability to meet the high requirements and aspirations of customers. The large network of branches spread throughout the country with millions of customers, the complicated variety of products and services provided, the diverse institutional framework-all this adds to the vastness and complexities of banking operations in India, which give spike to complaints about service deficiencies.	Customer Relationship Management (CRM) addresses the goal of maximizing business efficiency and improving shared value for the parties. It has the potential for maximizing marketing efficiency and effectiveness, which increases marketing productivity and generates shared values.	The analysis enabled the researcher to assess the consumer issue. Banks have seen remarkable development with economic transformations. In addition to providing a developing role in the economy, the banks, both public and private, are turning into profit-driven business associations. In an effort to be extra profitable, the banks have turned extremely competitive and customer oriented. This new orientation has pushed them to follow a much more realistic strategy to the operation of business. The CRM is one such tool that aims to fulfill the demands of the customer as per their evolving requirements.
Kachwala, T. (2018). Association Rule Mining Approach for Customer Relationship Management. CSEIT1833754| Received:20April2018| Accepted:30April 2018| March-April-2018 [(3)3: 1991-1995] International Journal of Scientific Research in Computer Science, Engineering and Information Technology, 3(3), 1991-1995. https://ijsrcseit.com	Rising rivalry and a competitive world, every business needs to recognize, predict and satisfy customers in order to optimize income. Association rule mining algorithms are used to figure out data mining issues in a fashionable way. Along with the broad range of current methods, it is continuously complicated to pick the best potential algorithm for rule-based mining tasks. Most of the current mining pattern proposals and pattern associations adopt optimization on the basis of objective quality metrics. This major appeal for this type of metric resides in the fact that pattern mining is primarily aimed at finding secret and previously undiscovered information from datasets, so there is no chance of contrasting the knowledge derived with the subjective knowledge given by the expert. In addition, the experience of two different users in a particular field can vary greatly, which triggers inaccuracy in the metrics. The purpose of this study is to explore frequent trends, impressive connections, and associations.	Association rule mining is a commonly utilized technique in data mining. Association rules are in a position to announce an exciting partnership in a wide database. The overwhelming quantity of data contained in the collection of association rules which could be used not only for the relationship in the database but also for the relationship between other types or groups of database instances. Assessing the best rules obtained by implementing the association rule collection of Apriori algorithm. It illustrates that monitoring overlapping rule, support and confidence in two comparable rule sets assists to assess the quality of algorithms or the measured accuracy of source datasets.	Association rules play a crucial role in numerous data mining applications, attempting to discover impressive trends in databases. Apriori is the easiest algorithm used to remove frequent patterns from the database. Apriori algorithm utilizes a broad item set property, which is simple to implement.
X. Zhou, G. Bargshady, M. Abdar, X. Tao, R. Gururajan and K. C. Chan, "A Case Study of Predicting Banking Customers Behaviour by Using Data Mining," 2019 6th International Conference on Behavioral, Economic and Socio-Cultural Computing (BESC), Beijing, China, 2019, pp. 1-6, doi: 10.1109/BESC48373.2019.8963436.	The Customer Knowledge Management model has lately gained interest in the integration of both technology and data aspects relationship management and a human related approach in knowledge management. Information technology tools and explosions in bank consumer data have strengthened and modified the bond between banks and their consumers. Data mining could be used for decision making and forecasting in banks and finance organizations. Classification is able to forecast patterns and trends of consumer behaviour. The prediction is made by classifying the database records in many predefined groups on the basis of certain parameters.	The Customer Knowledge Management framework is suggested. The neural networks and association rules mining are considered classification techniques. Moreover, applications from the banking line, customer segmentation, prospecting and absorption, protection, revenue growth, risk analysis, weaknesses and strengths were taken into account in order to demonstrate the effectiveness of classification models. K-fold cross-validation methodology utilized to validate the predictive approaches. Performance collation of the neural perceptron network and association rule classifiers is made in consideration of effectivity and efficiency.	A powerful CKM-data mining platform for predicting customer behaviour has been proposed. Neural Network and Association Rules have been implemented in the new customer knowledge management system for foretelling banking consumers' behaviour in the real world. The findings suggest that both models achieve good predictive efficiency and that data mining techniques could be used to enhance consumer behaviour, comprehension and predictability. The best model to achieve high predictive efficiency was the Neural Network with a 97% accuracy score. However, the neural network requires a lot of time to prepare the algorithm. In the future, neuro-fuzzy classifiers and ensemble of classifiers can be used to boost predictive ability.

Continued on following page

Table Continued

Citation	Problem(s) identified	Suggested Solution(s)	Significance (e.g. evaluated as performance)
Bezabeh, Belete. (2017). The application of data mining techniques to support customer relationship management: the case of ethiopian revenue and customs authority.	ERCA uses statistical analysis and evaluation methods to distinguish potential and low valued customers. The Authority shall verify whether or not the customers fulfill their duty in the revenue database while at the similar time cross-check the customs database whether the products are imported/exported with the payment of the appropriate tax by means of evaluation and statistical analysis techniques. However, these methods are not reliable and productive and have taken a significant duration of time to handle consumers as per their characteristics. These strategies are also less effective in regulating taxpayers who refuse to disclose their real income in order to minimize their tax bill and the income of the federal government.	Classification is a data mining learning model that intends to create a model for predicting future consumer actions by categorizing database records in a set of preselected groups depending on some parameters. The most popular classification techniques, like decision tree and neural network classification techniques, were evaluated in this research; the decision tree J48 algorithm and the neural network Multilayer-Perceptron algorithm were explored. The decision tree of the J48 offered a descriptive cluster classification model, allowing the investigation and identification of the characteristics of each cluster.	Classification models were developed with the J48 Decision Tree and Multilayer Perceptron Neural Net Algorithm. Out of these two classification models, the best classification model was chosen by comparing the cumulative accuracy of the classification of high-value consumers and the accuracy of the classification of low-value consumers. The decision tree model has proven valuable in all of these assessment parameters; the researcher assumes that the decision tree classification model has the required methodology for this CRM study. In general, the research findings were promising. It was practical to aggregate consumer data using data mining techniques that enabled business sense. To this result, the relevant literature on data mining techniques, CRM and customer segmentation has been studied.
Kumar, Dr. (2013). A Case Study of Customer Relationship Management using Data Mining Techniques. INTERNATIONAL JOURNAL OF TECHNOLOGICAL EXPLORATION AND LEARNING (IJTEL). 2. 275-280.	The main determinant of a successful business is the Value Creation for customers. Customer loyalty ensures the long-term success of companies. Customer bases established over a period of time have proven to be of considerable help in increasing the scope of the product or service of a specific company. However, the recent rise in business operating costs has made it more compelling for companies to increase loyalty among current customers when seeking to attract new ones. Processes through which a company generates value for the customer are often referred to as Customer Relationship Management (CRM).	Data mining techniques in CRM are based on a study of the literature on data mining techniques in CRM. Critical analysis of CRM data mining literature has helped to define significant CRM dimensions and data mining techniques for the implementation of CRM data mining techniques. It defines the dimensions of CRM as Customer Recognition, Customer Attraction, Customer Retention and Customer Growth. In conjunction, the forms of data mining models categorized as Association, Forecasting, Classification, Regression, Clustering, Sequence Discovery and Visualization.	Customer relationship management is important in order to compete successfully on the market today. The more efficiently you could use your consumer knowledge to serve their needs, the more successful you will be. It can be concluded that the operational CRM requires analytical CRM with predictive data mining models at its heart. The path to effective business means understanding your consumers and their demands, thus data mining is a key reference.

Continued on following page

Table Continued

Citation	Problem(s) identified	Suggested Solution(s)	Significance (e.g. evaluated as performance)
Maraghi, M., Adibi, M., Mehdizadeh, E. (2020). Using RFM Model and Market Basket Analysis for Segmenting Customers and Assigning Marketing Strategies to Resulted Segments. Journal of Applied Intelligent Systems and Information Sciences, 1(1), 35-43. doi: 10.22034/jaisis.2020.102488	Client Relationship Management (CRM) in grocery stores can draw in with customers in an adequate way with the end goal of building great connections and, subsequently, making full benefits. Clients are various gatherings of individuals who have various longings, styles and yearnings. Advertising the board of the fragments of the store to the clients to react accurately to their different necessities. Another huge issue of senior supervisors is the recognizable proof of beneficial clients. These clients make the primary benefit of the business and the protection of these benefits guarantees the endurance of the store. This exploration gives a model that supplements the CRM cycle from shopper attention to promoting plan task. Productive clients will be recognized because of the right comprehension, all things considered. Client Relationship Manager (CRM) divide in store divisions of customers to more readily comprehend and improve income and the viability of advertising recommendations. The more painstakingly fragmented customers, the more fruitful the promoting efforts relegated to them are.	Dataset with recency, recurrence and money related (RFM) measures is developed and grouped utilizing K-implies calculation. Six sections of clients are distinguished depending on the consequences of bunching. All fragments are completely investigated and promoting systems for them are portrayed in stage two. Item things are proposed to clients who bought forerunner items. Along these lines, devoted advertising recommendations are produced for some unique clients. Recency, recurrence and money related (RFM) model is quite possibly the most normally utilized methods lately for portioning clients. This model uses three measures including recency, recurrence and money related estimation of buys. The coordinated CRM model as a blend of RFM investigation and grouping and market crate examination is another issue concentrated here. This model finishes the cycle of CRM from client comprehension to appointing their showcasing system. This half breed model distinguishes clients who are probably going to do shopping later on from the start and afterward recognizes the items that will be bought by these clients. Furthermore, to streamline advertising recommendations and mine more excellent affiliation rules, ABC investigation of item stock administration is propelled, which is likewise another issue. Accordingly, a two-way enhancement is planned and actualized to recognize the more alluring clients.	Clients were scored utilizing the RFM model and the quarters of recency (R), recurrence (F), and money related (M) standards and afterward bunched utilizing the K-implies calculation. In view of past investigations, the quantity of groups was viewed as eight. To be more certain, Kohonen networks were utilized to decide the ideal number of bunches and the most noteworthy Silhouette was accomplished for eight groups. The ramifications of applying the consequences of this exploration for stores incorporate expanding productivity, brand advancement and improving consumer loyalty coming about because of an extensive comprehension of clients' requirements and desires and allotting proper advertising systems to them. For future examination, the idea of client lifetime esteem as RFM can be utilized to rank various portions of clients. What's more, the system of the proposed half breed model can be applied to different settings, for example, bank clients, protection, etc. Also, thinking about various loads for R, F and M models and utilizing AHP for setting them may deliver more attractive outcomes. At long last, the bunching quality (normal Silhouette) of the current strategy can be expanded utilizing other grouping calculations, for example, two-venture bunching.

DATA MINING RELATED SOLUTION(S)

Research Framework

Figure 1.

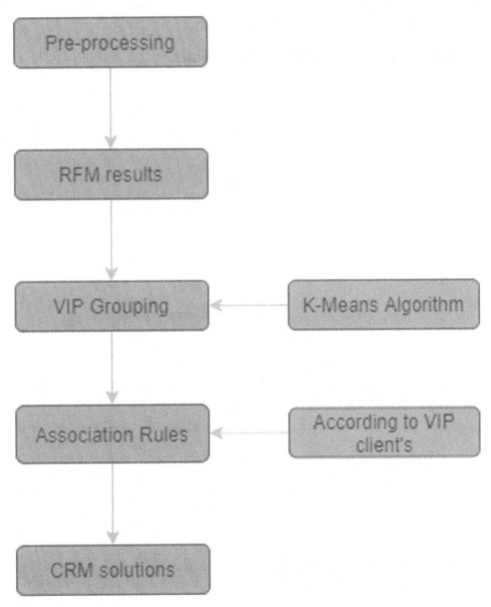

The structure of this research is provided as above. We pre-handled the client's information by erasing the copied records and those with many missing qualities or off base qualities. At that point, we characterized the RFM estimations of every client dependent on the information given by the target firm. Besides, before the model development, it is important to characterize the RFM scores and distinguish circulation highlights of these RFM scores. At that point, normalized RFM scores to assemble bunching models. Thirdly, utilizing K-implies calculation to assess and pick the best grouping model to decide the VIPs by choosing the significant factors and drop those with low connections with the objective factors. It is critical to recognize and choose factors which are taken on a decent informative force. Fourthly, utilizing the market crate examination strategy to distinguish affiliation rules from the exchanges of VIPs which are dictated by the best anticipated model. At last, utilized these examples to propose CRM procedures for administrators to give better assistance.

Market Basket Analysis for Generating CRM Rules

Figure 2.

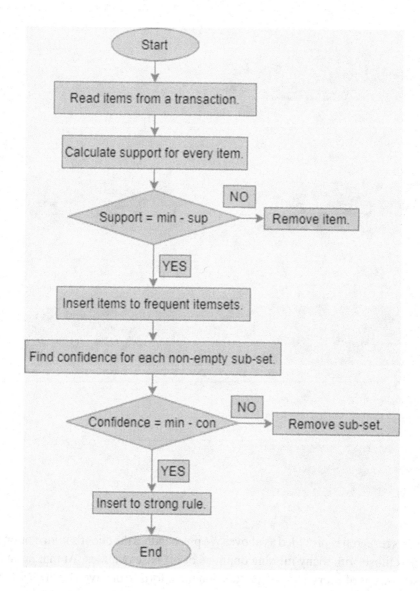

Market basket analysis is an assortment of things bought by a client is an itemset. The arrangement of things on the left-hand side is the precursor of the standard, while the one to the privilege is the resulting. The likelihood that the precursor occasion will happen is the help of the standard. That basically alludes to the relative recurrence that an itemset shows up in exchanges. In a QSR, the help of a thing or thing blend assists with recognizing cornerstone items. Thus, on the off chance that regular thing sets have high help, at that point they can be estimated to draw in individuals to the store. As per market container examination, a client that arranges this feast would be bound to arrange a beverage. The likelihood that a client will buy a beverage on the state of buying a sandwich and treats is alluded to as the certainty of the standard. Certainty can be utilized for item situation technique and expanding productivity. Setting high edge things close to related high certainty things can build the general edge on buys. The lift of the standard is the proportion of the help of the left-hand side of the standard co-happening with the right-hand side, partitioned by the likelihood that the left-hand side and right-hand side co-happen if the two are autonomous.

Market Basket Analysis ETL Process Flow Diagram

Figure 3.

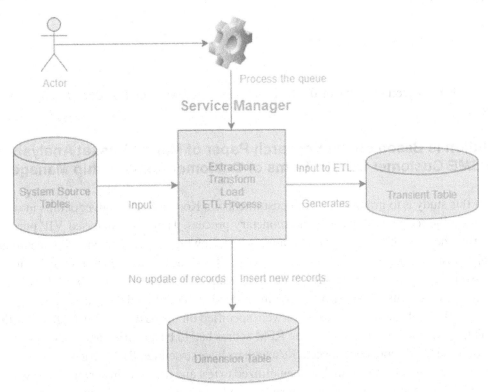

Market Basket Analysis is an information mining strategy that yields connections between different things in a client's container. Market Basket Analysis reports are utilized to comprehend what sells with what and incorporates the likelihood and productivity of market bushels. Such a report can be utilized to design advancements, streamline item arrangement, and backing store planogram choices. These reports assist you with understanding the measurable connection between deals for various products. The ETL Service changes the tables into various coordinated tables of MBA for effective information mining. MBA dwells in a different arrangement of information base patterns and uses ETL contents to move input information from Retail Analytics into the MBA tables. ETL programs are added to change Retail Analytics tables into various coordinated tables for association and product progression, client sections, and schedule for effective queries or accumulations for information mining. The ETL contents might be run week by week or during a planned time period to revive the information in the MBA composition prior to running reports as a component of the week after week clump measure. The consequences of any reports are ready for utilization by a different outbound ETL measure that distributes results to foreordained yield tables and emerging perspectives and tidies up any middle information in the MBA tables. The ETL cycle ought to be run preceding running some other administrations to assemble data for the MBA required tables. The accompanying schedules play out this assignment:

To set up the ETL administration in the line:

```
start
rse_srvc_mgr.batch_pre_process('CORE_ETL','MBA');
end;
To handle the line:
start
rse_srvc_mgr.process_queue('CORE_ETL');
end;
```

A survey of the objective tables of the ETL program rundown might be done to approve the ETL information.

The Solution is Based on the Research Paper of Market Basket Analysis of Korean SME Customer Data in Terms of Customer Relationship Management

The goal of this study is to group the overall customers of a Korean SME's buyer details into VIPs and non-VIPs using the recency, frequency, and monetary process. They then analysed VIP purchase data in order to find the most powerful rules and patterns by following different data mining techniques, like the K-means clustering algorithm and association rules. This study aims to facilitate SMEs to maintain the loyal customers or VIPs by providing effective rules and patterns (Lee et al., 2018). First researchers pre-processed buyers' transaction data by removing redundant records and those null or incorrect values. After that, the RFM values of every customer were specified on the basis of the data given by the target business. Due to the model development, the RFM scores had to be specified and the distribution characteristics of these RFM scores established. Researchers have structured RFM scores to create clustering models. Then, the K-means algorithm has been utilized to test and the best clustering model was selected to assess the VIPs by choosing the relevant variables and removing those with low correlations with the target variables. The market basket analysis approach has been utilized to define the association rules for VIP transactions that are calculated by the best predictive model. Ultimately, researchers used these patterns to develop CRM methods for managers to deliver better support.

DESCRIPTIVE OR VISUAL INTERPRETATION OF DATASET

Table 1. Descriptive or visual interpretation of dataset

Customer ID	Branch Code	Age
Gender	First Purchase Date	Final Purchase Date
Total Purchase Frequency	Total Purchase Amount	Average Purchase Amount
Average purchase cycle	Membership card	Online membership
Customer Relationship Type	Membership Card Join Date	Membership Self-Identification
Card Classification	Foreign Expensive Good Preference	Normal Good Preference Type
Most Expensive Good Preference	Multi-Branch Visit	Sensitive to Discount
Main Branch	Biz-Area	
Food	Goods	Cosmetic
Women Wear	Men Wear	Kids
Home Product	Home Appliance	Furniture
Total Purchase Category	Main Item Category	

The data set utilized in this case study was taken from a shopping centre in Korea that was pleased to give proper transaction data of its buyers. The shopping centre primarily sells products such as food, products, cosmetics, and garments through online or in the physical store. This whole data contains basic buyers' details and data on their purchasing activity from June 2015 to June 2016. This case study collected 2,000 buyers and 51,080 transaction data as basic records from the target shop customer segmentation. Above is a preview of the data collection in Table 1 used by the researchers.

Preprocessing on Data

For preprocessing the data set, duplicated and missing or incorrect values have been removed and left with 1866 buyers. The descriptive details of 1866 buyers are shown in Table 2, which was analyzed by SPSS 21. Based on the table used by the researchers, females are higher than male, and most of the buyers are middle-aged. For buyers grouping, RFM values had to be specified as shown in Table 3 consisting of partial data only, while Table 4 shows the descriptive statistical analysis of RFM values. To standardize the RFM score, the Analytical Hierarchy Process (AHP) and the Expert Consultation System (Delphi) have been utilized to assess the weight of each predictor by performing an AHP survey and inviting 15 CRM expert professors. After pairwise comparison is completed, the consistency ratio was evaluated and the geometric mean of the quantity of pairwise comparisons was calculated. The RFM weight was then computed and the RFM score was divided by five scaling points. The final RFM scores obtained are shown in Table 5.

Table 2. Descriptive or visual interpretation of dataset

Variables	Groups	Frequency	Percentage %
Gender	Male	332	12.80
	Female	1534	82.20
	Total	1866	100
Age	<=20	1	0.10
	21~30	151	8.10
	31~40	538	28.80
	41~50	662	35.50
	51~60	370	19.80
	61~70	105	5.60
	71~80	35	1.90
	81~90	4	0.20
	Total	1866	100

ID	Recency	Frequency	Monetary
6141	129	6	261.57
9937	49	14	270.08
19053	7	30	888.60
22258	24	30	765.28
28214	12	75	2447.44
...	...	...	...

** Unit: 10 Thousand Won (KRW)

RFM Indicators	Minimum	Maximum	Mean	Std.
R	1	366	48.07	67.54
F	2	227	26.25	20.60
M	188.63	6915.51	659.68	655.57

Scaling	ID	R Score	F Score	M Score	RFM Score
5 Score	6141	1	1	1	100
4 Score	9937	2	2	1	130
3 Score	19053	4	4	5	470
2 Score	22258	3	4	4	389
1 Score	28214	4	5	5	489

Descriptive or Visual Interpretation of Outcomes or Results

The study tested three different clustering approaches before selecting the best which are Kohonen, K-means and Two Step algorithms. The process started by normalizing RFM values and then obtaining the Z-score for RFM by SPSS 21. The RFM Z-scores are utilized as input variables for clustering analysis. The number of clusters is calculated to be 4 and the best segmentation result is the K-means of the SPSS Modeler 14.1. Figure 3 from the paper illustrates the difference curve with the usage of C5.0 algorithm.

Figure 4.

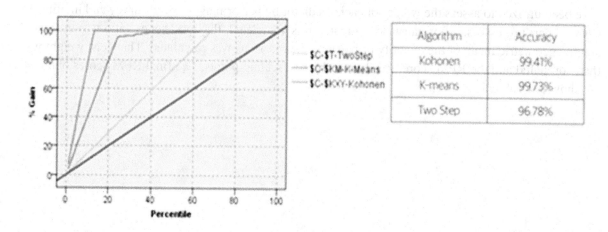

Algorithm	Accuracy
Kohonen	99.41%
K-means	99.73%
Two Step	96.78%

Since K-means has achieved the best curve, K-means clustering algorithm model has been developed by researchers. The cluster features were analyzed and, based on the results, cluster 3 had been selected as VIPs because the value of R (15.586) was lower than the average of R, whereas the values of F(62.132) and M(1093.515) were higher than the average of F and M. Cluster 3 could therefore offer greater economic advantages than other clusters. It was therefore concluded that the VIPs were 227 of the 1,866 customers. In addition to the RFM variables, it was important to identify other variables that would impact the segmentation of customers. As a result, 12 variables were evaluated by implementing the Gain Ratio attribute evaluator based on the ranked search tool. In the research paper, Table 8 sets out the descending order of significance of the 12 variables and Table 9 sets out the sense of each variable.

Table 3. Descriptive interpretation of outcomes

Items	Entropy	Gain Ratio
M	0	0.534
F	0.178	0.356
R	0.442	0.092
MEMBERSHIP_SELF_IDEN	0.506	0.028
AGE	0.514	0.020
BRANCH	0.522	0.012
GENDER	0.522	0.012
ZZ_CARD_FLG	0.524	0.010
MULTI_BRANCH_VISIT_YN	0.526	0.008
MOST_EXP_GOOD_PREF_YN	0.529	0.005
ENURI_SENSI_YN	0.530	0.004
FOREIGN_EXP_GOOD_PREF_YN	0.531	0.003

Attribute	Meaning
R	The time interval between the last purchasing behavior and current
F	The number of transactions over a certain period of time
M	The amount of money spent on products or services
MEMBERSHIP_SELF_IDEN	Membership Self-identify (For their own use=0, for family members use=1)
AGE	Customer's age
BRANCH	Gangnam=1, Konkuk=2, Gwanak=3, Gwangju Outlet=4, Gwangju=5, GimhaeOutlet=6, Nowon=7, Daegu=8, Daejeon=9, Dongnae=10, Myeong-dong=11, Mia=12, BusanJung-gu =13, Busanjin-gu=14, Bupyeong=15, Bundang=16, Sangin=17, Anyang=18, Yeongdeungpo=19, Ulsan=20, Incheon=21, Ilsan=22, Jamsil=23, Jeonju=24, Changwon=25, Cheongnyangni =26, Pohang=27, Haeundae=28
GENDER	The gender of customers (Male=0, Female=1)
ZZ_CARD_FLG	The classification of card (No card=0, Public card=1, holiday card=2, Ordinary card=3)
MULTI_BRANCH_VISIT_YN	Whether customers will go to multi-branches shopping (Single shop=1, Multiple shop=2)
MOST_EXP_GOOD_PREF_YN	Whether customers prefer to buy most expensive goods (No=0, Yes=1)
ENURI_SENSI_YN	Whether customers will be sensitive to discount (No=0, Yes=1)
FOREIGN_EXP_GOOD_PREF_YN	Whether customers prefer to buy foreign expensive goods (No=0, Discount sensitive =1, Prefer to buy overseas brand-name goods=2, Buy overseas brand-name goods once=3, Overseas brand-name freak =4)

Next, is market basket analysis which will discover association rules by using the apriori algorithm. Researchers have made several tries with various parameter values, and then specify the minimum support to 10%, minimum confidence to 85% and maximum number of antecedents to 3. Ultimately, it was appropriate to evaluate 14,104 transaction data for VIPs (227 customers) depending on the results of the classification. An interpretation of the transaction data is provided in the research paper in Table 10. The maximum number of rules that have been obtained is 29 and all of the rules have reached both the minimum support (10%) and the minimum confidence (85%) as defined previously. While confidence of 10 rules are greater than 90% have been chosen and shown in Table 11 in the paper.

Table 4. Association rules by using the Apriori Algorithm

Order	Consequent	Antecedent	Instances	Support %	Confid-ence %	Rule Support %	Lift
1	FOOD	HOME_PRODUCT GOODS	6292	44.611	96.249	42.938	1.243
2	FOOD	WOMEN_WEAR HOME_PRODUCT	6350	45.023	96.220	43.321	1.242
3	WOMEN WEAR	KIDS COSMETIC	8337	59.111	93.175	55.077	1.203
4	WOMEN WEAR	MEN_WEAR KIDS GOODS	8205	58.174	92.724	53.942	1.197
5	FOOD	MEN_WEAR HOME_PRODUCT	8229	58.345	92.575	54.013	1.195
6	FOOD	WOMEN_WEAR HOME_PRODUCT COSMETIC	3079	21.831	92.043	20.094	1.188
7	FOOD	MEN_WEAR HOME_PRODUCT GOODS	2791	19.789	91.652	18.137	1.257
8	FOOD	HOME_PRODUCT KIDS WOMEN_WEAR	2791	19.789	91.508	18.108	1.182
9	COSMETIC	HOME_PRODUCT KIDS WOMEN_WEAR	8043	57.026	91.172	51.992	1.177
10	COSMETIC	KIDS WOMEN_WEAR FOOD	3451	24.268	90.351	22.107	1.239

Customer ID	Purchase Date	Purchase Items	Amount
9937	2015/06/06	(Food, Women Wear, Home Appliance)	186,742
	2015/07/03	(Cosmetic, Men Wear)	170,300
	2015/07/31	(Food, Men Wear)	154,325
	2015/08/04	(Food)	68,942
	2015/08/13	(Food, Cosmetic, Men Wear)	154,700
	2015/09/07	(Food, Women Wear, Men Wear)	196,980
	2015/09/23	(Cosmetic, Women Wear, Home Appliance)	175,490
	2015/10/17	(Food, Men Wear, Home Appliance)	287,843
	2015/10/23	(Food)	49,826
	2015/11/11	(Food, Cosmetic, Women Wear, Men Wear, Home Appliance)	435,630
	2015/12/26	(Food, Cosmetic, Women Wear, Men Wear, Home Appliance)	214,878
	2016/01/08	(Food, Home Appliance)	39,593
	2016/02/23	(Food, Home Appliance)	305,738
	2016/04/13	(Cosmetic, Women Wear, Men Wear)	259,853
Total	Purchasing Frequency: 14. The Number of Item Category 5		2,700,840

Rule 1 has the support of 44.611% and the lift of 1.243 implies that this rule is highly reliable. Rule 3 has the highest level of support with a confidence level of 93.175%, which means that consumers who buy children's goods and cosmetics commonly prefer to purchase women's clothing at the same time simultaneously with each of them. Since the target company's VIPs are predominantly middle-aged women, they are normally concerned and buy these products, such as women's wear, kid's products and cosmetics. Researchers have re-organized the relationship between the variables and the implications of the 10 rules, as seen in Table 12. Centered on the table, food is often grouped with other categories such as home products, women wear, men wear, goods, kid's products, and cosmetics. Correspondingly, women's wear is also contrasted with other groups, such as kid's products, men's wear, goods and cosmetics. Cosmetic is often commonly associated with other categories, such as women's clothes, kid's products, household products, and food. This finding also indirectly points out that the buyers of the goal shop are predominantly female customers. This research examined the key purchasing item categories of VIPs to provide better comprehension of the rules of association.

Table 5. Usage Frequency of Confidence Top 10 Rules

Variables	Frequency	Consequent
Home Product	6	Food
Women Wear	3	Food
Men Wear	2	Food
Goods	2	Food
KIDs	1	Food
Cosmetic	1	Food
KIDs	2	Women Wear
Men Wear	1	Women Wear
Cosmetic	1	Women Wear
Goods	1	Women Wear
KIDs	2	Cosmetic
Women Wear	2	Cosmetic
Home Product	1	Cosmetic
Food	1	Cosmetic

CONCLUSION

As indicated by the outcomes, we characterized VIPs based on RFM esteems, and created division models that order clients into VIPs or Non-VIPs gatherings. All the while, we propose a dynamic system which joins a pattern forecast model. At that point, we removed 29 affiliation rules from the exchange information of VIP clients and chose to think of CRM methodologies against VIP clients for the objective CRM as follows. Most importantly, the characterization model can be promoted to recognize VIPs so the objective shopping center can practice advertising exercises against them. We can plan explicit special exercises for VIP clients and give them more exact administrations to accomplish long haul improvement of undertakings. For instance, endeavors can distinguish VIPs and plan explicit assistance projects, for example, sending limited time exercises and related item indexes through messages, which is helpful for diminishing costs and proficiently assigning assets. Also, the outcomes show that ladies wear is exceptionally connected with the classifications of men wear, children's item, corrective, it is suggested that the objective CRM overhauls their site so the page related with ladies wear can be a single tick away from men wear, children's item, restorative, and so forth Set up pertinent promoting exercises as per the substance of the affiliation rules. It will improve clients' fulfillment, which is useful for client maintenance. Thirdly, the objective shopping center can set suggestions on the gateway and give an inclination to item publicizing as indicated by clients' inquiry propensities. The flow watchwords of the objective CRM in significant hunt destinations will propose and incite clients to buy item liked. In this manner, it is suggested that the objective CRM resets current watchwords would assist with drawing in more likely clients. Last however not the least. Since VIPs are basically moderately aged ladies who favored beautifiers, ladies wear, home item, it is insightful to urge clients to purchase items effectively and set explicit advertising exercises for these VIPs. For instance, when buying 50$ of an item, the objective shop will introduce 5$ B item coupons as a blessing. The objective firm can improve consumer loyalty and faithfulness by offering a customized assistance, for example, by requesting clients a 20% rebate on a specific item on the date of their birthday. Ideally, the above recommendations could likewise be useful to SMEs that try to expand their piece of the pie.

REFERENCES

Akhmedov, R. (2017). *Implementation of CRM Strategies to Increase Customer Loyalty, Case of Kazakhstan Companies.* Academic Press.

Ariffin, N. H. M. (2013). The Development of a Strategic CRM-i Framework: Case Study in Public Institutions of Higher Learning. *Procedia: Social and Behavioral Sciences, 75*, 36–43. doi:10.1016/j.sbspro.2013.04.005

Awan, A. G., & Azhar, S. (2014). Customer Relationship Management System: A case study of Floor Mills Bahawalpur District. *British Journal of Marketing Studies, 2*(7), 1–13.

Bagherighadikolaei, S., Ghousi, R., & Haeri, A. (2020). A Data Mining approach for forecasting failure root causes: A case study in an Automated Teller Machine (ATM) manufacturing company. *Journal of Optimization in Industrial Engineering, 13*(2), 101–121. doi:10.22094/joie.2020.1863364.1630

Bezabeh, B. B. (2017). The application of data mining techniques to support customer relationship management: the case of ethiopian revenue and customs authority. *arXiv preprint arXiv:1706.10050.*

Cavique, L. (2005). *Next-item discovery in the market basket analysis.* Paper presented at the 2005 portuguese conference on artificial intelligence. 10.1109/EPIA.2005.341294

Gil-Gomez, H., Guerola-Navarro, V., Oltra-Badenes, R., & Lozano-Quilis, J. A. (2020). Customer relationship management: Digital transformation and sustainable business model innovation. *Economic Research-Ekonomska Istraživanja, 33*(1), 2733–2750. doi:10.1080/1331677X.2019.1676283

Gurudath, S. (2020). *Market Basket Analysis & Recommendation System Using Association Rules.* Academic Press.

Kumar, D. (2013). A Case Study of Customer Relationship Management using Data Mining Techniques. *International Journal of Technological Exploration and Learning*, 2, 275–280.

Liu, R.-Q., Lee, Y.-C., & Mu, H. (2018). Customer Classification and Market Basket Analysis Using K-Means Clustering and Association Rules. *Evidence from Distribution Big Data of Korean Retailing Company.*, *19*, 59–76. doi:10.15813/kmr.2018.19.4.004

Loshin, D., & Reifer, A. (2013). Customer Data Analytics. In D. Loshin & A. Reifer (Eds.), *Using Information to Develop a Culture of Customer Centricity* (pp. 68–78). Morgan Kaufmann. doi:10.1016/B978-0-12-410543-0.00009-3

Maraghi, M., Adibi, M. A., & Mehdizadeh, E. (2020). Using RFM Model and Market Basket Analysis for Segmenting Customers and Assigning Marketing Strategies to Resulted Segments. *Journal of Applied Intelligent Systems and Information Sciences*, *1*(1), 35–43. doi:10.22034/jaisis.2020.102488

Menzel, C. M., & Reiners, T. (2014). Customer Relationship Management System a Case Study on Small-Medium-Sized Companies in North Germany. In J. Devos, H. van Landeghem, & D. Deschoolmeester (Eds.), *Information Systems for Small and Medium-sized Enterprises: State of Art of IS Research in SMEs* (pp. 169–197). Springer Berlin Heidelberg. doi:10.1007/978-3-642-38244-4_9

Ngai, E. W. T., Xiu, L., & Chau, D. C. K. (2009). Application of data mining techniques in customer relationship management: A literature review and classification. *Expert Systems with Applications*, *36*(2, Part 2), 2592–2602. doi:10.1016/j.eswa.2008.02.021

Nugroho, A., Suharmanto, A., & Masugino. (2018). Customer relationship management implementation in the small and medium enterprise. *AIP Conference Proceedings*, *1941*(1), 020018. doi:10.1063/1.5028076

Paliouras, K., & Siakas, K. (2017). Social Customer Relationship Management: A Case Study. *International Journal of Entrepreneurial Knowledge*, *5*(1), 20–34. Advance online publication. doi:10.1515/ijek-2017-0002

Rayen, L. P., & Sreeranganachiyar, T. (2017). A Study on Problems Faced by the Customer in Relation to Customer Relationship Management Practices. *Sumedha Journal of Management*, *6*(3), 24–38.

Rygielski, C., Wang, J., & Yen, D. C. (2002). Data mining techniques for customer relationship management. *Technology in Society*, *24*(4), 483–502. doi:10.1016/S0160-791X(02)00038-6

Salah, M., & Abou-Shouk, M. (2019). *The effect of customer relationship management practices on airline customer loyalty*. Academic Press.

Singh, P., Yadav, S., & Pall, L. (2018). Sales Trend Analysis of Products Using Customer Relationship Management Tool. *International Journal of Scientific Research in Computer Science, Engineering and Information Technology*, 289-294. doi:10.32628/CSEIT183888

Tuzunkan, D. (2018). Customer relationship management in business-to-business marketing: Example of tourism sector. *Geo Journal of Tourism and Geosites*, *22*(1), 329. doi:10.30892/gtg.22204-291

Zhou, X., Bargshady, G., Abdar, M., Tao, X., Gururajan, R., & Chan, K. C. (2019). *A Case Study of Predicting Banking Customers Behaviour by Using Data Mining*. Paper presented at the 2019 6th International Conference on Behavioral, Economic and Socio-Cultural Computing (BESC).

Chapter 8
Machine Intelligence as a Foundation of Self–Driving Automotive (SDA) Systems

Goh Bian Chiat
Universiti Malaya, Malaysia

Muneer Ahmad
National University of Sciences and Technology, Pakistan

N. Z. Jhanjhi
https://orcid.org/0000-0001-8116-4733
Taylor's University, Malaysia

Yasir Malik
New York Institute of Technology, Vancouver, Canada

ABSTRACT

Machine intelligence is a backbone of self-driving automotive (SDA) systems. Presently, ResNet, DenseNet, and ShuffleNet V2 are excellent convolution choices, whereas object detection focuses on YOLO and F-RCNN design. This study discovers the uniqueness of methods and argues the suitability of using each design in SDA technology. Real-time object detection is imperative in SDA technology, for CNN, as well as to object detection algorithms, an architecture that is a balance between speed and accuracy is important. The most favorable architecture in the scope of this case study would be ShuffleNetV2 and YOLO since both are networks that prioritize speed. But the drawback of speed prioritization is that they suffer from slight inaccuracies. One way to overcome this is to replace the neural network with a more accurate (albeit slower) model. The other solution is to use reinforcement learning to find the best architecture, basically using neural networks to create neural networks. Both approaches are resource-intensive in the sense of capital, talent, and computational budget.

DOI: 10.4018/978-1-7998-9201-4.ch008

INTRODUCTION

SElf-Driving Automotive (SDA), or Autonomous Vehicle is a feature which allowed automotive (mainly car, trucks, vehicle with four wheels or more) to drive itself. According to SAE (Society of Automotive Engineers) International Standard J3016, there are 6 Level of Autonomous. Starting at Level Zero, which is no automation, up till Level 5, full vehicle autonomy (Shuttleworth, 2019). Many automotive and software company are currently racing to develop their own SDA technology, such as Tesla, Waymo, Toyota, Honda, BMW, etc.

The rational for auto maker to pursuit this technology comes in 3 folds: Satisfaction, Safety and Survival. The first two rationales combined to become a strong selling point and ultimately boost the company images and profit. When a car is able to drive itself, the task of driving becomes optional. This meant that driver choose to drive, and not had to drive. This generates a lot of satisfaction for the driver, especially when the driver must drive on the traffic-dense road, with multiple stop-start traffic, every day to work. On a long highway, driver now able to sit back and enjoy the view, or having conversation with passenger without the distraction of driving, the trip can become very enjoyable. Safety is a main consideration for auto maker, function such as Vehicle Stabilization Control, ABS exist in modern car. Because SDA technology can driver itself, it can also apply brake when there is a sudden object appeared in front of the car. This is possible by means of tuning the camera and increasing the image quality during the night, as well as tuning the SDA Chip to have a high refresh rate. The average driver reaction time for an unexpected (like animal jumping in front) are 2.3s (McGehee, Mazzae, & Baldwin, 2000). If the SDA Chip is tuned to ingest images at a rate of 10 fps, taking account the reaction time for the SDA Chip to apply brakes is 500ms, and required 2 frame to validate that the object capture was not a lens flare, the total respond time for the SDA from image acquisition to processing to action will took 700ms. This is a 3x improvement, comparing SDA with average human driver. In addition, as the hardware begins to improve overtime, a customized hardware and a software accelerator will only bring the respond time down even further. Finally, company that does not innovates according to the market trend will find their sales dwindled and eventually washed away by those who innovates accordingly, one good example would be Nokia (Troianovski & Grundberg, 2012). Innovation based on market trend or create a blue-ocean market is one of the cores for the long-term benefits and survival of any company. When market demands function such as SDA, company that refuse to innovate in this realm will not receive support from the market.

At this point, majority of automotive manufacturer are still in SAE Level 2, which means provides support for both steering and throttle/braking function ("The State of the Self-Driving Car Race 2020," 2020). Support meaning the AI will help or assist the driver in steer and brake, such as Lane Keep Assistance from Honda. Waymo and Tesla are few of the companies that really focuses on developing SDA. Their technology has reached SAE Level 3, and in some area, Level 4. Which means, the car can drive itself without the intervention of driver in some of the roads and drive act as supervision only. Waymo has started their RoboTaxi in some part of US, where cars can drive itself ferrying passengers without having a driver present (Ohnsman, 2020).

Both Waymo and Tesla both approach the SDA with a very different solution. Waymo uses Lidar as their primary sensor and uses high definition map to navigate. Whereas Tesla Autopilot uses combination of camera, ultrasonic sensors and radar, with the aid of normal map for navigation. Lidar is a hardware that uses photon to measure the distance and build a 3-D representation internally. Tesla engineer argued that human does not shoot laser from the eyes. Instead, human relies primarily on visual to navigate and drive. Tesla approach to reach SAE Level 5 is to solve computer vision (Musk, Karpathy, & Bannon, 2019).

In our case study, I will focus on Tesla side of Self-Driving Automotive Technology (aka Autopilot ®). More specifically, I will discuss in-depth on the image recognition part of the SDA, which is the core technology in Autopilot®.

Data Acquisition

One drawback in training neural network is that they required a large amount of data that are also diverse. There are three way to solve the issue of acquiring training samples: Simulation, dedicated team or self-report. Simulation done by create a computer simulation of traffic condition and train the SDA system on it. According to Elon Musk, founder of Tesla Inc., this is equivalent to "grading your own homework" (Musk, Karpathy, & Bannon, 2019). The next solution is dedicated team. This is where the engineer hired a bunch of drivers to drive around certain city to collected data (Sun et al., 2020). Data are then processed and feed into the SDA training system. Using this approach to collect data are slow, expensive and could only captured a portion of long tail real-world data. Therefore, Tesla uses self-report system. When Autopilot® is engaging, if and when the driver took over the control of the vehicle, the car will capture the driving scenario and sent the dataset back to Tesla Training Center (aka Mothership) for training. This is one of the self-reporting triggers, the reason behind this trigger is that, when the driver does not feel comfortable or confidence about the driving of Autopilot®, this signals to Tesla that the SDA system are either underperforming (does not recognize the risk and did not slow down) or overperforming (overestimate the risk and slows down beyond needs). By using self-report system, Tesla can rely on the fleet to capture massive amount of real-world data that is very valuable for training CNN; as well as indicator for the AI engineer where is the ability gap of the Autopilot®. As of 2019, Tesla is reported to have the largest and most diverse repository of real-world driving because of that report system (Musk, Karpathy, & Bannon, 2019). This case study will focus only on the image recognition and will not go in-depth into data acquisition and storage due to limited time.

Future Implication

If restricted the implication of solving computer vision to automotive related industry. SDA technology will have major impact of how vehicles are operated on the road. Your car becomes your taxi, and passenger can simply rest in the car during commute. There is no need for parking space, especially for areas where real estates are at its premium (places like downtown KL, New York, Tokyo, Beijing). You can even lease out your car as a taxi service for other passenger and earn some side income. Buses and trucks that does long haul will be safer and faster. Because SDA eliminates the risk of driver getting too tired to focus and causes accidents, driver stops for rest, changing driver for long distance drive. This is, to the limits of my imagination, the implication of fully developed SDA, but they are still many possible utilizations for SDA that is currently not obvious for me.

LITERATURE REVIEW

Convolutional Neural Networks

Convolutional neural networks (CNN) are a branch of deep learning that focuses on vision. Circa year 2010, CNN model was unable to reach great depth (more than 20 layers) and their accuracy was poor. In year 2011, the discovery of AlexNet (Krizhevsky, Sutskever, & Hinton, 2012) sparked a wide interest in computer vision, they won ILSVR (Russakovsky et al., 2015) in year 2012, with Top-5 error rate 1-.8% lower than the runner up. Leading to that, in 2014, Google develop their first ever CNN model, dubbed GoogLeNet, or Inception V1 (Szegedy et al., 2015), that was able to reach 27 layers. The won the ILSVRC that year.

A landmark paper by (He et al., 2015) revolutionized the CNN architecture by introducing Residual Network (ResNet). ResNet was the first of its kind, it had a unique skip connection, where the gradient, of feature channels can "jump" layer and merge with convolution layer below. This skip/residual connection has two unique benefits: the element-wise addition (aka merge) means adding very minute computation overhead; it allowed the model to overcome depth issue and go ultra deep. This is the first time when a CNN can reach more than 50 layer without suffering from accuracy degradation. In ILSVRC, ResNet achieve top 5 error rate of 3.57% on ImageNet dataset, which was about 50% more accurate than previous best model, which was GoogLeNet. ResNet was also able to reach more than 1000 convolution layers with only small drop in accuracy, which is an indication of overfitting. The authors argued that, it is very easy for a large CNN model to overfit, an additional intervention such as maxout, dropout, image augmentation, in general, will improve the performance of CNN model. To prove this point, (Huang, Sun, Liu, Sedra, & Weinberger, 2016) modified the existing ResNet and by randomly dropping layers and had better accuracy than the unmodified ResNet.

Following the discovery of residual connection, (Huang, Liu, van der Maaten, & Weinberger, 2018) designed an architecture that had extensive feature reuse, called DenseNet. The convolutional layers in DenseNet are grouped into several Dense Block by the resolution dimension of its feature channels. Meaning all convolution layer feature channel that has resolution of 56x56 will be grouped together into Dense Block. In between Dense Block are transition layer. Transition layer is a layer where the feature channels are downsized before entering another Dense Block. What is unique about Dense Block are the "extreme" feature reuse. The feature channels from earlier conv. layer are shared with all conv. layer below. This means all convolution kernel can "see" the feature channel from every layer before it. The kernel can "see" the previous feature channels because the operation of skip (jump) connection is not by element-wise summation, but concatenation. This means the original value of the previous feature channel are not altered, therefore this type of skip connection is called feature reuse. Because of this extensive feature reuse, a 40-layer DenseNet with 1 million parameters is able to achieve better error rate than a 110-layer ResNet with 1.7 million parameters in CIFAR100 data (24.24% vs 27.22% with standard data augmentation).

ShuffleNet V2 builds on ShuffleNet V1 (Zhang, Zhou, Lin, & Sun, 2017), with has special emphasis on speed. The word shuffle in the names comes of the operation "Channel Shuffle" that is originated from version 1. The authors argued that using FLOPs to measure the computation cost will only yield and approximation (aka indirect metric) but does not shed information on actually how fast each batch is processed. They verified and introduce four general guidelines when manually building CNN block that emphasis on efficiency while retail good accuracy. Those guidelines are: 1) Equal channel (input and output) width minimizes MAC; 2) Too many group conv. increases MAC; 3) Parallelism decrease when network become increasingly fragmented; 4) Cost of element-wise operation is nontrivial. Based on the four finding, authors modified parts of ShuffleNet V1. Point-wise group convolution is replaced with normal convolution. Element-wise addition ops for residual connection is replace with concatenation. Authors also design the flow such as feature channel concatenation; channel shuffle and channel split are "bundled" into a single step. Comparing between ShuffleNet and DenseNet when both are on similar number of parameters, ShuffleNet V2 2x has a Top-1 error rate of 25.1% while DenseNet 2x was 34.6%. But a 201-layer DenseNet was able to achieve 22.6% Top -1 error rate with identical single crop in validation that uses about 3.8 Giga FLOPs (very similar to ResNet-50 FLOP value); 264-layer DenseNet achieve 22.15% Top-1 error. 50-layer ShuffleNet V2 requires about 2.3 Giga FLOPs to run, achieves 22.8% Top-1 error. This showed that, ShuffleNet V2 is highly computation efficient and suitable to run on lightweight processor. If computation resources are abundant due to factor such as dedicated neural hardware accelerator, then DenseNet can yield better result. This small improvement might seem menial, but when a active SDA system making thousands of prediction every second on the road, a 0.65% improvement can have a very huge cumulative benefits.

Table 1. Accuracy and FLOPs value of three different CNN. ResNet serves as a benchmark for ShuffleNet V2 and DenseNet. Data in the table are aggregated from respective paper.

CNN Architecture	Top-1 ImageNet	FLOPs (Giga)
ShuffleNet V2 2x	25.1	-
DenseNet 2x	34.6	-
DenseNet-201	22.6	3.8
DenseNet-264	22.15	15.1
ResNet-50	24.6	4.0
ShuffleNet V2-50	22.8	2.3

Object Detection

In this paper, the focus will be multi object detection. As of this writing, there are generally two neural network design for object detection: Two-Stage and One-Stage detectors (Zou, Shi, Guo, & Ye, 2019). Table 2 showed the accuracy between One-Stage and Two-Stage detector. One shared design in all object detection algorithm is the backbone. They all use convolution neural network (CNN) as their backbone.

Two-stage Detectors

The second coming of CNN has a cascading effect on the development of detector, which has been relatively stagnant since discovery of DPM (Deformable Part-base Model) (P. F. Felzenszwalb, Girshick, & McAllester, 2010; P. F. Felzenszwalb, Girshick, McAllester, & Ramanan, 2010; P. Felzenszwalb, McAllester, & Ramanan, 2008), 2014 saw a renew discovery of object detection algorithm called Regions with CNN feature (RCNN). RCNN (Girshick, Donahue, Darrell, & Malik, 2014) is comprised from two parts. A CNN for computer vision and object detection. The object detection has three main components:

regional proposal, feature extraction and linear SVM classifier. Regional proposal highlights a region with bounding boxes in the input images that it thinks there is an object, say candidate A. Feature extractor will extract feature from candidate A, the feature will then run through a classifier to assign a class. For each candidate, R-CNN will generate five value. One for class prediction and four values for coordinates of bounding box, (x,y) coordinate of the upper right corner of bounding box plus (w, h) which are size of the box. Although R-CNN achieve good accuracy versus the previous best mode, which is DPM, its drawback is slow. The regional proposal algorithm called Selective Search can generate about 2,000 regions through the images, but each proposed region (prediction) are fed into CNN independently. That means the CNN has to perform 2,000 independent forwards and backward pass for each image. A single image can took up to 14s to detect even on GPU. [1]Faster-R-CNN was proposed in 2015, with an accelerated detection speed to nearly real time. To achieve that, the task of regional proposal Selective Search was replaced by Region Proposal Network (RPN). RPN more desirable than Selective Search is because RPN sits on GPU, while Selective Search sits on CPU. This design allowed two neural networks (CNN and RPN) to train jointly, which produce four groups of loss value. Object identification (output Yes/ No) and region regress box from RPN; object classification score and final bounding box coordinate. This means Faster-R-CNN are made up of two neural networks, RPN, which is the detector, sits above the CNN. Feature Pyramid Networks (FPN) leverage on the feature channel formed by CNN, using a lateral connection in CNN layer to build multiple different scales of high-level semantics. Which means the detector is now sitting "side-by-side" with CNN; instead siting on-top design. This design leverage on the forward propagation feature channel pyramid to build their detector, resulting in fast and more accurate model than Faster-R-CNN.

Single-stage Detector

YOLO (You Only Look Once) (Redmon, Divvala, Girshick, & Farhadi, 2016) emphasis on speed but suffer from accuracy degradation. In a simpler term, single-stage detector combines RPN and CNN into a single architecture that are category specific. Each image is divided into a grid (m x m), at each grid, multiple bounding box and class prediction are generated in each grid. Because the prediction for each grid runs simultaneously, it gives a great boost to the speed. But YOLO still suffer from "small object inaccuracy" due to its design, meaning, its object localization algorithm does not perform well when the object is small. YOLO detector "sits" on top of the network, which makes it prone to "small object inaccuracy". This means, YOLO object localization power is weaker when the object in the image is small. E.g., YOLO can detect a man sitting in a café drinking coffee, but it may have difficulty identify the espresso cup he is holding, especially if the cup is very small. To overcome this, Single Shot MultiBox Detector (SSD) (Liu et al., 2016) place detector along the different convolution layer. With this placement, SSD can detect object of different scales, which improve the small object inaccuracy. But single-stage network still loses out to two-stage in terms of accuracy. RetinaNet (Lin, Goyal, Girshick, He, & Dollár, 2018) introduce a new loss term called "Focal Loss". Focal loss emphasis on hard example and ignores background object that are not of interest. In two-stage object detection, this task was managed by Selective Search, where it will ignore background objects that are not in training list. In RetinaNet, a hyperparameter called *focusing* is added to cross entropy loss that penalizes out-of-focus object, or background object. Over the year, there are several versions of YOLO, each version made a slight improvement of the classification accuracy while keeping the speed fast. The latest version of YOLO, YOLO V4 (Bochkovskiy, Wang, & Liao, 2020) finally able to achieve better accuracy than Two-Stage detector in MS COCO dataset (Lin et al., 2014).

Table 2. Accuracy comparison between popular object detection neural network for both single and two stage detectors

Object detector	Faster-R-CNN	Feature Pyramid Net	YOLO	YOLO V4	Retina Net
COCO mAP@.5	42.7	59.1	-	65.7	59.1
COCO mAP@[.5,.95]	21.9	36.2	-	43.5	39.1
VOC07 mAP	73.2	-	63.4		-
VOC12 mAP	70.4	-	57.9		-

Table 2 YOLO V4 was the latest release (April 2020) in YOLO family that are both fast and accurate. Since we are looking at YOLO, I believe it would be interesting to see the development of YOLO over the year. Citation for: VOC07 (Everingham, Gool, Winn, & Zisserman, n.d.-a) and VOC12 (Everingham, Gool, Winn, & Zisserman, n.d.-b). Data are aggregated from respective paper.

Modern Perspective in CNN Object Detection

Since 2018, there has been an extensive discovery of new CNN architecture that are parameter efficient. CNN design such as ShuffleNet V2 that is by-itself fast, or DenseNet (Huang, Liu, Van Der Maaten, & Weinberger, 2017) that have higher accuracy but more computational expensive.

These two modern CNN designs, in my opinion, are both equally capable to be used as CNN backbone for the SDA. Both architecture is more parameter efficient and more accurate than ResNet. While one focus speed and other focus on accuracy, selection of CNN model comes down to the business strategy of the automaker or AI company that produces SDA system: 1) Faster rollout or custom system. Using parameter efficient model like ShuffleNet V2 enable company to use off-the-shelf GPU chip and hardware that is less optimized (Nvidia PX-Drive, Intel Mobileye, etc.). This approach allowed SDA to be rolled out quicker and with lesser cost and sacrifices bits of accuracy. 2) In the other hand, company can also use a more accurate CNN architecture such as DenseNet. (G. Huang, Liu, Pleiss, Van Der Maaten, & Weinberger, 2019) introduces an optimization algorithm to manage and accelerate the model. But because DenseNet computation is fundamentally heavy, a custom hardware is needed so that it can run as fast as ShuffleNet V2 in an off-the-shelf hardware. This approach will drive up the cost and slowed the roll-out time, but have a benefit of highly accurate model that are running at very high frame-per-second, which is a strong selling point by itself.

Year 2020 see the design of YOLO V4 (Bochkovskiy, Wang, & Liao, 2020) that are both as accurate as two-stage detector, and at the same time retain the speed of single-stage. This discovery means that for the first time, SDA engineer do not need to compromise between the two most important requirement in SDA system.

APPROACHES FOR SDA SYSTEM (IN TESLA)

Engineering Consideration of SDA System

SDA comprise of three main components: object identification, vehicle operation and navigation. This case study focuses on object identification.

There are three engineering consideration when designing an SDA hardware: 1) SDA software must sits in the vehicle itself because it required real-time detection. 2) SDA software run much faster in GPU than in CPU, but draws more power from the vehicle. 3) High bandwidth memory access to avoid data transfer bottleneck. Higher processing speed will draw more energy from the vehicle battery pack and stresses the alternator, but generate better accuracy, vice versa. This means, the main thing SDA engineer has to balance is between energy consumption, model accuracy and processing speed.

Balance between Accuracy and Speed

Took ResNet as an example, using ImageNet data, Top-1 accuracy for ResNet-101 is 23.40% at 8 GFLOPs, accuracy of ResNet-152 is 23.00% at 11 GFLOPs. Using a larger model (in the case of ResNet, more layer) can improve the accuracy of the prediction. But by using 3GFLOPs more calculation, the accuracy only improves by 0.40%. This means, pursuing high accuracy at a cost of exponential computational expenses as well as energy consumption. Pursuit highly accurate model by increasing the model's depth w.r.t ResNet, has another undesirable effect in the perspective of SDA: heavy in memory consumption and slow in image processing speed. ResNet-101 requires 170MB to store its parameters, but ResNet-152 requires 219MB, which is 28.8% more memory and 37.5% more FLOPs. Therefore, balance between accuracy, speed and memory consumption are something SDA engineer has to take into consideration and strike a balance give-and-take when designing.

Divide and Conquer

In each image, or instance, a typical SDA system (such as Tesla Autopilot) may produce up to 1,000 predictions (Karpathy, 2019). A single CNN may not able to generate this much of prediction with road-safety level accuracy, especially when the ingestion image has resolution of 800x600 pixels (or higher, such as 1280x960), or tensor shape (3, 800, 600). In order to improve error rate, what Tesla AI engineer has done, was divide-and-conquer. All the predictions are spread out into different category and sub-categories, and one CNN model only responsible for prediction in one of those sub-categories. By doing so, the prediction tasks are spread out into multiple CNN, as they are all "looking" at the same image. This is a unique solution by Tesla Engineers to solve the single CNN model capacity limitation, they called this design: HydraNet. Another simile would be a growing tree. The root (or base) is the input image, and the final branch is the prediction output.

HydraNet

The entire HydraNet are looking at single image in each timesteps. Early convolution layer has more general feature, or more shared features. As the convolution layer goes deeper towards output, the feature now becomes more distinct. Because of that, the number of CNN "bodies" or "backbone" can be reduced. By leveraging on transfer learning, Tesla Engineer can freeze the earlier convolution layer, and only train deeper layer when only new information is discovered.

Using multiple CNN to share the prediction burden so that the error rate remains low. In SDA, because the system is operating a vehicle, having very low accuracy is literally a matter of life-and-death. Such as, misclassifying a human as a debris will spill significant disaster in all shape and forms.

This meant, the SDA software need to be accurate, but the hardware needs to be relatively lightweight because it is an on-board computation. In SDA algorithm, one of the ways to reduce computational complexity while maintaining the accuracy is by using shared CNN backbone. Because the CNN models are all looking at same image, it can use "common" or "shared" convolution layer at earlier stage. This can be done because earlier convolution stage convolves non-distinct features. As the convolution goes deeper, the convolved feature combined, beginning to form more distinct features. That is why Tesla engineers was able to use shared backbone, then at the later layer, begin to add different "branches" or "prediction heads" for each categories of sub-categories (hence the name HydraNet). Take an example, if 20 CNN backbone is adequate in SDA system, using shared backbone, four CNN backbone can be fused into one shared backbone. This means the in earlier layer, SDA hardware only needs to perform five convolution computation, instead of 20 convolutions. Reduce computation means the SDA can run faster, and uses lesser energy, without compromising on the accuracy.

Backpropagation on HydraNet

Backpropagation is the fundamental means of learning in any neural network design. With the shared backbone and multiple prediction head, at the joints (or branching points), feature channels are duplicated according to number of prediction heads. During backpropagation, the shared backbone will receive gradient flow from above in an element-wise summation because it is a "duplication/copy joints (gate)" (Li, Johnson, & Karpathy, 2016). Because of the summation of gradient, its backpropagation, it might elongate the training time because the earlier layer (aka shared backbone) has to take into accounts the gradient from multiple prediction heads, but during testing and production, the computation duration will be less.

Figure 1.

- (3, 960, 1280) input images
- "ResNet-50 like" dilated backbones
- FPN / DeepLabV3 / UNet -like heads
- ~15 tasks => "prototypes framework"

One great feature in CNN is transfer learning. Using an example of the red "Stop" sign. There are normal stop sign, occluded, Stop sign with condition, Stop sign that is inactive, Stop sign with illumination. Similar to real world, Stop sign has multiple, varied datasets and it occupy its own prediction head (Karpathy, 2020). When Tesla fleet encounter new form of Stop sign, it has to be trained, and the way to train it is by using transfer learning. This means the entire pillar of CNN and its heads are frozen (meaning no active training), and the only convolution layer that was actively trained are the Stop sign prediction head. Using transfer training, Tesla engineer can just train any prediction head without needing to training the entire network from ground up each time. Transfer learning with respect to HydraNet allowed new variation of same types of data to be incorporated to the SDA algorithm faster, therefore, expanding the capability of SDA. This is preferable for both the company and vehicle owner, because driver can enjoy faster update that expand the range of SDA, which creates a good reputation for the company.

Figure 2.

Figure 3.

Figure 4.

Figure 5.

FLEET LEARNING

In terms of SDA system as a whole, building a CNN object detector is consider to be fundamental but not difficult, as many of the CNN architecture are open sourced. Therefore, the real challenge in any CNN training is to get data, or data acquisition.

As mention in Part One of the case study, there are three means of data acquisition for SDA: simulation, hire someone to drive, or fleet learning. Tesla took the approach of Fleet Learning to obtain the volume and variation of data. It is been establish that volume (large amount of data) and variation (in different shade, form, factor, etc.) are two most important criteria when training a neural network (Musk, Karpathy, & Bannon, 2019). The way to gather this through constant internet communication between the Tesla fleet and its training center. Tesla engineer can "ask" the fleet to upload any form of images simply by providing small sample image to the detector. In the example of Stop sign, Tesla engineer can insert small training sample for "Stop Except Right Turn". The instruction sets are uploaded to the Tesla fleet via Over-The-Air update, and each time Tesla Autopilot® "see" visually similar images, it will capture a short clip (or images, depending on setting), store in its on-board memory system and upload it back to the training center.

Figure 6. Fleet learning allowed Tesla engineer to "ask" the fleet for visually similar images, and the fleet will capture, and send it back to the Tesla training center (Karpathy, 2020)

Fleet learning is the unique property of Tesla Autopilot system, as Tesla driver uses the car, their system learns from the driver to constantly makes it better. "Real world are messy and complex", a quote from Elon Musk. Tesla engineer uses fleet learning and shadow mode to capture the long tail data that are varied, strange, rare but important for the SDA, the idea is very similar to crowd sourcing, or wisdom of the crowd.

Figure 7. Fleet learning and Shadow mode contribute to varied data of the same catogories. Both func-
tion allowed Tesla engineer to train their SDA on long-tailed end of data that are both rare but critical
Musk, Karpathy, & Bannon, 2019)

DATA PREPROCESSING

Before images are annotated, images from fleet learning will be screened by a team of annotators. This
step is part of the image preprocessing step to delete: 1) mis-captured images, 2) similar images that
already exists in the repository (repetition).

Image Annotation

As of this writing, Tesla is running their Autopilot® SDA system based on supervised learning. News
and rumors from Tesla that they are working on a system called "Project Dojo" that required lesser image annotation and it is a semi supervised learning, but no concrete detail so far, therefore, this case
study will focus on supervised learning. This technique require human to annotate the images uploaded
by the fleet. According to Elon Musk, there is not much processing in terms of modifying the contrast
of the images, the only preprocessing they focus on it's the image annotation.

Image annotation is a process of building a bounding box with categorical and sub categorical label-
ing. Each bounding box has a corresponding pixel coordinates that indicate the location of "focal object"
or "foreground object".

In the context of SDA, image annotations process is extremely important, as example below, image
annotation will highlight, or box important environmental tags. Those environmental tags are divided into
categories and subcategories. Example of category are "Drivable Lane" or "Moving Objects", within,
subcategories can be "Human", "Cyclist", "Motorcyclist", Trucks", etc. Image annotation allowed SDA
algorithm to "pay attention" to what is important on the road, road marking, drivable lane and ignore
background object like building and playground. Example, "Human with baby stroller" (new discovered

data) is in the subcategory of "Human", when training, only "Human" subcategories will receive active training, while the rest of the "Moving Object" shared backbone are frozen.

Any mislabeling at this stage will affect the neural network training later. Mislabeling "Human with baby stroller" into "Static Object" category have tremendous effect, on the weight layer (kernel weight and feature channels) in the CNN as well as real world implication. Because priority reaction for "Moving Object" may be higher than "Static Object", such mislabeling may tell the SDA that "Human with baby stroller" is a low priority response, or required lesser degree of evasion.

Figure 8. Using a highly simplified demonstration, images sent from the fleet need to be preprocessed, one of the most important steps is to annotate all the important environmental tag, especially those that are novel and never or rarely seen by the neural network. Those annotation can be store in multiple format, including CSV and JSON and will be fed together into the training with the related images. (credit image on top left: Karpathy, 2019)

Flow of data from the fleet to the Tesla training center. The path showed the process of data annotation, storage, neural network training tracker etc. This flow chart is not exhaustive.

Figure 9. Data process flow in Tesla Training Center called "The Mothership". This chart is not from Tesla, rather, this chart is created based on information gathered from Tesla presentation such as: Tesla Autonomous Day, Tesla Shareholder Meeting, Elon Musk's presentation

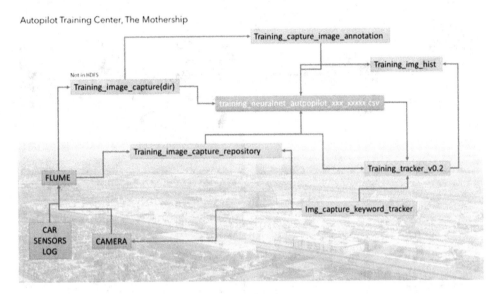

CNN ARCHITECTURE SELECTION FOR TESLA

Based on presentation by Tesla, they are using "ResNet-50-like" shared (dilated) backbone with "Feature PyramidNet-like" prediction head (Karpathy, 2019). This means they opted for slower object detector that are more accurate. One of the reason Tesla opted for ResNet and FPN was because it has the lowest error rate compare to another model back in 2016 (or earlier), when Tesla decided to ditch Nvidia and build their in-house SDA chip that have built-in neural engine. It is to be believe that Tesla Engineer opted for ResNet-50-like instead of a deeper model such as ResNet-101 was because of a small 0.4% gain in accuracy, at a much more computation and memory overhead. ResNet-50 has the benefits of smaller computation (ResNet-101 require 2x FLOPs compared to ResNet-50), which can generate independent prediction faster, smaller capacity of ResNet-50 is overcome by using more prediction heads or shared backbone, which is a unique design from Tesla called: HydraNet.

When Autopilot® training is completed and reaches satisfactory (according to Tesla) accuracy level during test-time, their CNN object detection weight will be upload back to the fleet.

Final verification of model accuracy will be completed by the Tesla driver themselves. In terms of neural network training, this is the most optimal approaches, where the model is tested on real world, using actual road user, in a validation software called "Shadow Mode". During a drive, Shadow model will actively generate prediction while comparing the validation with the driver's respond and produce a prediction error. When prediction error reaches threshold (predetermined by Tesla team), it passed the final validation test and therefore, ready for actual road use.

Figure 10. "Shadow Mode" is the final real-world validation for their Autopilot. This GUI is only accessible to Tesla Autopilot engineering team. (Karpathy, 2019; Karpathy, 2020)

Priority for any SDA design should be accuracy, because the consequences of mis-classification is much severe than slow respond. In addition, computation speed is largely depending on the hardware, whereas model accuracy is solely a software challenge. Therefore, Tesla move to develop their own SDA Chip and hardware package called "Full Self Driving" or "Hardware 3.0" that run faster than previous off-the-shelf chip. Currently Tesla Hardware 3.0 can achieve 2,300 fps, while using less than 100 watts. It's SOC (system-on-chip) design included 12 ARM CPU, GPU and custom TPU in a 14nm FinFET design. AI researcher also discovered CNN architecture that are both fast and accurate such as ShuffleNet V2 (Ma, Zhang, Zheng, & Sun, 2018). With research into modern hardware undergoing as a response to growth of neural network architecture, new chips design has incorporated neural network processor (or neural engine or TPU, Tensor Processing Unit) into it. Example are Apple A14 Chip (Apple Inc., 2020), HiSilicone Kirin 980 Processor (Huawei, 2018), Google Tensor Processing Unit (TPU) (Jouppi et al., 2017).

Training a SDA

Using Tesla Autopilot as an example, they uses a "ResNet-50-like" CNN shared/dilated backbone with "Feature Pyramid Network-like" prediction head. According to Tesla, they have 48 CNN, generate 1,000 distinct prediction for their SDA system. From ground-up, it took them 70,000 GPU hour to train, using SOTA GPU in 2016 such as Nvidia P100 (Pascal) or V100 (Tesla) (Karpathy, 2019).

CONCLUSION

The objective of this case study was to showcase one of the most advance SDA in the market, and its decision-making process to achieve Autonomy Level 3, and what are some other CNN and object detection algorithm that is available yet suitable to be used in SDA. In a nutshell: there are many CNN architectures that are parameter efficient, fast or both, which means they are able to achieve good accuracy in a reasonable amount of time using reasonable amount of computation thanks to continuous active research in this field; training a CNN is relatively straightforward as the only requirement is good hardware, especially GPU. The hardest part in building a SDA neural network that can safely navigate in the real world is getting a large volume of different/varied dataset, especially those that are rare (such as a car turn turtle).

REFERENCES

Inc. (2020, September 16). *Apple Keynote*. Retrieved from www.apple.com/newsroom/2020/09/apple-unveils-all-new-ipad-air-with-a14-bionic-apples-most-advanced-chip/

Bochkovskiy, A., Wang, C.-Y., & Liao, H.-Y. M. (2020). *YOLOv4: Optimal Speed and Accuracy of Object Detection*. Retrieved from https://arxiv.org/abs/2004.10934

Everingham, M., Gool, V., Williams, Winn, J., & Zisserman, A. (n.d.a). *The PASCAL Visual Object Classes Challenge 2007 (VOC2007) Results*. Academic Press.

Everingham, M., Gool, V., Williams, Winn, J., & Zisserman, A. (n.d.b). *The PASCAL Visual Object Classes Challenge 2012 (VOC2012) Results*. Academic Press.

Felzenszwalb, P. F., Girshick, R. B., & McAllester, D. (2010, June 1). *Cascade object detection with deformable part models*. doi:10.1109/CVPR.2010.5539906

Felzenszwalb, P. F., Girshick, R. B., McAllester, D., & Ramanan, D. (2010). Object Detection with Discriminatively Trained Part-Based Models. *IEEE Transactions on Pattern Analysis and Machine Intelligence*, 32(9), 1627–1645. doi:10.1109/TPAMI.2009.167 PMID:20634557

Felzenszwalb, P., McAllester, D., & Ramanan, D. (2008, June 1). *A discriminatively trained, multiscale, deformable part model*. doi:10.1109/CVPR.2008.4587597

Girshick, R., Donahue, J., Darrell, T., & Malik, J. (2014). Rich Feature Hierarchies for Accurate Object Detection and Semantic Segmentation. *2014 IEEE Conference on Computer Vision and Pattern Recognition*. 10.1109/CVPR.2014.81

Huang, G., Liu, Z., Van Der Maaten, L., & Weinberger, K. Q. (2017). Densely Connected Convolutional Networks. *2017 IEEE Conference on Computer Vision and Pattern Recognition (CVPR)*. 10.1109/CVPR.2017.243

Huang, G., Sun, Y., Liu, Z., Sedra, D., & Weinberger, K. (2016). Deep Networks with Stochastic Depth. doi:10.1007/978-3-319-46493-0_39

He, K., Zhang, X., Ren, S., & Sun, J. (2015). *Deep residual learning for image recognition.* arXiv preprint arXiv:1512.03385.

Krizhevsky, A., Sutskever, I., & Hinton, G. E. (2012). *ImageNet Classification with Deep Convolutional Neural Networks.* Retrieved from https://papers.nips.cc/paper/4824-imagenet-classification-with-deep-convolutional-neural-networks.pdf

Lin, T.-Y., Dollar, P., Girshick, R., He, K., Hariharan, B., & Belongie, S. (2017). Feature Pyramid Networks for Object Detection. *2017 IEEE Conference on Computer Vision and Pattern Recognition (CVPR).* 10.1109/CVPR.2017.106

Lin, T.-Y., Goyal, P., Girshick, R., He, K., & Dollár, P. (2018). *Focal Loss for Dense Object Detection.* Retrieved from https://arxiv.org/abs/1708.02002

Lin, T.-Y., Maire, M., Belongie, S., Hays, J., Perona, P., Ramanan, D., ... Zitnick, C. L. (2014). Microsoft COCO: Common Objects in Context. *Computer Vision – ECCV 2014*, 740–755. doi:10.1007/978-3-319-10602-1_48

Liu, W., Anguelov, D., Erhan, D., Szegedy, C., Reed, S., Fu, C.-Y., & Berg, A. C. (2016). SSD: Single Shot MultiBox Detector. *Computer Vision – ECCV 2016*, 21–37. doi:10.1007/978-3-319-46448-0_2

Ma, N., Zhang, X., Zheng, H.-T., & Sun, J. (2018). ShuffleNet V2: Practical Guidelines for Efficient CNN Architecture Design. *Computer Vision – ECCV 2018*, 122–138. doi:10.1007/978-3-030-01264-9_8

McGehee, D. V., Mazzae, E. N., & Baldwin, G. H. S. (2000). Driver Reaction Time in Crash Avoidance Research: Validation of a Driving Simulator Study on a Test Track. *Proceedings of the Human Factors and Ergonomics Society Annual Meeting, 44*(20). 10.1177/154193120004402026

Musk, E., Karpathy, A., & Bannon, P. (2019). *Tesla Autonomy Day* [YouTube Video]. Retrieved from https://www.youtube.com/watch?v=Ucp0TTmvqOE

Ohnsman, A. (2020, October 8). *Waymo Restarts Robotaxi Service Without Human Safety Drivers.* Retrieved December 22, 2020, from Forbes website: https://www.forbes.com/sites/alanohnsman/2020/10/08/waymo-restarts-robotaxi-service-without-human-safety-drivers/?sh=6ab6c85d69d8

Redmon, J., Divvala, S., Girshick, R., & Farhadi, A. (2016). You Only Look Once: Unified, Real-Time Object Detection. *2016 IEEE Conference on Computer Vision and Pattern Recognition (CVPR).* 10.1109/CVPR.2016.91

Ren, S., He, K., Girshick, R., & Sun, J. (2015). *Faster R-CNN: Towards Real-Time Object Detection with Region Proposal Networks.* Retrieved from arXiv.org website: https://arxiv.org/abs/1506.01497

Russakovsky, O., Deng, J., Su, H., Krause, J., Satheesh, S., Ma, S., Huang, Z., Karpathy, A., Khosla, A., Bernstein, M., Berg, A. C., & Fei-Fei, L. (2015). ImageNet Large Scale Visual Recognition Challenge. *International Journal of Computer Vision, 115*(3), 211–252. doi:10.100711263-015-0816-y

Shuttleworth, J. (2019, January 7). *SAE J3016 automated-driving graphic.* Retrieved December 22, 2020, from Sae.org website: https://www.sae.org/news/2019/01/sae-updates-j3016-automated-driving-graphic

Sun, P., Kretzschmar, H., Dotiwalla, X., Chouard, A., Patnaik, V., Tsui, P., … Caine, B. (2020). *Scalability in perception for autonomous driving: Waymo open dataset*. Academic Press.

Szegedy, C., Ioffe, S., Vanhoucke, V., & Alemi, A. A. (2017). *Inception-v4, inception-resnet and the impact of residual connections on learning*. AAAI Press.

Szegedy, C., Liu, W., Jia, Y., Sermanet, P., & Reed, S. (2015). *Going deeper with convolutions*. Retrieved from https://arxiv.org/abs/1409.4842

Troianovski, A., & Grundberg, S. (2012, July 19). Nokia's Bad Call on Smartphones. *Wall Street Journal*. Retrieved from https://www.wsj.com/articles/SB10001424052702304388004577531002591315494

Zou, Z., Shi, Z., Guo, Y., & Ye, J. (2019). *Object Detection in 20 Years: A Survey*. Retrieved from https://ui.adsabs.harvard.edu/abs/2019arXiv190505055Z/abstract

Huawei. (2018, August 31). *Kirin 980, the World's First 7nm Process Mobile AI*. Retrieved from https://consumer.huawei.com/en/campaign/kirin980/

Jouppi, N., Young, C., Patil, N., Patterson, D., Agrawal, G., Bajwa, R., … Yoon, D. (2017). *In-Datacenter Performance Analysis of a Tensor Processing Unit TM*. Retrieved from https://arxiv.org/ftp/arxiv/papers/1704/1704.04760.pdf

Karpathy, A. (2019). *PyTorch at Tesla - Andrej Karpathy, Tesla* [YouTube Video]. Retrieved from https://www.youtube.com/watch?v=oBklltKXtDE&t=553s

Karpathy, A. (2020). *Tesla Andrej Karpathy in CVPR 2020: Scalability in Autonomous Driving Workshop*. Retrieved from https://www.youtube.com/watch?v=X2CpuabzRaY

Li, F.-F., Johnson, J., & Karpathy, A. (2016, March). *CS231n Convolutional Neural Networks for Visual Recognition*. Retrieved from CS231n: Convolutional Neural Networks for Visual Recognition website: http://cs231n.stanford.edu/2016/

Szegedy, C., Ioffe, S., Vanhoucke, V., & Alemi, A. A. (2017). *Inception-v4, inception-resnet and the impact of residual connections on learning*. AAAI Press.

The State of the Self-Driving Car Race. (2020). *Bloomberg Hyperdrive*. Retrieved from https://www.bloomberg.com/features/2020-self-driving-car-race/

ENDNOTE

[1]   Between R-CNN and Faster-R-CNN, there is another architecture called Fast-R-CNN.

Chapter 9
Machine Learning–Based Wearable Devices for Smart Healthcare Application With Risk Factor Monitoring

Suja A. Alex
 https://orcid.org/0000-0003-4429-6715
St. Xavier's Catholic College of Engineering, India

Ponkamali S.
Cognizant Technology Solutions, USA

Andrew T. R.

Intuit Inc., USA

N. Z. Jhanjhi
 https://orcid.org/0000-0001-8116-4733
Taylor's University, Malaysia

Muhammad Tayyab
Taylor's University, Malaysia

ABSTRACT

The stroke is an important health burden around the world that occurs due to the block of blood supply to the brain. The interruption of blood supply depends on either the sudden blood supply interruption to the brain or a blood vessel leak in tissues. It is tricky to treat stroke-affected patients because the accurate time of stroke is unknown. Internet of things (IoT) is an active field and plays a major role in stroke prediction. Many machines learning (ML) techniques have been used to automate the process and enable many machines to detect the prediction rate of stroke and analyze the risk factor. The ML-based wearable device plays a significant role in making real-time decisions that benefit stroke patients. The parameters such as risk factors associated with stroke and wearable sensors and machine learning techniques for stroke prediction are discussed.

DOI: 10.4018/978-1-7998-9201-4.ch009

INTRODUCTION

Stroke has been considered as one of the most dangerous causes of all-cause mortality and morbidity throughout the world. Although the world has improved the vascular risk factor management, stroke systems of care, which has helped to mitigate the effects of stroke from number three cause to number five cause, but it is still a dangerous disease for human life (Fugate et al., 2014). In a survey and intensive investigation, 30% of stroke cases remains unsolvable and hidden due to lack of awareness (Bilal et al., 2020; Garkowski et al., 2015). To investigate and determine the correct disease on time, there requires a complex, critical, and detailed examination under micro-observation during clinical phase which can then detect the systematic symptoms of stroke (Bersano et al., 2020; Shulman & Cervantes-Arslanian, 2019). Such detailed observation become more complicated by the involvement phenotype heterogeneity as it is observed the stroke can also be primary cause for a particular body. When the blood supply disturbance happened in the brain then stroke occurs. It is a dangerous disease because it is hard to predict the precise time of occurrence. The result of brain stroke is oxygen decrease in brain cells that leads to death of brain cell (Hong et al., 2013). Some of the symptoms of stroke are bleeds, clots, and transient ischemic attack. The most common symptoms are arm weakness, difficult to speak, difficult to walk, blurred vision, tiredness, memory loss, stiffness in muscles and co-ordination issue (Subramaniyam et al., 2017) (World Health Organization, 2004). The success of treatment says that it should be treated within a few hours (4.5 hours) of occurrence to prevent the long-term damage from stroke. The development in imaging technique suggests that whether the stroke initiated within 4.5 hours. Further, the lesion data extracted from MRI image helps to predict the patients' speech production skills (Hope et al., 2013).

Wearable technology supports brain stroke prediction because it performs real-time monitoring of stroke-related physiological parameters. Based on the wearable characteristics, such as weight, accessibility, frequency of use, data continuity, and response time, the wearables have margins in reporting high-precision prediction outcomes. The trend of integration of wearables into the internet of things (IoT), electronic health records (EHRs) and machine learning (ML) algorithms to launch a stroke risk prediction system (Chen & Sawan, 2021; Usmani et al., 2021). Moreover, the risk factors that is measured with the help smart devices or wearable devices helped to observe the health of patient. While histories of other diseases with these wearable devices can also be maintained to further treatment, medication, and other medical purposes in different available EHRs (Sirsat et al., 2020; Tayyab, Marjani, Jhanjhi, & Hashem, 2021). Due to abundant real-time dataset that is recorded using machines and different EHRs, ML techniques (Tayyab, Marjani, Jhanjhi, Hashim et al, 2021) have been used to analyze the association and order them in specific risk factor to predict the correct brain stroke rate (Goldstein et al., 2017; Vashistha et al., 2019).

Brain Stroke

Brain stroke is a critical cerebrovascular disease in adult humans. It is one of the leading causes of death in India (Joshi et al., 2006). Brain stroke makes the body to rest 20% of the oxygen and 2% of glucose of the entire body weight is constituted by the brain. The blood flow in the brain increases when recreation of the neurons takes palace in certain parts of the brain, and it is done by the internal carotid and vertebral arteries. The blood is then changed from brain to heart through the internal veins (Wang et al., 2020). The loss of the blood might be observed in two scenarios in which the flow of blood among the blood tissues decreases results in the ischemic stroke whereas if internal bleeding occurs among the brain tissues known as hemorrhagic stroke (Harrar et al., 2020) and the types of strokes are demonstrated in Figure 1. The flow of the blood from arteries among the brain tissues leads to blockage or becomes slender (Sun et al., 2020; Wilkinson et al., 2021). This blocking might also happen when the tiny parts of plaque that are caused through atherosclerosis damage thereby creating a clot in the blood vessel (Lattanzi & Silvestrini, 2016; Rasmussen et al., 2020). Hemorrhagic stroke referred to as a severe stroke in which the blocked artery might damage resulting in the bleeding or explosion of the artery (Boukobza et al., 2019; Verma et al., 2020). When the blood is divulged and leaked it spreads out causing pressure on the brain (Uppal et al., 2020).

Figure 1. Classification of stroke type (Vamsi Badi, 2020)

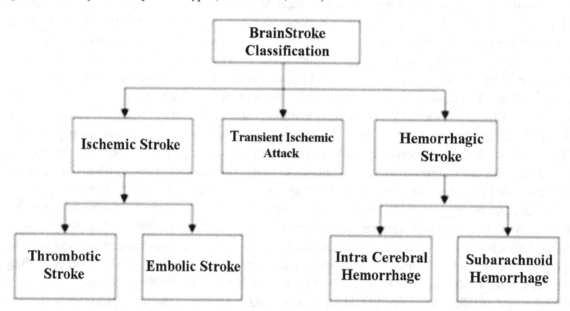

Machine Learning

It is defined as the branch of AI which can focuses on the use of data and algorithms to create a mimic of human learning and it can improve gradually with its accuracy and precision. It has ability to controls the data to generate the output values within the limit (Convertino et al., 2011). The algorithms are developed for the machine that used data sets. These algorithms are implemented through supervised learning and unsupervised learning algorithms. Supervised learning takes data with labelled data and produced output. But unsupervised learning did not use the labelled data by allowing it to find the output based upon the input data.

Internet of Things (IoT) Wearable Devices

Wearable devices are now considered as the most significant ubiquitous technology introduced by IoT in modern life. To have the accurate and efficient dataset for achieving high results, model IoT based wearable devices are recommendable with consumer-oriented services. Various important and very common devices like smart wrist watches, smart clothes, and smart medical equipment that are enabled with most AI and IoT based technologies have become most attractive for healthcare systems. Model wearable devices have been updated with health, safety, interaction, geolocation, identity, fitness etc. Such technology has brought the world under the umbrella of IoT by using physical and digital contents.

IoT consists of various devices that generate data continuously communicating among them. Wearable Technology is a wonderful innovation. The area of wearable IoT classifies the wearable devices into four groups:

1. Health
2. Sports and daily activity
3. Tracking and localization
4. safety.

Numerous kinds of wearable devices can be used to monitor the risk factors in diseases like heart attack, brain stroke etc. Such behavior dependent devices are based on modern AI and ML models like DNN, CNN and SVM to predict the accurate prediction rate for calculating the risk factor of brain stroke, heart attack, kidney failure and more about the health risk parameters (Shehab et al., 2017). The modern devices which have AI enabled, have shown a exponential growth of popularity in modern world which has changed the entire environment (Poongodi et al., 2020).

Applications of ML Based Wearable Devices

ML has been increasingly acknowledged by many modern life applications as the most important and useful tool in healthcare system. The ML with wearable technology plays a vital role in medical applications. Nowadays, E-health has been upgraded with numbers of new innovative features that not only design but also develop wearable devices. These devices can have ability to detect many diseases using sensors embedded into it. For example, the remote monitoring of patients with wearable devices can allow emergency care and hence reduce hospitalizations. The cardiac function of a patient is recorded by the wearable devices and ML algorithms analyze the cardiac response and can evaluate the compensated and decompensated heart failure (HF) states (Inan et al., 2018). The status of patients with HF also evaluated by this ML algorithm can also be implemented in a real-time monitoring system. The another application of wearable devices and ML techniques integration become popular to analyze the large quantities of longitudinal data streamed from these devices (Mannini & Sabatini, 2010). The ML algorithms identify non-trivial and complex patterns in long-term continuous biological signals (Witten et al., 2005). The IoT has modified and improved the lifestyle by introducing many features that have ML and deep learning (DL) models embedded to decide many run time decisions which are related to human health system (Shehab et al., 2017). Many researchers have introduced IoT devices which are based on ML to facilitate modern human life and reduce the doctor to patient rate (Al-Turjman & Baali, 2019).One of the most important application despite the brain stroke, is the application used in epilepsy based on ML techniques.

The wearable biosensor and ML integration provides a remote monitoring platform for managing hospitalized COVID-19 patients. It showed strong correlation between the parameters such as heart rate and oxygen saturation that are captured by the wearable biosensors and manual measurements. In addition, the COVID-19 patients were classified with moderate/high viral load (Un et al., 2021).

RELATED WORKS

Brain stroke is considered as one of the top leading death causes throughout the world, particularly in densely populated countries in America, Europe, and in Asia (Vos et al., 2020). China, Germany, and United States are the leading countries where stroke has high morbidity and mortality rate. However, it is reported that 80% risk can be reduced only if the risk factor is controlled and monitored with modern ML based devices (Johnson et al., 2019). Hypertension, hyperglycemia, hyperlipidemia, obesity, diabetes, atrial fibrillation, smoking, heavy drinking, sedentary lifestyle, and most importantly unhealthy diet are the most common key factors and causes of stroke in modern countries where the health systems are very strong (Benjamin et al., 2019; Wu et al., 2019). There have been many prevention measures taken to mitigate the risk of having stroke. World-wide many ML based wearable devices have been launched by different IT companies to reduce the risk of having stroke among the population. Furthermore, those who have high risk of stroke are required to have high-risk prevention strategies so that high-risk can be removed (Spence, 2020).

Wearable Devices for Stroke Prediction

Some of the devices such as blood pressure, pulse pressure, and continuous blood flow monitoring can be used predict stroke risk. The following table 1 shows various devices used for different purposes.

Table 1. Devices for Risk Factors Monitoring

Device	Purpose
Biobeat BB-613	Blood Pressure Monitoring (El-Hajj & Kyriacou, 2020; Hosanee et al., 2020)
Carotid Neckband	Blood Flow Dynamics Monitoring (Song et al., 2019)
Kardia Mobile 6L	ECG Monitoring (Isakadze & Martin, 2020)
Smart Watches	Pulse Rate, Blood Pressure, and workout (Lin, Lan, Hsu et al, 2018)
Glucose Tracker (Beta Bionic)	Diagnose the sugar level in blood (Lin, Lan, Hsu et al, 2018)
Wearable for heart attack (Zoll LifeVest)	To normalize the heartbeat of patient (Lin, Lan, Hsu et al, 2018)
Brain Tumor (Q-Collar)	To support the brain around the neck (Chen & Sawan, 2021).
Electromyography (EMG)	A sensor used to measure the muscle movements that help the rehabilitation of prosthetic patient and stroke patients (Chen & Sawan, 2021).
Asthma Monitoring	To detect the oxygen level and respiratory system (Chen & Sawan, 2021)
Movement Disorder	Used for Parkinson's patients (Chen & Sawan, 2021)
Coagulation Monitoring	Blood coagulation is monitored (Chen & Sawan, 2021)
Depression Monitor	Depression is monitored using an application namely, Major Depression Disorder (MDD) (Chen & Sawan, 2021)
Cancer Treatment	Used to treat the cancer patient (Chen & Sawan, 2021)

The device Biobeat BB-613 can be used to monitor blood pressure (El-Hajj & Kyriacou, 2020; Hosanee et al., 2020). Another device called Carotid neckband monitors continuous blood flow and KardiaMobile 6L monitors ECG level.

Machine Learning Based Stroke Prediction

Machine Learning and EEG can be used to make stroke monitoring. Here the essential attributes were extracted from the EEG data received from six channels. The author Yoon-A Choi proposed a system that collects data and predict stroke early. Since it collects various types of data, data preprocessing method called Fast Fourier Transform (FFT), was applied further machine learning models where used to predict and analyze the stroke precursors (Choi et al., 2021). Here the Random Forest algorithm was used for prediction of stroke and experimentally proved with F1-score of 92.51%.

Support vector machine (SVM) is one of the most popular ML model that has been used in this task. In most recent, deep neural network (DNN) models have gained the attention (Joo et al., 2020). These models that is based on neural network (NN) layers have modified and updated the ability to study the relationship between features which are used for input and output through hidden layers (Block et al., 2020). Age, family history, disease history, hypertension, heart attack rate and diabetes are considered as primary and key features that have been selected by ML techniques based on extensive clinal studies (Chen & Sawan, 2021; Hung et al., 2019). Furthermore, some of ML techniques has suggested other features like creatinine, glucose and Hemoglobin as a key factor that has causes the stroke. In the recent technologies, it is worthfully important to have knowledge of convolutional neural network (CNN) model that have ability to integrate the EHRs and 24 hours data (Liu et al., 2020). Such dataset is recorded with the help of IoT base wearable devices like electrocardiogram (ECG), Electromyography (EMG), bloop pressure (BP), and heartbeat rate to predict the stroke risk rate, these are not important and key factor in other predicting models (Miotto et al., 2018). During the training process of CNN model, the streaming dataset is converted into graph. This enhanced features selection and model prediction using CNN based have recorded accurate prediction and precision rate since real time datasets have physiological parameters that can enhance and improve the predicting rate as compared only with EHRs based systems (Lin, Ye, Wang et al, 2018).

MOBILE APP BASED STROKE PREDICTION

There have been many mobile applications which are AI enabled active systems. These devices now have been using to detect the stroke rate and mitigate the factors that are responsible for creaking stroke. Stroke prediction is considered as essential element to save the health as well as to secure the proactive measures which can detect and minimize the stroke rate. There are various prediction models that have been proposed in modern EHRs to provide the as high as accurate risk rate of stroke. For example, the stroke health-education mobile app (SHEMA) helps the stroke patients to know about stroke risk factors and health-related quality of life in patients with stroke (Kang et al., 2019). The timely information helps stroke patient with easy access. The American Heart Association and American Stroke Association provided recommendations for risk factor evaluation in online in order to prevent recurrent stroke (Kernan et al., 2014).

MACHINE LEARNING AND MOBILE APP BASED STROKE PREDICTION

In recent past, there are few studies existing on stroke that examined and predict the most dangerous cardiovascular/non cardiovascular morbidity and mortality rate and relationship with stroke. Most of the predicting applications are based on ML or apparently AI which can analyze the dataset which is available. Based on large dataset and medical history, ML algorithms predict the risk rate of stroke in human life. Few of modern predicting systems are based on CNN to achieve high rate of accuracy despite using SVM. For example, the combined approach of Bayesian model and Mobile application classifies and diagnose stroke to take some measures to avoid unnecessary damages. The bayesian algorithm was integrated in mobile application program to improve usability of prediction algorithm. After that, the mobile application was installed in any smart phone. So this method helps the person to know whether healthy or stroke (Alex & Nayahi, 2018; Subramaniyam et al., 2020). In table 1, some of the modern health prediction systems are given which are based on model ML algorithms like CNN and DNN to get the high accuracy (Lip et al., 2021). It is observed in various predicting system that complex relationship of different comorbidities discovered many approached based on ML and multimorbidity have major reservations for predicting stroke rate.

CONCLUSION

The number of stroke patients is increasing day by day in the whole world. The stroke prediction in smart healthcare predicts the stroke earlier. IoT applications can detect blood pressure that can reduce the risk factor. A sensor connected to the device sends real-time blood pressure to the cloud, and doctors can check that data anywhere from the world. The machine learning algorithm analyzes data in the mobile and connected to the internet. Hence the use of IoT healthcare solutions reduces cost and drives the treatment to home-centric.

REFERENCES

Al-Turjman, F., & Baali, I. (2019). Machine learning for wearable IoT-based applications: A survey. *Transactions on Emerging Telecommunications Technologies*, e3635. doi:10.1002/ett.3635

Alex, S. A., & Nayahi, J. J. V. (2018). Deep Incremental Learning for Big Data Stream Analytics. In *International conference on Computer Networks, Big data and IoT*. Springer.

Benjamin, E. J., Muntner, P., Alonso, A., Bittencourt, M. S., Callaway, C. W., Carson, A. P., Chamberlain, A. M., Chang, A. R., Cheng, S., Das, S. R., Delling, F. N., Djousse, L., Elkind, M. S. V., Ferguson, J. F., Fornage, M., Jordan, L. C., Khan, S. S., Kissela, B. M., Knutson, K. L., ... Virani, S. S. (2019). Heart disease and stroke statistics—2019 update: A report from the American Heart Association. *Circulation*, *139*(10), e56–e528. doi:10.1161/CIR.0000000000000659 PMID:30700139

Bersano, A., Kraemer, M., Burlina, A., Mancuso, M., & Finsterer, J. (2020). Heritable and non-heritable uncommon causes of stroke. *Journal of Neurology*, 1–28. PMID:32318851

Bilal, M., Usmani, R. S. A., Tayyab, M., Mahmoud, A. A., & Abdalla, R. M. (2020). Smart cities data: Framework, applications, and challenges. Handbook of Smart Cities, 1-29.

Block, L., El-Merhi, A., Liljencrantz, J., Naredi, S., Staron, M., & Odenstedt Hergès, H. (2020). Cerebral ischemia detection using artificial intelligence (CIDAI)—A study protocol. *Acta Anaesthesiologica Scandinavica, 64*(9), 1335–1342. doi:10.1111/aas.13657 PMID:32533722

Boukobza, M., Nahmani, S., Deschamps, L., & Laissy, J.-P. (2019). Brain abscess complicating ischemic embolic stroke in a patient with cardiac papillary fibroelastoma–Case report and literature review. *Journal of Clinical Neuroscience, 66*, 277–279. doi:10.1016/j.jocn.2019.03.041 PMID:31097380

Chen, Y.-H., & Sawan, M. (2021). Trends and Challenges of Wearable Multimodal Technologies for Stroke Risk Prediction. *Sensors (Basel), 21*(2), 460. doi:10.339021020460 PMID:33440697

Choi, Y.-A., Park, S., Jun, J.-A., Ho, C. M. B., Pyo, C.-S., Lee, H., & Yu, J. (2021). Machine-Learning-Based Elderly Stroke Monitoring System Using Electroencephalography Vital Signals. *Applied Sciences (Basel, Switzerland), 11*(4), 1761. doi:10.3390/app11041761

Convertino, V. A., Moulton, S. L., Grudic, G. Z., Rickards, C. A., Hinojosa-Laborde, C., Gerhardt, R. T., Blackbourne, L. H., & Ryan, K. L. (2011). Use of advanced machine-learning techniques for non-invasive monitoring of hemorrhage. *The Journal of Trauma and Acute Care Surgery, 71*(1), S25–S32. doi:10.1097/TA.0b013e3182211601 PMID:21795890

El-Hajj, C., & Kyriacou, P. A. (2020). A review of machine learning techniques in photoplethysmography for the non-invasive cuff-less measurement of blood pressure. *Biomedical Signal Processing and Control, 58*, 101870. doi:10.1016/j.bspc.2020.101870

Fugate, J. E., Lyons, J. L., Thakur, K. T., Smith, B. R., Hedley-Whyte, E. T., & Mateen, F. J. (2014). Infectious causes of stroke. *The Lancet. Infectious Diseases, 14*(9), 869–880. doi:10.1016/S1473-3099(14)70755-8 PMID:24881525

Garkowski, A., Zajkowska, J., Moniuszko, A., Czupryna, P., & Pancewicz, S. (2015). Infectious causes of stroke. *The Lancet. Infectious Diseases, 15*(6), 632. doi:10.1016/S1473-3099(15)00020-1 PMID:26008834

Goldstein, B. A., Navar, A. M., Pencina, M. J., & Ioannidis, J. (2017). Opportunities and challenges in developing risk prediction models with electronic health records data: A systematic review. *Journal of the American Medical Informatics Association: JAMIA, 24*(1), 198–208. doi:10.1093/jamia/ocw042 PMID:27189013

Harrar, D. B., Salussolia, C. L., Kapur, K., Danehy, A., & Kleinman, M. E. (2020). A stroke alert protocol decreases the time to diagnosis of brain attack symptoms in a pediatric emergency department. The Journal of Pediatrics, 216, 136-141. doi:10.1016/j.jpeds.2019.09.027

Hong, K.-S., Bang, O. Y., Kang, D.-W., Yu, K.-H., Bae, H.-J., Lee, J. S., Heo, J. H., Kwon, S. U., Oh, C. W., Lee, B.-C., Kim, J. S., & Yoon, B.-W. (2013). Stroke statistics in Korea: part I. Epidemiology and risk factors: a report from the korean stroke society and clinical research center for stroke. *Journal of Stroke, 15*(1), 2. doi:10.5853/jos.2013.15.1.2 PMID:24324935

Hope, T. M., Seghier, M. L., Leff, A. P., & Price, C. J. (2013). Predicting outcome and recovery after stroke with lesions extracted from MRI images. *NeuroImage. Clinical*, *2*, 424–433. doi:10.1016/j. nicl.2013.03.005 PMID:24179796

Hosanee, M., Chan, G., Welykholowa, K., Cooper, R., Kyriacou, P. A., Zheng, D., Allen, J., Abbott, D., Menon, C., Lovell, N. H., Howard, N., Chan, W.-S., Lim, K., Fletcher, R., Ward, R., & Elgendi, M. (2020). Cuffless single-site photoplethysmography for blood pressure monitoring. *Journal of Clinical Medicine*, *9*(3), 723. doi:10.3390/jcm9030723 PMID:32155976

Hung, C.-Y., Lin, C.-H., Lan, T.-H., Peng, G.-S., & Lee, C.-C. (2019). Development of an intelligent decision support system for ischemic stroke risk assessment in a population-based electronic health record database. *PLoS One*, *14*(3), e0213007. doi:10.1371/journal.pone.0213007 PMID:30865675

Inan, O. T., Baran Pouyan, M., Javaid, A. Q., Dowling, S., Etemadi, M., Dorier, A., Heller, J. A., Bicen, A. O., Roy, S., De Marco, T., & Klein, L. (2018). Novel wearable seismocardiography and machine learning algorithms can assess clinical status of heart failure patients. *Circulation: Heart Failure*, *11*(1), e004313. doi:10.1161/CIRCHEARTFAILURE.117.004313 PMID:29330154

Isakadze, N., & Martin, S. S. (2020). How useful is the smartwatch ECG? *Trends in Cardiovascular Medicine*, *30*(7), 442–448. doi:10.1016/j.tcm.2019.10.010 PMID:31706789

Johnson, C. O., Nguyen, M., Roth, G. A., Nichols, E., Alam, T., Abate, D., Abd-Allah, F., Abdelalim, A., Abraha, H. N., Abu-Rmeileh, N. M. E., Adebayo, O. M., Adeoye, A. M., Agarwal, G., Agrawal, S., Aichour, A. N., Aichour, I., Aichour, M. T. E., Alahdab, F., Ali, R., ... Murray, C. J. L. (2019). Global, regional, and national burden of stroke, 1990–2016: A systematic analysis for the Global Burden of Disease Study 2016. *Lancet Neurology*, *18*(5), 439–458. doi:10.1016/S1474-4422(19)30034-1 PMID:30871944

Joo, G., Song, Y., Im, H., & Park, J. (2020). H. Im and J. Park, "Clinical implication of machine learning in predicting the occurrence of cardiovascular disease using big data (Nationwide Cohort Data in Korea). *IEEE Access: Practical Innovations, Open Solutions*, *8*, 157643–157653. doi:10.1109/ACCESS.2020.3015757

Joshi, R., Cardona, M., Iyengar, S., Sukumar, A., Raju, C. R., Raju, K. R., Raju, K., Reddy, K. S., Lopez, A., & Neal, B. (2006). Chronic diseases now a leading cause of death in rural India—Mortality data from the Andhra Pradesh Rural Health Initiative. *International Journal of Epidemiology*, *35*(6), 1522–1529. doi:10.1093/ije/dyl168 PMID:16997852

Kang, Y.-N., Shen, H.-N., Lin, C.-Y., Elwyn, G., Huang, S.-C., Wu, T.-F., & Hou, W.-H. (2019). Does a Mobile app improve patients' knowledge of stroke risk factors and health-related quality of life in patients with stroke? A randomized controlled trial. *BMC Medical Informatics and Decision Making*, *19*(1), 1–9. doi:10.118612911-019-1000-z PMID:31864348

Kernan, W. N., Ovbiagele, B., Black, H. R., Bravata, D. M., Chimowitz, M. I., Ezekowitz, M. D., Fang, M. C., Fisher, M., Furie, K. L., Heck, D. V., Johnston, S. C. C., Kasner, S. E., Kittner, S. J., Mitchell, P. H., Rich, M. W., Richardson, D. J., Schwamm, L. H., & Wilson, J. A. (2014). Guidelines for the prevention of stroke in patients with stroke and transient ischemic attack: A guideline for healthcare professionals from the American Heart Association/American Stroke Association. *Stroke*, *45*(7), 2160–2236. doi:10.1161/STR.0000000000000024 PMID:24788967

Lattanzi, S., & Silvestrini, M. (2016). Blood pressure in acute intra-cerebral hemorrhage. *Annals of Translational Medicine, 4*(16), 320. doi:10.21037/atm.2016.08.04 PMID:27668240

Lin, R., Ye, Z., Wang, H., & Wu, B. (2018). Chronic diseases and health monitoring big data: A survey. *IEEE Reviews in Biomedical Engineering, 11*, 275–288. doi:10.1109/RBME.2018.2829704 PMID:29993699

Lin, S.-S., Lan, C.-W., Hsu, H.-Y., & Chen, S.-T. (2018). Data analytics of a wearable device for heat stroke detection. *Sensors (Basel), 18*(12), 4347. doi:10.339018124347 PMID:30544887

Lip, G. Y., Genaidy, A., Tran, G., Marroquin, P., & Estes, C. (2021). Improving Stroke Risk Prediction in the General Population: A Comparative Assessment of Common Clinical Rules, a New Multimorbid Index, and Machine-Learning-Based Algorithms. *Thrombosis and Haemostasis*. PMID:33765685

Liu, Y., Yin, B., & Cong, Y. (2020). The Probability of Ischaemic Stroke Prediction with a Multi-Neural-Network Model. *Sensors (Basel), 20*(17), 4995. doi:10.339020174995 PMID:32899242

Mannini, A., & Sabatini, A. M. (2010). Machine learning methods for classifying human physical activity from on-body accelerometers. *Sensors (Basel), 10*(2), 1154–1175. doi:10.3390100201154 PMID:22205862

Miotto, R., Wang, F., Wang, S., Jiang, X., & Dudley, J. T. (2018). Deep learning for healthcare: Review, opportunities and challenges. *Briefings in Bioinformatics, 19*(6), 1236–1246. doi:10.1093/bib/bbx044 PMID:28481991

Poongodi, T., Krishnamurthi, R., Indrakumari, R., Suresh, P., & Balusamy, B. (2020). Wearable devices and IoT. In *A handbook of Internet of Things in biomedical and cyber physical system* (pp. 245–273). Springer. doi:10.1007/978-3-030-23983-1_10

Rasmussen, M., Valentin, J. B., & Simonsen, C. Z. (2020). Blood Pressure Thresholds During Endovascular Therapy in Ischemic Stroke—Reply. *JAMA Neurology, 77*(12), 1579–1580. doi:10.1001/jamaneurol.2020.3819 PMID:33044508

Shehab, A., Ismail, A., Osman, L., Elhoseny, M., & El-Henawy, I. M. (2017). Quantified self using IoT wearable devices. In *International conference on advanced intelligent systems and informatics*. Springer.

Shulman, J. G., & Cervantes-Arslanian, A. M. (2019). Infectious etiologies of stroke. Seminars in Neurology, 39(4), 482-494. doi:10.1055-0039-1687915

Sirsat, M. S., Fermé, E., & Câmara, J. (2020). Machine learning for brain stroke: A review. *Journal of Stroke and Cerebrovascular Diseases, 29*(10), 105162. doi:10.1016/j.jstrokecerebrovasdis.2020.105162 PMID:32912543

Song, I., Yoon, J., Kang, J., Kim, M., Jang, W. S., Shin, N.-Y., & Yoo, Y. (2019). Design and implementation of a new wireless carotid neckband doppler system with wearable ultrasound sensors: Preliminary results. *Applied Sciences (Basel, Switzerland), 9*(11), 2202. doi:10.3390/app9112202

Spence, J. D. (2020). Uses of ultrasound in stroke prevention. *Cardiovascular Diagnosis and Therapy, 10*(4), 955–964. doi:10.21037/cdt.2019.12.12 PMID:32968653

Subramaniyam, M., Hong, S. H., Yu, J., & Park, S. J. (2017). Wake-Up Stroke Prediction through IoT and Its Possibilities. *2017 International Conference on Platform Technology and Service (PlatCon)*, 1-5. 10.1109/PlatCon.2017.7883738

Subramaniyam, M., Lee, K.-S., Park, S. J., & Min, S. N. (2020). Development of Mobile Application Program for Stroke Prediction Using Machine Learning with Voice Onset Time Data. In *International Conference on Human-Computer Interaction*. Springer. 10.1007/978-3-030-50726-8_87

Sun, F., Liu, H., Fu, H.-x., Li, C.-b., & Geng, X.-j. (2020). *Predictive factors of hemorrhage after thrombolysis in patients with acute ischemic stroke* (Vol. 11). Frontiers in Neurology.

Tayyab, M., Marjani, M., Jhanjhi, N., & Hashem, I. A. T. (2021). A Light-weight Watermarking-Based Framework on Dataset Using Deep Learning Algorithms. *2021 National Computing Colleges Conference (NCCC)*, 1-6. 10.1109/NCCC49330.2021.9428845

Tayyab, M., Marjani, M., Jhanjhi, N., Hashim, I. A. T., & Almazroi, A. A (2021). Cryptographic Based Secure Model on Dataset for Deep Learning Algorithms. *Computers Materials & Continua*, 69(1), 1183–1200. doi:10.32604/cmc.2021.017199

Un, K.-C., Wong, C.-K., Lau, Y.-M., Lee, J. C.-Y., Tam, F. C.-C., Lai, W.-H., Lau, Y.-M., Chen, H., Wibowo, S., Zhang, X., Yan, M., Wu, E., Chan, S.-C., Lee, S.-M., Chow, A., Tong, R. C.-F., Majmudar, M. D., Rajput, K. S., Hung, I. F.-N., & Siu, C.-W. (2021). Observational study on wearable biosensors and machine learning-based remote monitoring of COVID-19 patients. *Scientific Reports*, *11*(1), 1–9. doi:10.103841598-021-82771-7 PMID:33623096

Uppal, S., Goel, S., Randhawa, B., & Maheshwary, A. (2020). Autoimmune-Associated Vasculitis Presenting as Ischemic Stroke With Hemorrhagic Transformation: A Case Report and Literature Review. *Cureus*, *12*(9). Advance online publication. doi:10.7759/cureus.10403 PMID:33062521

Usmani, R. S. A., Saeed, A., & Tayyab, M. (2021). Role of ICT for Community in Education During COVID-19. In *ICT Solutions for Improving Smart Communities in Asia* (pp. 125–150). IGI Global. doi:10.4018/978-1-7998-7114-9.ch006

Vamsi Badi, D. B. (2020). *Prediction of Brain Stroke Severity Using Machine Learning. International Information and Engineering Technology Association*. IIETA. doi:10.18280/ria.340609

Vashistha, R., Yadav, D., Chhabra, D., & Shukla, P. (2019). Artificial intelligence integration for neurodegenerative disorders. In *Leveraging Biomedical and Healthcare Data* (pp. 77–89). Elsevier. doi:10.1016/B978-0-12-809556-0.00005-8

Verma, A., Jaiswal, S., & Sheikh, W. R. (2020). Acute thrombotic occlusion of subclavian artery presenting as a stroke mimic. *Journal of the American College of Emergency Physicians Open*, *1*(5), 932–934. doi:10.1002/emp2.12085 PMID:33145542

Vos, T., Lim, S. S., Abbafati, C., Abbas, K. M., Abbasi, M., Abbasifard, M., Abbasi-Kangevari, M., Abbastabar, H., Abd-Allah, F., Abdelalim, A., Abdollahi, M., Abdollahpour, I., Abolhassani, H., Aboyans, V., Abrams, E. M., Abreu, L. G., Abrigo, M. R. M., Abu-Raddad, L. J., Abushouk, A. I., ... Murray, C. J. L. (2020). Global burden of 369 diseases and injuries in 204 countries and territories, 1990–2019: A systematic analysis for the Global Burden of Disease Study 2019. *Lancet*, *396*(10258), 1204–1222. doi:10.1016/S0140-6736(20)30925-9 PMID:33069326

Wang, S., Li, Y., Tian, J., Peng, X., Yi, L., Du, C., Feng, C., Liu, C., Deng, R., & Liang, X. (2020). A randomized controlled trial of brain and heart health manager-led mHealth secondary stroke prevention. *Cardiovascular Diagnosis and Therapy*, *10*(5), 1192–1199. doi:10.21037/cdt-20-423 PMID:33224743

Wilkinson, D. A., Daou, B. J., Nadel, J. L., Chaudhary, N., Gemmete, J. J., Thompson, B. G., & Pandey, A. S. (2021). Abdominal aortic aneurysm is associated with subarachnoid hemorrhage. *Journal of Neurointerventional Surgery*, *13*(8), 716–721. doi:10.1136/neurintsurg-2020-016757 PMID:33158992

Witten, Frank, Hall, & Pal. (2005). Practical machine learning tools and techniques. *Data Mining*, *2*, 4.

World Health Organization. (2004). *The atlas of heart disease and stroke.* Author.

Wu, S., Wu, B., Liu, M., Chen, Z., Wang, W., Anderson, C. S., Sandercock, P., Wang, Y., Huang, Y., Cui, L., Pu, C., Jia, J., Zhang, T., Liu, X., Zhang, S., Xie, P., Fan, D., Ji, X., Wong, K.-S. L., ... Zhang, S. (2019). Stroke in China: Advances and challenges in epidemiology, prevention, and management. *Lancet Neurology*, *18*(4), 394–405. doi:10.1016/S1474-4422(18)30500-3 PMID:30878104

Chapter 10
Predicting the Early Stage of Diabetes and Finding the Association of the Symptoms

Tasnim Mohamad Naim
Universiti Malaya, Malaysia

Siti Nurnabila Abdul Rashid
Universiti Malaya, Malaysia

Muneer Ahmad
National University of Sciences and Technology, Pakistan

N. Z. Jhanjhi
https://orcid.org/0000-0001-8116-4733
Taylor's University, Malaysia

ABSTRACT

Diabetes has become a growing global public health issue. This illness can become chronic since there are no early symptoms and it can lead to several negative health impacts. This study focuses on the early identification of diabetes based on the symptoms of the disease and finds the relation between the symptoms and the diagnosis. Successful early diagnosis of this disease could boost a person to a better treatment plan before it worsens the health to a critical stage. This study exploits recent classification algorithms including Naïve Bayes, Logistic Regression, REPTree, J48, and Random Forest. The association rules mining using the apriori algorithm is used. Further, the 10-fold cross-validation with split criteria was adopted. The authors adopted several evaluation metrics including accuracy, precision, recall, F-measure, and area under the ROC curve. The research findings revealed the Random Forest to be the best classification algorithm as compared to other classifiers.

DOI: 10.4018/978-1-7998-9201-4.ch010

INTRODUCTION

This case study focuses on the health sector in particular focusing specifically on the early stage of a disease called diabetes. Diabetes is a condition that affects most of the populations in the world. Early diabetes or also referred to as pre-diabetes is the early sign before the diabetes is developed in a person. As the year goes by, diabetes has become more and more of a bigger public health concern around the world as it affects a large group of the population. Globally, it is estimated that diabetes affects approximately 415 million people worldwide and is expected to reach 642 million people by the year 2040 (Al-Lawati, 2017). If the number goes on to increase exponentially, the health sector might have to take on a huge toll on managing the increasing surge of patients with other extended health complications from diabetes.

Problems observed are that it was found that most of the time, people that have early diabetes often do not know about their conditions until either it has turned into diabetes or any health implications from this arises (Embong, 2019). This is as the symptoms may appear as mild or appear to have no correlation to diabetes at a glance. To get an accurate diagnosis, several stages of tests have to be performed. It is estimated that around 193 million people with diabetes worldwide don't know their condition (Al-Lawati, 2017). This could be fatal as when diabetes progresses, it can increase to the level in which it can bring on further serious health implications such as heart disease, kidney failure, stroke, nerve damage and so on (Olsen, 2020). If it is left untreated and undetected, it would be harder to manage a person's health condition.

The main stakeholders that are involved with this case study are people related to the healthcare industry which are the patients, doctors, and so on. As mentioned previously, early stages of diabetes can be present in a person with little to no noticeable symptoms at all which can cause the untreated high blood glucose to continually increase within time (HonorHealth, 2020). This can make diagnosing and spotting early diabetes signs to be tricky without rigorous testing for the doctor and at the cost of the patient's health worsening if it went further undiagnosed.

Possible solutions that can help in addressing these issues are by predicting the possibility of a patient having early diabetes with the symptoms that they have as well as studying the association present between the symptoms. Methods that can be used to predict is by using classifying algorithms while the association can use associate algorithms. This is as in many diseases, people with diabetes do have their own tell-tale symptoms that may not have been so obvious at early stages. Data of existing diabetes patient diagnosis based on their symptoms can be used as part of data training in order to determine the correct diagnosis based on a collection of the symptoms. Whereas a test data could be used to test the trained model to evaluate the accuracy of the diagnosis.

To the stakeholders, the positive impact from this issue if early diabetes can be predicted is that quick decision making can be taken swiftly to curb potential risks and avoid unwanted health implications. This is as it can be detected early and the symptoms that a patient may have can easily be associated with how early diabetes may be present in a person. The patient would be able to take early health precautions or start seeking further medicinal help while the doctor could plan out treatment plans quickly or shorten the time usually taken to diagnose a patient. Plus, the prediction can be done faster, on mass of people at once and accurately based on the data.

REVIEW OF LITERATURE FOR THE EXISTING SOLUTIONS

Diabetes is a disorder in which the blood sugar level of a person is above average 4.4-6.1 mmol/L (Saravananathan, K., & Velmurugan, T.). According to clinicians, diabetes happens when the organ known as the pancreas does not release a hormone called insulin in adequate amounts (Manjusree & Satees h Kumar, 2019). There are 3 types of diabetes which are Type 1 Diabetes also called insulin-dependent diabetes or juvenile-onset diabetes, Type 2 Diabetes is called insulin resistance diabetes or adult-onset diabetes and Type 3 Diabetes is called gestational diabetes it occurs only during pregnancy when insulin blocks the hormones (Rani, 2018). Other than that, nearly 80% of the people impacted from diabetes belong to middle and low-income countries which also can lead to death (Nagarajan & Chandrasekaran, 2015).

Figure 1.

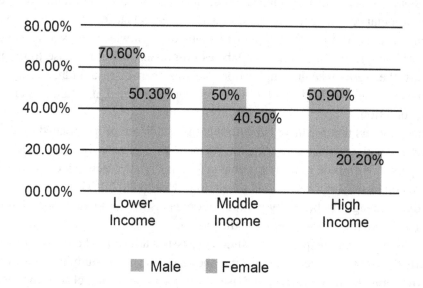

Figure 1 presents death rates among the different categories of people which include male and female from lower-income, middle income and high income. The survey showed that male lower income leads to the death rate rather than middle income and high income. Female high income showed that they get the least percentage which is 20.20% rather than other income. There are many factors that someone can get diabetes such as a lifestyle of someone and ages also play the main factor.

Based on most of the research, they suggest that early diagnosis can avoid any other diseases related to diabetes such as blindness, heart diseases and amputation. The early and accurate diagnosis can be achieved by using data mining tools such as the Naïve Bayes technique has been used for the prediction of diabetes (Mahboob Alam et al., 2019). There also has a collection of the review that performs a comprehensive analysis of the application of machine learning, data mining techniques and diabetes research tools by using Knowledge Discovery in Database.

Figure 2.

Class: class (Nom)
Visualize All

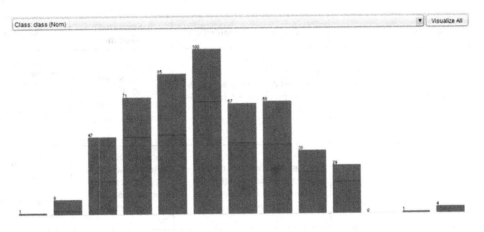

Figure 3.

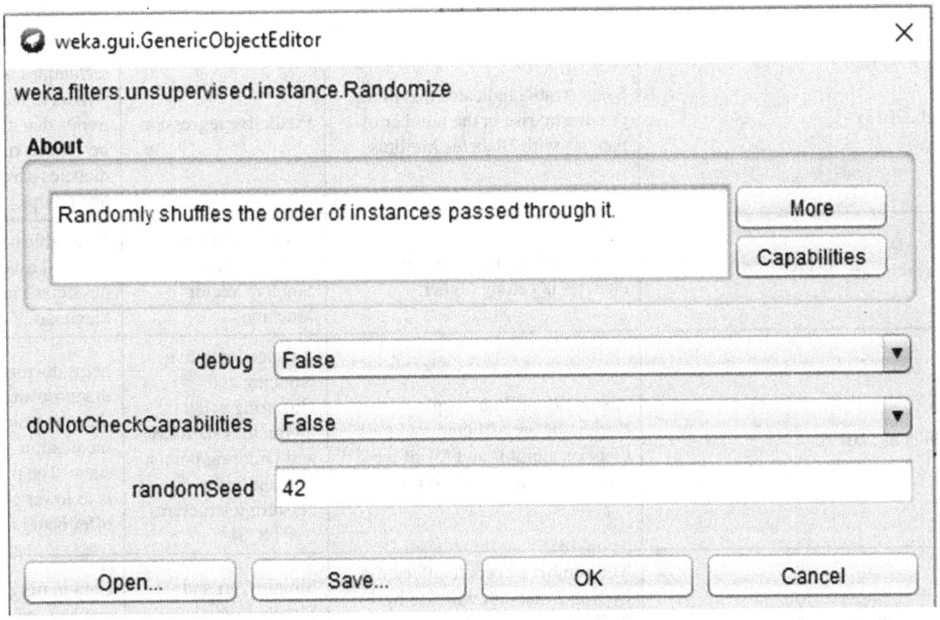

weka.gui.GenericObjectEditor
weka.filters.unsupervised.instance.Randomize
About
Randomly shuffles the order of instances passed through it.
More
Capabilities
debug
False
doNotCheckCapabilities
False
randomSeed
42
Open...
Save...
OK
Cancel

The entire selection process for the research is shown in a workflow in Figure 2, with Figure 3 indicating the number of publications each year. The main purpose of the collection review from all these articles is to analyse the use of machine learning and data mining techniques in the prediction of diabetes at an early stage over the year.

Table 1. An overview of machine intelligence solutions for cervical cancer identification

Citation	Problem(s) Identified	Suggested Solution(s)	Significance
(Mahboob Alam et al., 2019)	A diabetes diagnosis is known to be a complex topic for quantitative studies.	Artificial Neural Network, Random Forest and K-mean clustering	To assess important attribute collection, and for clustering, estimation, and correlation rule mining for diabetes, different methods are used
(Kazerouni et al., 2020)	The statistics show that the key non-genetic determinants of the condition are conditions such as obesity and physical inactivity	Logistic Regression, K-nearest neighbour, support vector machine, and artificial neural networks	These algorithms have a high ability to forecast and diagnose many diseases early, including T2DM,
(Manjusree & Sateesh Kumar, 2019)	The lack of reliable treatment at low expense, meaning that patients are accurately diagnosed	The predictive model and descriptive model	To predict the occurrence or recurrence of diabetes risks, an application using a class comparison data mining algorithm has been developed
(Aljumah et al., 2013)	Saudi Arabia indicates that there is a drastic rise in the number of patients with Diabetes Mellitus	Predictive regression	To use data mining techniques in Saudi Arabia to uncover trends that classify the best mode of care for diabetes across various age groups
(Sanakal & Jayakumari, 2014)	The possibility of women getting diabetes is getting higher.	Fuzzy C-means clustering and Support Vector Machine	To establish a device that will assist physicians in medical diagnosis.
(Mamatha Bai et al., 2019)	About 145 million people worldwide are afflicted by diabetes mellitus and 5% of the Indian population add to this rate.	Balanced iterative reducing and clustering using hierarchies (BIRCH) and Ordering points to identify the clustering structure (OPTICS)	helps doctors diagnose and have the recommended medication at an early stage. The primary goal is to lower costs and offer better care.
(Nagarajan & Chandrasekaran, 2015)	The number of people affected by type 2 diabetes will rise by 2025 and diabetes prevalence is decreased by 2.7% in rural areas in India.	Simple Cart and Simple Logistic. Naïve Bayes and REPTree	Data mining methods can be used to predict the nature and risk levels of diabetes.
(Balpande & Wajgi, 2017)	Diabetes induces many other illnesses, such as blindness, blood pressure, Heart disease, renal disease, and nerve injury.	Equivalence Class Clustering and bottoms up lattice traversal Algorithm (ECLAT)	To assess the predictability and incidence estimates of diabetes.

Continued on following page

Table 1. Continued

Citation	Problem(s) Identified	Suggested Solution(s)	Significance
(Iyer et al., 2015)	Effects of diabetes have been reported to have a more fatal and worsening effect on women than on men due to their poorer survival rate and reduced quality of life.	Decision Tree and Naïve Bayes	Detection of diabetes at an early stage is the key to recovery, which may help prevent complications.
(Sneha & Gangil, 2019)	The World Health Organization's study discusses diabetes and its complications that threaten families medically, financially and socially.	AdaBoost algorithm with Decision Stump be used as the base classifier for classification, Naïve Bayes, Support Vector Machine, and Decision Tree	Seeking to diagnose and avoid early-stage complications of diabetes by predictive analysis through developing classification techniques.
(Wu et al., 2018)	The key challenges they are attempting to solve are to increase the accuracy of the prediction algorithm and to make the model scalable to more than one dataset	K-mean cluster algorithm and Logistic Regression.	To develop an effective forecast model for the high-risk T2DM community.
(Kavakiotis et al., 2017)	The diabetes is the prognosis and diagnosis of human-threatening and/or disease-reducing quality of life	Machine learning, Knowledge Discovery in Database, Support VectOR Machine	Important work has been performed to date on nearly all areas of DM science and, in particular, biomarker recognition and prediction-diagnosis
(Sa'di et al., 2015)	Diabetes is one of the most common diseases in the world today with high death and morbidity rates, making it one of the biggest health issues in the world	RBF Network, Naïve Bayes and J48	Naive Bayes algorithm with an accuracy rate of 76.95% had the best accuracy compared to J48 and the RBF Network diagnosis.
(Perveen et al., 2016).	Studies have found that life expectancy for people with diabetes may be shortened by as many as 15 years.	J48 decision tree.	Comprehensive attempts are being made to enhance the performance of such structures using an ensemble classifier
(Saravananathan & Velmurugan, 2016)	There is a proportionate spike in complications associated with diabetes, and the condition has become the most lethal disease in the United States with no immediate solution in sight.	J48-Algorithm, REPTree, Support Vector Machine and k-Nearest Neighbour Algorithm.	The results show that the performance of the J48 technique is significantly higher than the other three techniques for the classification of diabetes data.
(Habibi et al., 2015)	The biggest difficulty in diabetes screening is the need to research and take blood samples from a variety of patients, which are costly in terms of financial and manpower resources	The Decision Tree and J48 algorithms	They built a model using the T2DM decision tree that does not include laboratory testing for the diagnosis of T2DM

Continued on following page

Table 1. Continued

Citation	Problem(s) Identified	Suggested Solution(s)	Significance
(Das et al., 2018)	According to the World Health Organisation (WHO) projection, diabetes will be one of the leading causes of death in 2030 and the death rate will double between 2005 and 2030	J48 Algorithm and Naïve Bayes	Data mining plays a crucial role in the diabetes dataset to reveal and discover the secret information of a vast volume of unused diabetes data that can greatly help to advance the level of care for patients with diabetes
(Saravananathan & Velmurugan, 2016)	The number of cases for diabetes increases exponentially day by day.	Decision Tree	Data mining tools can be a valuable advantage for diabetes researchers because they can discover and reveal secret information from a large array of diabetes-related data that dramatically increases the quality of health care for diabetes patients.
(Rani, 2018)	A vast volume of data relating to health is generated at different stages of the health system because of the scale of the data, it would be difficult to analyse the data and derive the analysis	Association clustering and Time Series	The patient's prediction system based on various diabetic factors is beneficial to the patient. This will serve as an early warning device and will alert people of the potential risk of diabetes.

METHODOLOGY

The objectives of this case study are to predict early-stage diabetes and finding the association rule from the symptoms using the WEKA tool. The study adopts a selected range of available classifiers and association namely Naïve Bayes, Logistic Regression, REPTree, J48 and Random Forest. An overview of the data-mining related solutions for this case study is as in Figure 4 below.

Figure 4.

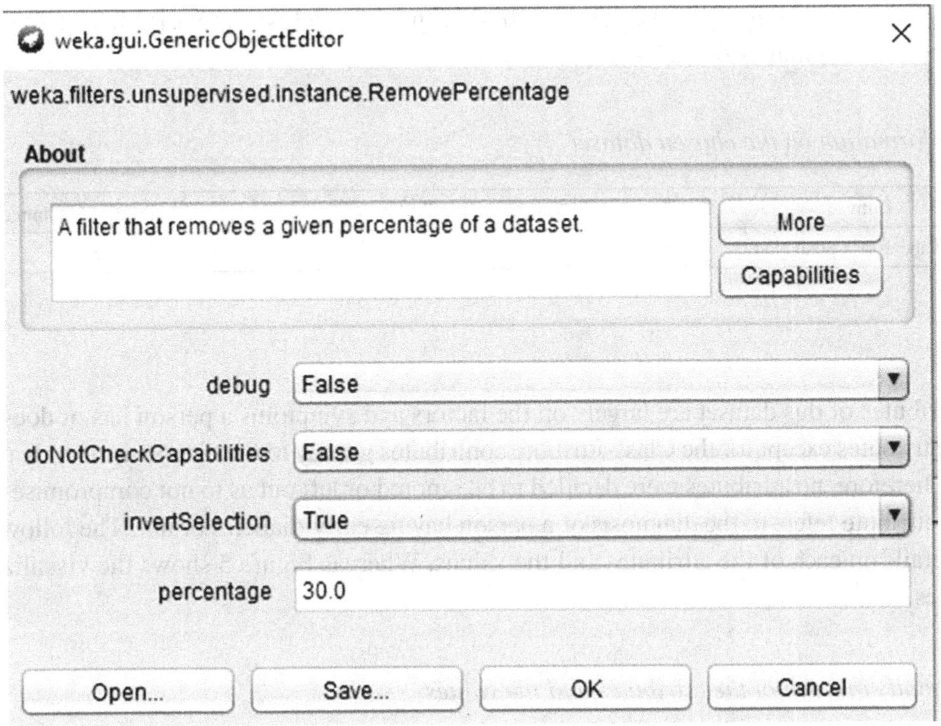

The experimental processes have all been achieved by using the WEKA toolkit. The purpose of using WEKA tool to retrieve the data for prediction and association of the symptoms for early-stage diabetes for running the related algorithms.

The data mining technique that was used to predict if a person is positive or negative for diabetes is classification. A few classification algorithms in WEKA were put to test as part of the solutions. The algorithms we chose to analyze the data are Naïve Bayes, Logistic Regression, REPTree, J48 and Random Forest. The best classifiers among these will be chosen through some evaluation metrics based on their performances of the classified instances.

While for associating the symptoms with the diabetes diagnosis, an association technique was used which is the Apriori Algorithm. The algorithm helped in finding the common relations between one or more symptoms that can be present in a person with early-stage diabetes which can help them in identifying possible symptoms and risks.

RESULTS AND DISCUSSION

Descriptive and Visual Interpretation of Dataset

The dataset used in this study is originally taken from the UCI Machine Learning Repository.

This model is designed to discover trends in the data and to consider the interaction between the data attributes. A descriptive model is the core function of the data and is summarized. The Table 2 below shows a brief description of the dataset.

Table 2. Information on the chosen dataset

Dataset	Number of attributes	Number of instances
Early stage diabetes risk prediction	17	520

The attributes of this dataset are largely on the factors and symptoms a person has or does not have. All of the attributes except for the Class attribute contributes greatly to the diagnosis of early diabetes in a person. Therefore, no attributes were decided to be ignored or left out as to not compromise the result. The Class attribute refers to the diagnosis of a person having early-diabetes or not. The following Table 3 below details on each of the attributes and the values. Whereas Figure 5 shows the visualizations for all attributes.

Table 3. Details on each of the attributes and the values

No	Attribute	Values
1	Age	Range from 16 - 90
2	Gender	1.Male 2. Female
3	Polyuria	1.Yes 2. No
4	Polydipsia	1.Yes 2. No
5	Sudden weight loss	1.Yes 2. No
6	Weakness	1.Yes 2. No
7	Polyphagia	1.Yes 2. No
8	Genital thrush	1.Yes 2. No
9	Visual blurring	1.Yes 2. No
10	Itching	1.Yes 2. No
11	Irritability	1.Yes 2. No
12	Delayed healing	1.Yes 2. No
13	Partial paresis	1.Yes 2. No
14	Muscle Stiffness	1.Yes 2. No
15	Alopecia	1.Yes 2. No
16	Obesity	1.Yes 2. No
17	Class	1. Positive 2. Negative

Figure 5.

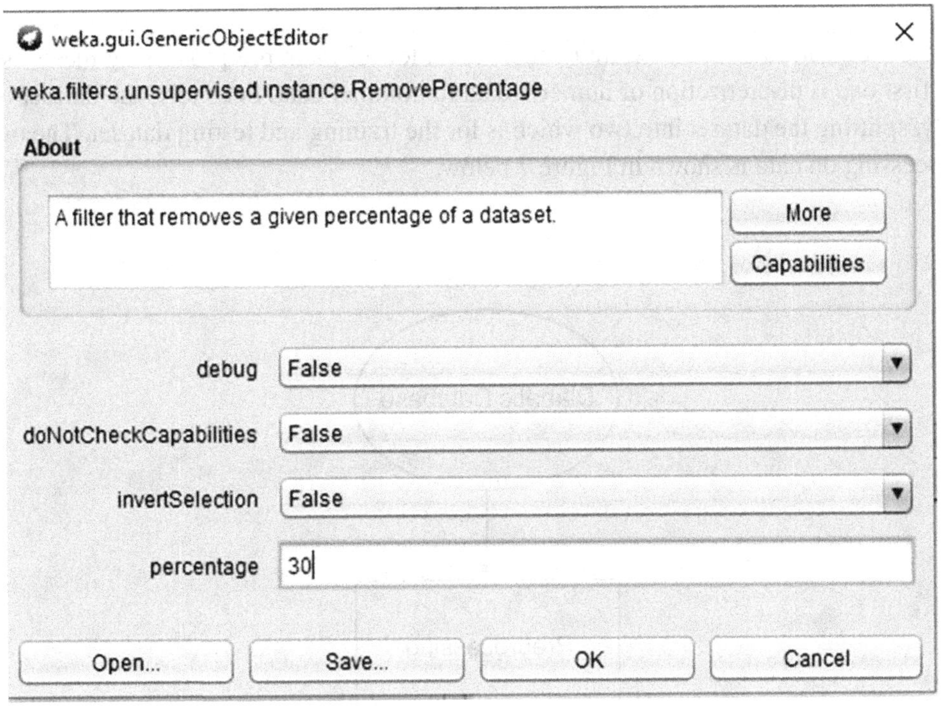

weka.gui.GenericObjectEditor
weka.filters.unsupervised.instance.RemovePercentage
About
A filter that removes a given percentage of a dataset.
More
Capabilities
debug False
doNotCheckCapabilities False
invertSelection False
percentage 30
Open...
Save...
OK
Cancel

To illustrate on the relation between the attributes with each other and also with the class, the following Figure 6 is drawn. The red color is the positive class whereas the blue color is the negative class.

Figure 6.

class
Negative
Positive
Obesity
Alopecia
delayed healing
muscle stiffness
partial paresis

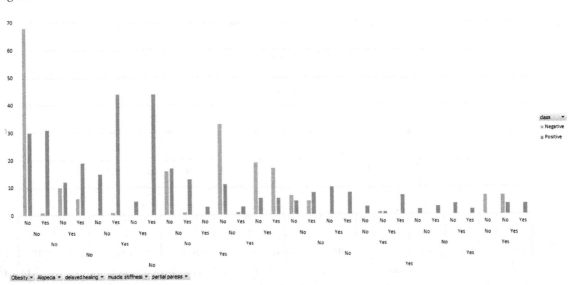

Preprocessing on Data

To enhance the outcomes of this case study, there are a few of main pre-processing methods that were done. The first one is discretization of numeric data to nominal data. Secondly, the dataset is randomized. Lastly, splitting the dataset into two which is for the training and testing dataset. The overview of the pre-processing on data is shown in Figure 7 below.

Figure 7.

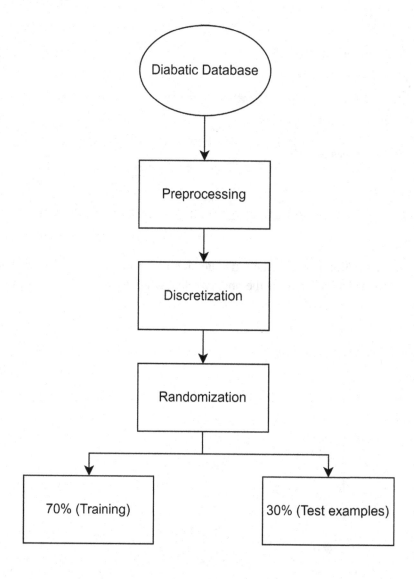

The discretization of numeric data to nominal data is performed on the "Age" class which is originally the only column for numeric data while the rest is nominal data. This process is done for the Apriori algorithm to be performed more accurately and efficiently without leaving out the "Age" attribute from consideration. The process is done with equal-width binning which sets the same gap interval between the ages. The discretization process is performed using a Discretize filter with the number of bins or interval chosen is 13.

Figure 8.

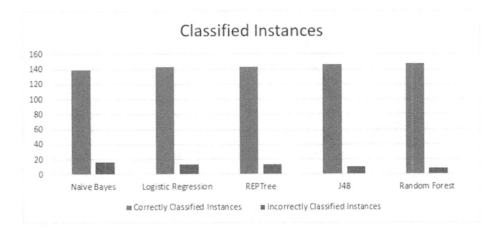

The following Figure 9 and Figure 10 are the visualized distribution of the Age class before and after the discretization process.

Figure 9.

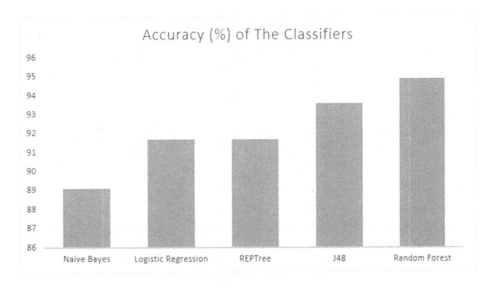

Figure 10.

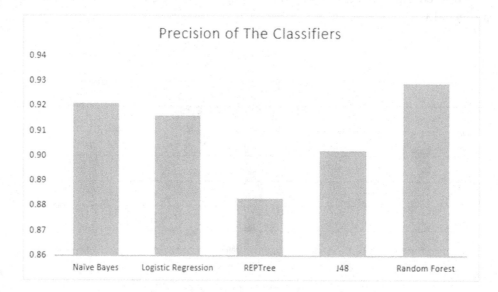

In the next pre-processing step, the dataset is split into two for training and testing datasets. Each with about 70% for training and 30% for testing. Before that, the dataset is randomized using the Randomize Filter. It is randomized first to ensure both split datasets can represent the original dataset.

Figure 11.

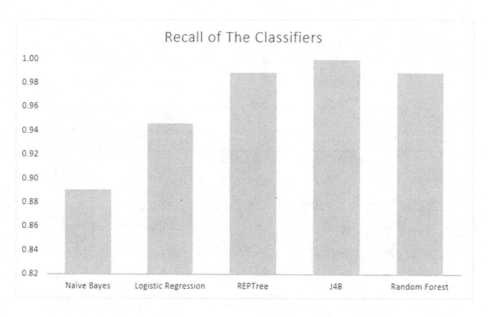

Then, 30% of the dataset is then removed from the original dataset to form the split testing dataset using RemovePercentage Filter. This will be used later as the supplied testing data to test on the training data.

Figure 12.

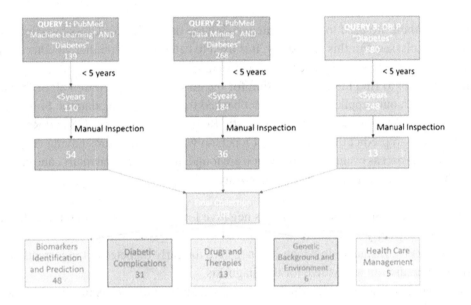

The previous step before is then undo to get back the original dataset. Another remaining 70% of the dataset is then also split to form a training dataset using RemovePercentage Filter again but with selected invertSelection as 'False' and percentage 30% which still will invert to the needed 70%. Lastly, the split datasets are then saved to be easily retrieved for later.

Descriptive and Visual Interpretation of Outcomes and Results

Figure 13.

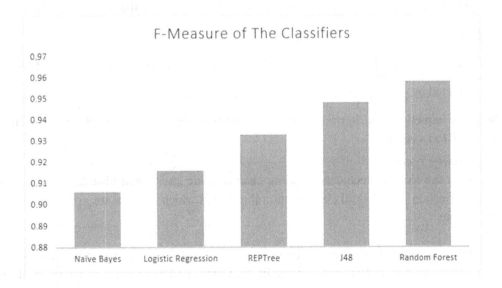

Classification

The classifier algorithms are used to help in categorizing the data value for Class by using the characteristics of the other classes. For this case study, it is used to categorize to predict whether the person is "Positive" or "Negative" for early diabetes with the existing symptoms.

In the initial attempt to find the best algorithm for the classifier, a few chosen algorithms were put into test in WEKA with our supplied testing dataset backed from the training dataset. Some of the algorithms that were put into test are Naïve Bayes, Logistic Regression, REPTree, J48 and Random Forest. These algorithms help to predict the likelihood of a person with the symptoms having early diabetes.

The metrics used as part as evaluation calculation are found in the confusion matrix and can calculated accordingly. The values used from confusion matrix are,

1. True Negative (TN): Instances in which it is negative for early diabetes and returns the classification prediction correctly.
2. False Positive (FP): Instances in which it is negative for early diabetes but incorrectly returns the classification prediction as positive for the diagnosis.
3. False Negative (FN): Instances in which it has early diabetes but returns the classification prediction incorrectly as negative for the diagnosis.
4. True Positive (TP): Instances in which it has early diabetes and returns the classification prediction correctly.

For this experiment, there are a few of the main performance evaluation metrics adopted from the confusion matrix above, which are,

1. Accuracy which measures the fraction of predictions on the instances that the model correctly classifies. It is calculated by $\dfrac{TP+TN}{TP+TN+FP+FN}$.
2. Precision which measures the fraction of positive classifications that were actually correct. It is calculated by $\dfrac{TP}{TP+FP}$.
3. Recall which measures the fraction of actual positive classifications that was identified correctly. It is calculated by $\dfrac{TP}{TP+FN}$.
4. F-Measure which is the harmonic mean of both precision and recall. It is calculated by $\dfrac{2 \times precision \times recall}{precision + recall}$.
5. Area under the ROC or receiver operating characteristic curve area visualizes the performance of a classification model at all classification thresholds which uses the true positive rate and true negative rate.

Among all the classifiers used, there are two classifiers in which it is tree based with nodes branching out the root node. Those classifiers can be visualized in Figure 14 and Figure 15 below to illustrate on how the classification trees make their decisions.

Figure 14.

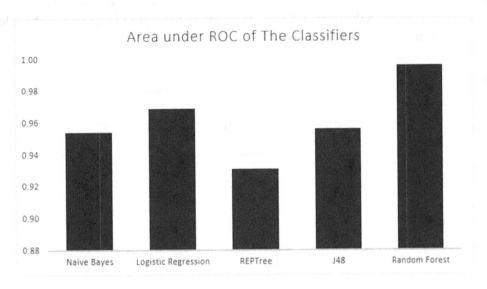

Figure 15.

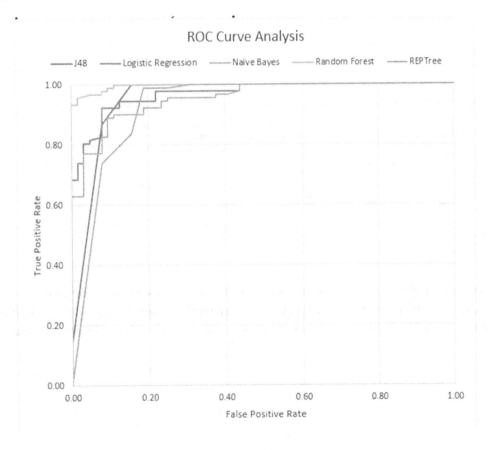

The following Table 4 below shows how many instances were correctly and incorrectly classified in the testing dataset based on the trained classifiers. Whereas the Figure 16 illustrates the values from the table in a bar chart.

Figure 16.

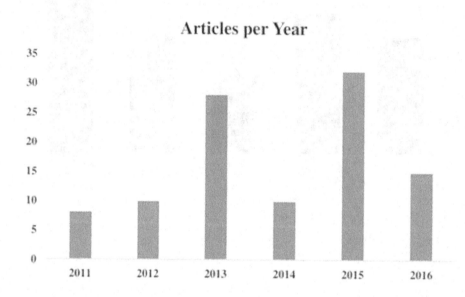

Table 4. *Correctly and incorrectly classified instances by the classification algorithm*

Classification Algorithms	Correctly Classified Instances	Incorrectly Classified Instances	Total Instances
Naïve Bayes	139	17	
Logistic Regression	143	13	
REPTree	143	13	156
J48	146	10	
Random Forest	148	8	

To observe and evaluate more on the classifiers used, other factors based on the confusion matrix is observed. The evaluation of the classifiers used are based on the accuracy, precision, recall, F-measure and area under the ROC. The results of these evaluation metrics are as shown below in Table 5.

Table 5. Evaluation metrics based on the classification algorithm

Classification Algorithms	Accuracy (%)	Precision	Recall	F-Measure	Area under ROC
Naïve Bayes	89.1	0.921	0.891	0.906	0.954
Logistic Regression	91.67	0.916	0.946	0.916	0.969
REPTree	91.67	0.883	0.989	0.933	0.931
J48	93.59	0.902	1	0.948	0.956
Random Forest	94.87	0.929	0.989	0.958	0.996

Looking at the first evaluation metric which is the accuracy as illustrated in Figure 17 below, we can see that the Random Forest algorithm has performed better than the rest. Notably, logistic regression and REPTree have the same accuracy whereas Naïve Bayes falls behind than the others. This shows that the diagnosis for early diabetes predicted by Random Forest have higher chances of being correct based on the classification.

Figure 17.

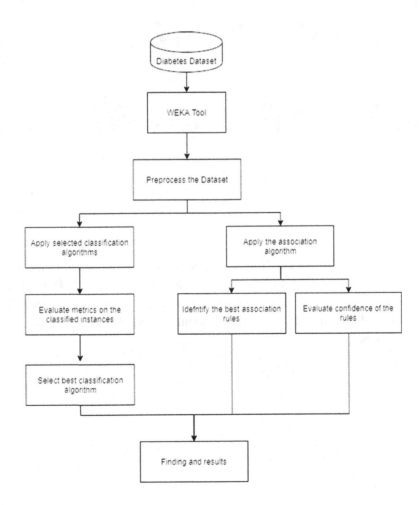

Whereas for the precision of the classifiers, as illustrated in Figure 18 below, Random Forest algorithm again have the highest precision score among the rest at 0.929. Despite having higher accuracy previously, J48 followed by REPTree scores the precision considerably lower than Naïve Bayes and Logistic Regression classifications. From this, it shows that Random Forest is able to predict that a person has early diabetes and being correct on it 92.9% of the time.

Figure 18.

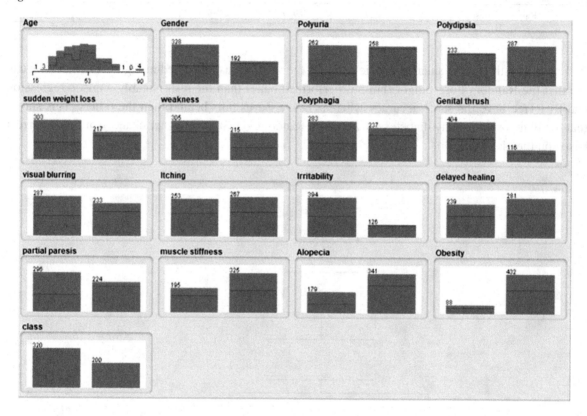

As for the recall evaluation of the classifiers, J48 turns out to be the one with the highest value at 1 as illustrated in Figure 19 below. This is followed by Random Forest and REPTree falling slightly by a little bit at 0.989. Whereas, Logistic Regression is at 0.946 and lastly is 0.891 which is Naïve Bayes. From this evaluation, it shows that J48 can correctly identifies of all the early-diabetes diagnosis more than the rest of classifiers.

Figure 19.

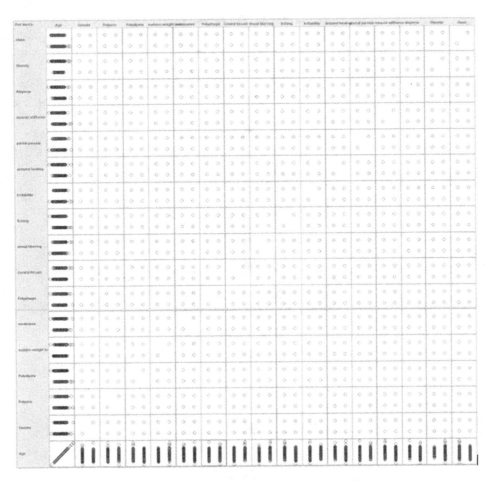

To look for the balance between precision and recall, F-measure can show on the harmony between them as shown in Figure 20. Random Forest scores the highest at 0.958 compared to the rest. In the next order falling descendingly is J48, REPTree, Logistic Regression and Naïve Bayes.

Figure 20.

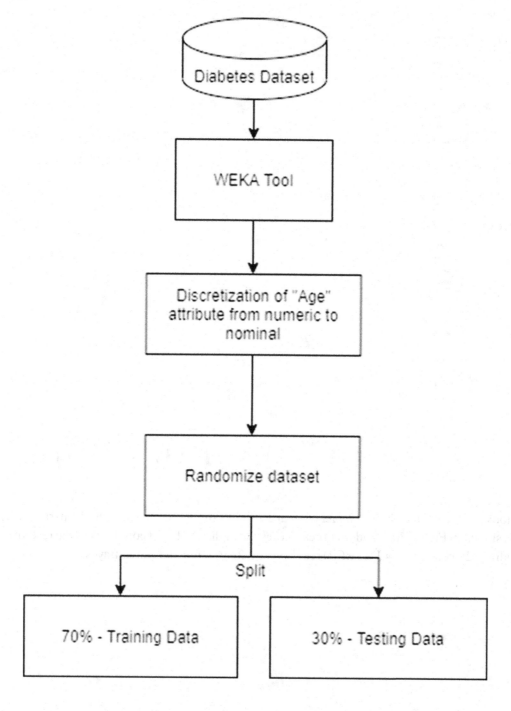

Figure 21 shows the ROC area of the classifiers. The Random Forest algorithm performs the best compared to the other at an area of 0.996. It is followed second by logistic regression at the area of 0.969. The third highest area is J48 followed by Naïve Bayes with little difference in the area. The classifier with the least ROC area is the REPTree.

Figure 21.

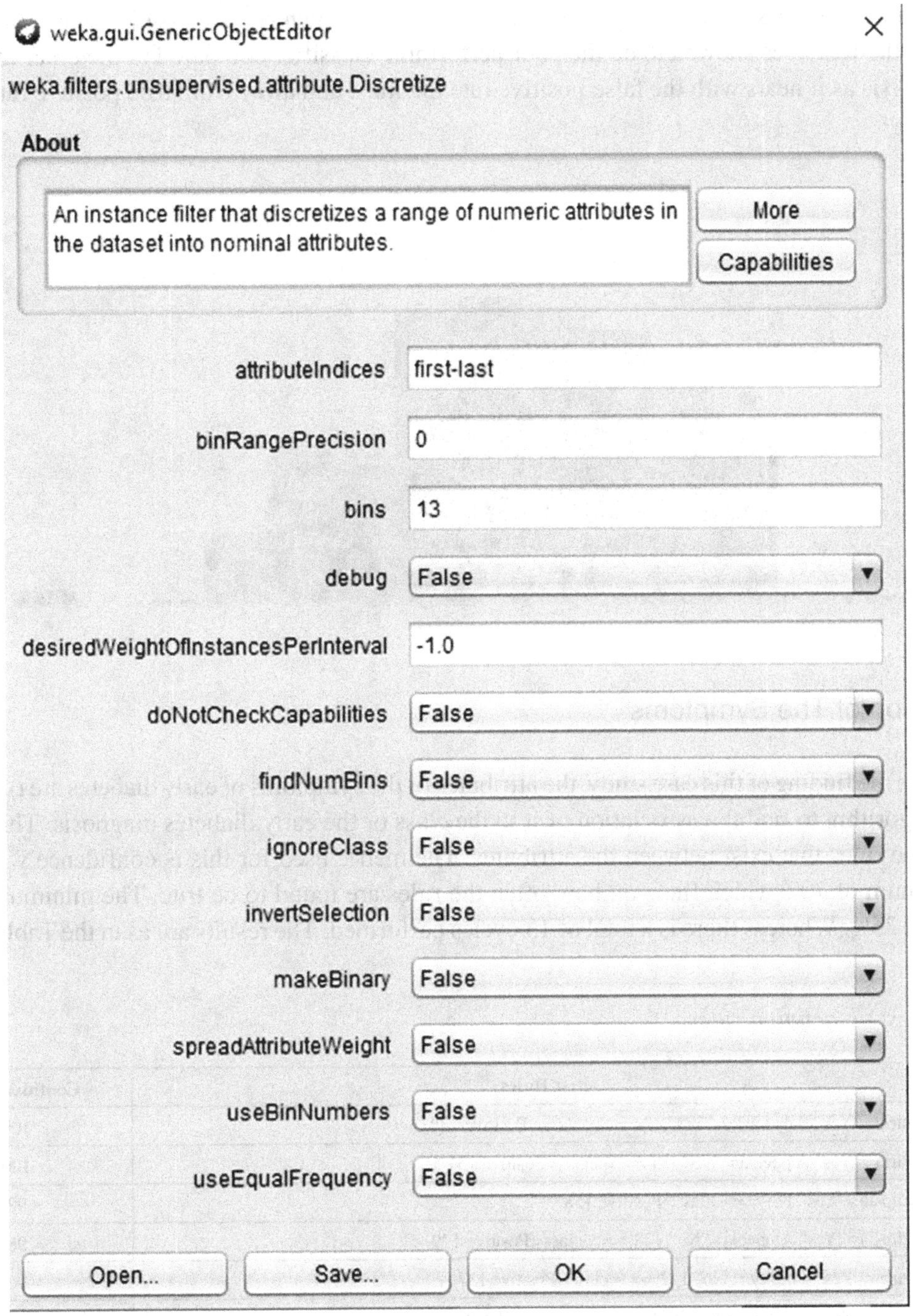

From the value of the ROC area, an analysis curve can be made based on the true positive rate and false positive rate. This is as part to visualize the evaluation of the model whether it is good or not. Figure 22 visualizes the ROC Curve Analysis of classifier for performance evaluation. It shows that Random Forest to be the best classifier following the metric as it is nearest with the true positive rate and thus at the optimal area under the curve compared to the others. It is followed by the descending order in ranking by Logistic Regression and J48 along with Naïve Bayes which relates closely to each other. In the last place, the REPTree showed to be the poor performing classifier compared to the other in the ROV curve analysis as it nears with the false positive rate the most and away from true positive rate.

Figure 22.

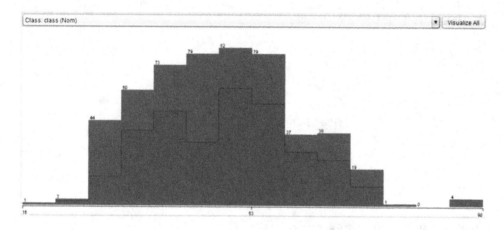

Association of The Symptoms

To enhance the finding of this case study, the attributes or the symptoms of early diabetes are run through Apriori algorithm to find the association of it to the class or the early diabetes diagnosis. This helps in drawing the rules that exist between the attributes. The metric used for this is confidence which is set to a minimum of 70% which finds on how often the rules are found to be true. The minimum support gathered is 35%. Whereas there is a total of 13 cycles performed. The results are as in the Table 6 below.

Table 6. Best association rules

No	Best Rules	Confidence (%)
1	Polyuria="Yes" Polydipsia="Yes" 134 ==> class=Positive 134	100
2	Polyuria="Yes" Alopecia="No" 131 ==> class=Positive 131	100
3	Polydipsia="Yes" 161 ==> class=Positive 158	98
4	Polydipsia="Yes" Alopecia="No" 132 ==> class=Positive 129	98
5	Polyuria="Yes" Obesity="No" 142 ==> class=Positive 137	96
6	Polyuria="Yes" Genital thrush="No" 138 ==> class=Positive 129	93
7	Polyuria="Yes" 183 ==> class=Positive 170	93
8	partial paresis="Yes" 147 ==> class=Positive 132	90
9	sudden weight loss="Yes" 157 ==> class=Positive 137	87
10	Polyphagia="Yes" 164 ==> class=Positive 130	79

Based on the best rules gained, it was indicated that all of the people who have polyuria and polydipsia all have diagnosis class as positive. It was also found that people who have polyuria and alopecia too are all positive for early diabetes. Whereas it is seen as often where people who have polydipsia or polydipsia without alopecia symptoms have early diabetes. The significance of this to the case study scenario is that the relations of the symptoms that are most prevalent especially with the presence of other symptoms alongside with it are able to be found out.

CONCLUSION

To conclude on this case study, we experimented on the prediction and association on the dataset of early-stage diabetes risk detection. The prediction part was found to be successful with the classification algorithms done and chosen. Whilst the association part too was successful in finding the relationship between the symptoms and the diagnosis of early diabetes in a person.

Therefore, from this finding, it was found that Random Forest is the most suitable algorithm when taking account of all of the metrics where not only the correctness matters, but the accuracy of the positive classes classified as well as calculated in precision, recall, F-measure and area under ROC. These metrics would be significant as a slight wrong in classification can cause the wrong diagnosis for a person.

From the association part that was conducted, we found the likelihood of early diabetes from someone with certain symptoms present. It was mostly found that instances with polyuria, polydipsia and without alopecia have very high likeliness to get a positive diagnosis. This draws out the likelihood of the person having a pattern of symptoms to affect the diagnosis.

Hence, detection of diabetes at an early stage is an important and significant breakthrough in the health sector as it can help to curb the disease that has affected so many people around the world to worsen. This is especially significant for patients who might be susceptible to diabetes or with symptoms that may not have been so obvious at first glance. Using data mining in the health sector would tremendously boost the effectiveness and the effectiveness of the system at a very short time and cost.

In future studies and the development of this case study, a bigger dataset on diabetes patients is recommended to improve accuracy. Not only that, but it is also recommended that the dataset have more attribute classes that can cover more factors of early diabetes such as gender, family background and so on as it may help a lot more on the classification and the association.

REFERENCES

Al-Lawati, J. A. (2017). Diabetes mellitus: A local and global public health emergency! *Oman Medical Journal*, *32*(3), 177–179. doi:10.5001/omj.2017.34 PMID:28584596

Aljumah, A. A., Ahamad, M. G., & Siddiqui, M. K. (2013). Application of data mining: Diabetes health care in young and old patients. *Journal of King Saud University - Computer and Information Sciences*, *25*(2), 127–136. doi:10.1016/j.jksuci.2012.10.003

Balpande, V. R., & Wajgi, R. D. (2017). Prediction and severity estimation of diabetes using data mining technique. *IEEE International Conference on Innovative Mechanisms for Industry Applications, ICIMIA 2017 - Proceedings, Icimia*, 576–580. 10.1109/ICIMIA.2017.7975526

Das, H., Naik, B., & Behera, H. S. (2018). *Classification of Diabetes Mellitus Disease (DMD): A Data Mining (DM) Approach.* doi:10.1007/978-981-10-7871-2_52

Embong, M. (2019, November 29). *What is prediabetes?* https://www.thestar.com.my/lifestyle/health/2013/11/14/what-is-prediabetes

Habibi, S., Ahmadi, M., & Alizadeh, S. (2015). Type 2 Diabetes Mellitus Screening and Risk Factors Using Decision Tree: Results of Data Mining. *Global Journal of Health Science, 7*(5), 304–310. doi:10.5539/gjhs.v7n5p304 PMID:26156928

HonorHealth. (2020). *Signs, Symptoms and Diagnosis of Diabetes.* https://www.honorhealth.com/medical-services/diabetes/signs-symptoms-diagnosis%0A

Iyer, A., S, J., & Sumbaly, R. (2015). Diagnosis of Diabetes Using Classification Mining Techniques. *International Journal of Data Mining & Knowledge Management Process, 5*(1), 1–14. doi:10.5121/ijdkp.2015.5101

Kavakiotis, I., Tsave, O., Salifoglou, A., Maglaveras, N., Vlahavas, I., & Chouvarda, I. (2017). Machine Learning and Data Mining Methods in Diabetes Research. In *Computational and Structural Biotechnology Journal* (Vol. 15, pp. 104–116). Elsevier B.V. doi:10.1016/j.csbj.2016.12.005

Kazerouni, F., Bayani, A., Asadi, F., Saeidi, L., Parvizi, N., & Mansoori, Z. (2020). Type2 diabetes mellitus prediction using data mining algorithms based on the long-noncoding RNAs expression: A comparison of four data mining approaches. *BMC Bioinformatics, 21*(1), 1–13. doi:10.118612859-020-03719-8 PMID:32854616

Mahboob Alam, T., Iqbal, M. A., Ali, Y., Wahab, A., Ijaz, S., Imtiaz Baig, T., Hussain, A., Malik, M. A., Raza, M. M., Ibrar, S., & Abbas, Z. (2019). A model for early prediction of diabetes. *Informatics in Medicine Unlocked, 16*, 100204. doi:10.1016/j.imu.2019.100204

Mamatha Bai, B. G., Nalini, B. M., & Majumdar, J. (2019). Analysis and Detection of Diabetes Using Data Mining Techniques: A Big Data Application in Health Care. In Advances in Intelligent Systems and Computing (Vol. 882). Springer Singapore. doi:10.1007/978-981-13-5953-8_37

Manjusree, M., & Sateesh Kumar, K. A. (2019). Diabetes prediction using data mining classification techniques. *International Journal of Recent Technology and Engineering, 8*(3), 5901–5905. doi:10.35940/ijrte.C4735.098319

Nagarajan, S., & Chandrasekaran, R. M. (2015). Design and implementation of expert clinical system for diagnosing diabetes using data mining techniques. *Indian Journal of Science and Technology, 8*(8), 771. Advance online publication. doi:10.17485/ijst/2015/v8i8/69272

Olsen, N. (2020). *The Effects of Diabetes on Your Body.* Academic Press.

Perveen, S., Shahbaz, M., Guergachi, A., & Keshavjee, K. (2016). Performance Analysis of Data Mining Classification Techniques to Predict Diabetes. *Procedia Computer Science, 82*, 115–121. doi:10.1016/j.procs.2016.04.016

Rani, S. (2018). mining in Continuous data for Diabetes Prediction. *2018 Second International Conference on Intelligent Computing and Control Systems (ICICCS), Iciccs*, 1209–1214. 10.1109/IC-CONS.2018.8662909

Sa'di, S., Maleki, A., Hashemi, R., Panbechi, Z., & Chalabi, K. (2015). Comparison of Data Mining Algorithms in the Diagnosis of Type Ii Diabetes. *International Journal on Computational Science & Applications*, 5(5), 1–12. doi:10.5121/ijcsa.2015.5501

Sanakal, R., & Jayakumari, S. T. (2014). Prognosis of Diabetes Using Data mining Approach-Fuzzy C Means Clustering and Support Vector Machine. *International Journal of Computer Trends and Technology*, 11(2), 94–98. doi:10.14445/22312803/IJCTT-V11P120

Saravananathan, K., & Velmurugan, T. (2016). Analyzing Diabetic Data using Classification Algorithms in Data Mining. *Indian Journal of Science and Technology*, 9(43). Advance online publication. doi:10.17485/ijst/2016/v9i43/93874

Sneha, N., & Gangil, T. (2019). Analysis of diabetes mellitus for early prediction using optimal features selection. *Journal of Big Data*, 6(1), 13. Advance online publication. doi:10.118640537-019-0175-6

Wu, H., Yang, S., Huang, Z., He, J., & Wang, X. (2018). Type 2 diabetes mellitus prediction model based on data mining. *Informatics in Medicine Unlocked*, 10, 100–107. doi:10.1016/j.imu.2017.12.006

Chapter 11
Role of Machine Learning in Handling the COVID–19 Pandemic

Sadia Aziz
Melbourne Institute of Technology, Australia

Qazi Mudassar Ilyas
ⓘ https://orcid.org/0000-0003-4238-8093
King Faisal University, Saudi Arabia

Abid Mehmood
King Faisal University, Saudi Arabia

Ashfaq Ahmad
Jazan University, Saudi Arabia

ABSTRACT

Since its appearance in late 2019, Severe Acute Respiratory Syndrome Coronavirus 2 (SARS-CoV-2) has become a significant threat to human health and public safety. Machine learning has been extensively exploited in the past to solve a range of problems in everyday life. It has also played its role in virtually all aspects of pandemic management, ranging from early detection and contact tracing to vaccine and drugs development and treatment. This chapter discusses some of the ways in which machine learning-based solutions have helped. In this regard, computer vision approaches have been used for the early detection of disease. Contact tracing has been enhanced by machine learning models to improve distance estimation techniques. Similarly, machine learning techniques have been used to accurately predict mortality rates to optimize resource management. These techniques have also helped in the otherwise tedious processes of vaccine and drugs development in numerous ways, such as providing insights into drug target interactions and possibilities of repurposing the existing drugs.

DOI: 10.4018/978-1-7998-9201-4.ch011

INTRODUCTION

The COVID-19 outbreak caused by the SARS-CoV-2 virus has shaken the entire world. Ordinary individuals were gripped with fear, uncertainty, and doubts as the number of infections ballooned in most countries. The health systems of many countries collapsed or reached the brink of collapse. Health professionals needed time to understand the disease. Medical doctors tried to suggest medicines to reduce the harmful effects of the disease as much as possible. Epidemiologists analyzed the spread of the virus to understand spread patterns and infections and devise strategies to contain the damage. Virologists strived to understand the virus DNA to develop the vaccine. Statisticians provided excellent support in collecting the data, making it publicly accessible to all, and identifying trends from this data. IT experts also came to the fore to support these efforts. The recent advancements in artificial intelligence and machine learning technologies enabled IT professionals to play an active role in these tasks. The massive amount of data helped machine learning professionals develop models for various issues related to the pandemic and propose valuable solutions. Some critical aspects of the pandemic in which machine learning played a key role included:

- Early detection
- Contact tracing
- Prediction of mortality
- Drugs and vaccination
- Screening and treatment
- Prediction of spread
- Prediction of number of infections
- Prediction of number of critical cases
- Prognosis prediction
- Prevention of the disease
- Management of Critical COVID-19 Patients
- Reducing the workload of healthcare workers
- Hospital Resource Management

Although each one of these topics deserves a thorough treatment, we focus on the first four aspects and discuss machine learning-based solutions to these issues.

Easly detection of the disease can help in better treatment, improving the chances of recovery of the infected individual. Such individuals can also isolate themselves to stop the further spread of the virus. Chest x-rays and CT scans are frequently used for the early detection of COVID-19 disease. In machine learning, computer vision and image classification techniques have been used successfully on these images by several studies for early detection of disease. Several classification techniques such as decision trees, random forests, support vector machines, and convolutional neural networks have been employed in these studies for the early detection of COVID-19. The role of machine learning in contact tracing is the second topic of this chapter.

Contact tracing has proven to be very useful in identifying the suspected infections and controlling the spread of the virus. Bluetooth low energy-based methodologies have been commonly used for contact tracing. However, estimation of the distance between two individuals – the most critical aspect of contact tracing – suffers from low accuracy with conventional BLE techniques as they are based on

Received Signal Strength Indicator (RSSI) measurement only. Machine learning techniques augment RSSI value with several parameters from smartphone sensors such as gyroscope, linear accelerometer, magnetic, and gravity sensors and feed them into various machine learning models to improve distance estimation accuracy.

Many countries of the world have seen several waves of the spread of the virus. During the peak spread time of these ways, it was critical to predict mortality rate and the number of serious infections to utilize the limited resources such as hospital beds, ICU (intensive care unit) rooms, oxygen, and ventilators. The chapter gives a brief exposé of machine learning techniques for COVID-19 mortality prediction. The classic susceptible, infected, removed/recovered (SIR)-based model was enhanced by machine learning techniques. Researchers also used the classic logistic growth model and advanced versions like the Prophet model proposed by Facebook Inc. However, the most commonly used models for this purpose include the Autoregressive Integrated Moving Average (ARIMA), various variations of ARIMA, and deep learning models.

Drug and vaccine development are the first things that come to our minds when we think of a new infectious disease. Scientists have been struggling to develop drugs and vaccines to reduce the pandemic spread and help cure the already infected individuals. As drug development is a lengthy process, health professionals have been trying to repurpose drugs developed for other diseases. Repurpose of several medicines, e.g., Chloroquine, Remdesivir, Baricitinib, and Dexamethasone, have proven effective in treating critical patients. Machine learning techniques have been used in drug and vaccine development for drug target interaction, drug reprofiling, modeling the molecular interaction between drugs and the target proteins, Gene Set Enrichment Analysis, epitope prediction, and precision medicine. Machine learning techniques used for these issues include multimodal restricted Boltzmann machine, random walk propagation algorithm, support vector machine, convolutional neural networks, long short-term memory networks, and BioBERT.

MACHINE LEARNING IN SARS-COV-2 EARLY DETECTION

The Coronavirus Disease (COVID-19) is an infectious and critical viral disease that spreads exponentially by close contact with infected persons or touching such items. It is caused by Severe Acute Respiratory Syndrome Coronavirus 2 (SARS-CoV-2), and its pandemic has turned out as the most prominent global health concern. It has variant symptoms from gentle to high fever, lung infection causing breathing difficulty, and chest pain. Severe cases may face loss of speech and smell and sometimes loss of movement. Its timely detection and diagnosis help control the infection. Starting early treatment of the COVID-19 disease needs efficient diagnostic methods. An infected person goes in quarantine to prevent the spread of the virus infection.

Positive diagnosis helps in rapid decisions on treatment and isolation. The conclusion requires concentrating on the affected individual's appearances, manifestations, findings, clinical history, and comorbidities. According to recent research, medical imaging like X-rays and CT scans, etc., help detect COVID-19 infection. The detection of disease from radiological images is not possible by the naked eye. Artificial Intelligence, particularly machine learning techniques, is beneficial in accurately diagnosing many diseases from radiology images.

Although all diagnostic methods have limitations, Polymerase Chain Reaction (PCR) and antibodies are the most popular testing methods. The PCR testing method is susceptible to high false-negative results whereas, in contrast, the antibodies testing is likely to have a high false-positive ratio. To reduce the examination span and increase accuracy, motivated researchers are investigating alternative automated methods. Medical images like Chest X-Ray (CXR) and Computerized Tomography (CT) scans help measure the intensity of COVID-19 infection.

COVID-19 Diagnosis Methods

Various factors may influence the COVID-19 diagnosis results. Some of the vital features include a suitable selection of biomarkers and specimen defile. A bioindicator scales the exposure to a specific virus to indicate the illness caused by that virus. The four prominent biomarkers of COVID-19 are listed in Table 1. The antibodies such as IgG/IgM are not apt for early detection of SARS due to the time required to emerge. The IgM antibody requires 3-6 days to be detected, whereas IgG could be detected after eight days (Zhuoyue et al., 2003). Table 1 depicts various biomarkers and their limitations.

Table 1. Bio indicators with limitations (Cui & Zhou, 2020)

Bio indicator	Remarks
RNA	susceptible to false-negative results and large computing time
Virus/Antigen	Requires ultra-sensitive techniques
Antibodies (IgG/IgM)	susceptible to false-positive ratio
Lungs' Assessment	Difficult to differentiate COVID-19 infection from other lungs infections such as pneumonia.

1. ***RNA:*** The single-stranded RNA genome of CoV-2 is an imperative diagnosis biomarker containing ample data amount nearly 30KB in size encoding 9860 amino acids (Cui & Zhou, 2020). Due to the limitations of high false-negative results and big read-out time of the linked techniques with RNA are not well reliable for virus eminent detection. The RT-PCR is a well-reputed method for reporting results in a reasonably lesser time of 1-2 days. The Clustered Regularly Interspaced Short Palindromic Repeats (CRISPR) is an improved method in reducing the false-negative results and big computing time. Table 2 shows different diagnosis procedures used along with RNA.

Table 2. RNA-based COVID-19 diagnostic methods (Cui & Zhou, 2020)

Diagnosis Method	Computing Time	Sensitivity	Specificity
Quantitative RT-qPCR (J et al., 2020)	More than 4 hours	45% - 60%	-
Digital PCR RT-dPCR (Dong et al., 2020)	-	90%	100%
Loop-Mediated Isothermal Amplification (LAMP) (Yu et al., 2020)	30 mins	97.6%	-
CRISPR (Broughton et al., 2020)	45 mins	90%	100%

2. **Antigen:** Most presumably, the SARS-CoV-2 has twenty-eight proteins (Wu et al., 2020), including the spike (S) glycoprotein and nucleocapsid (N) proteins. COVID-19 diagnosis probably can use these proteins as antigens. S and N proteins are likely the uttermost relevant antigen bioindicators for COVID-19 determination, as multiple diagnosis methods preferred these proteins for diagnosing SARS (PC et al., 2005; XY et al., 2004). A proteome microarray was built with eighteen out of the twenty-eight prognosticated proteins to examine the antibody acknowledgments. The outcomes reveal that the subjects at the recovering stage have categorical antibody counter-response to the proteome, notably to protein N, S1, ORF9b, and NSP5 (Jiang et al., 2020).

3. **Antibodies:** In contrast to RNA-based diagnoses, Antibodies-based diagnoses are susceptible to false-positive results. This serological-based technique is generally used to counter-check the RT-PCR negative results. The COVID-19 related antibodies are developed due to exposure to the coronavirus. Antibodies such as IgM reveal the recent exposure of the coronavirus, whereas IgG is known to show the earlier exposure. Due to this reason, the antibodies-based test may diagnose a person as infected with positive results after recovering from illness (Zhuoyue et al., 2003). Huang et al. (PC et al., 2005) has developed Lateral Flow Immuno-Chromatographic Strip (LFICS) for detecting antibodies in human blood promptly within 15 minutes, having overall sensitivity of 88.66% and specificity of 90.63% (Z et al., 2020).

4. **Lungs Assessment Based Diagnosis:** To detect severe symptoms of COVID-19 infection, image-based lung assessment diagnoses, including CXR, chest CT scans, chest ultrasound, etc., are truly helpful. These symptoms may include breathing difficulty, pneumonia, and infection in the upper breathing airways (Burhan et al., 2020).

COVID-19 Diagnosis by Image-Based Methods

The Chest Ultrasound (CUS) method results in the fast and reliable diagnosis of COVID-19 as compared to the other relevant image-based methods such as CXR and chest CT scan. The following section briefly illustrates the characteristics of various image-based methods for COVID-19 detection.

Chest X-Ray (CXR)

Although CXR demands expert knowledge and experience in analyzing X-ray images (Jaiswal et al., 2019), it is a pretty popular and quick method to identify various lung infections. CXR is used to apprehend images of the chest, airways, heart, lungs, blood vessels, and spine.

CXR images illustrate patchy reticular–nodular opacifications for detecting COVID-19 infection. As per the researchers(Jaiswal et al., 2019; Qin et al., 2018), CXR identifies oddities including solidification, i.e., the amassment of liquid or tissue in respiratory alveoli hindering air transposition, Ground Glass Opacity (GGO), and spot on the lung, essentially striking surface and lower ranges of the lung (Manych, 2020).

The CXR method's reliability depends on the gap between the manifestation of early signs and the accuracy of the adopted method. Some researchers believe in the invisibility of signs of infection in CXR images within three days after starting fever and cough. It can only be effectively detected after 8-10 days(Wong et al., 2020). For achieving a precise diagnosis, the number of medical images matters (Chetouani et al., 2012; Qureshi et al., 2017). Figure 1 portrays four sample images (a) Normal CXR, (b) CXR of COVID-19 affected individual with mild lung tissue reflection, (c) CXR of COVID-19 patient suffering moderate to critical lung tissue reflection, and (d) CXR of COVID-19 patient experiencing critical lung tissue reflection (N et al., 2021).

Figure 1. CXR images (a) Normal (b) Mild (c) Moderate to severe, and (d) Severe Infected (N et al. 2021)

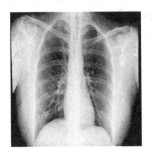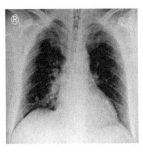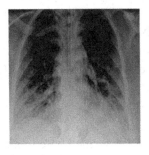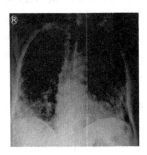

Chest-Ultrasound (CUS)

CUS is considered more superior to CXR in the case of pneumonia diagnosis as well as it has proved its significance in the detection of COVID-19 (E et al., 2020). This imaging method has its pros and cons. CXR needs ionizing radiations, whereas CUS has no such requirement (QY et al., 2020). Nevertheless, in the case of CUS physician, due to close contact with the patient, is vulnerable to the risk of infection, which is significant to consider. CUS is useful for monitoring a patient's condition during hospitalization. CUS could prove beneficial in finding lungs disorder at the onset of the disease (Kanne et al., 2020). Due to the benefits including, portability, relatively inexpensiveness, and no exposure of a patient to radiations, CUS can be used repeatedly for ICU patients (Clevert, 2020).

Chest CT Scan (CCT)

A chest CT scan integrates numerous chest X-rays captured from specific angles and employs advanced algorithms to provide transection reflections of the chest. CCT is a source of detailed information on internal organs, blood vessels, soft tissues, and bones using low ionizing radiations compared to the CXR. Because of a precise internal analysis of the body, CCT becomes a more effective imaging method that can be useful in detecting lung infections at early stages. However, the excessive use of this method is not encouraged as exposure to radiation raises the potential menace of cancer (*CAT Scan (CT) - Chest*, n.d.).

Assistance of Machine Learning (ML) Algorithms in Image-Based Diagnosis of COVID-19

COVID-19 infection spreads exponentially due to close contact with infected patients. The best-recommended method to avoid this spread is to quarantine the carriers from the rest of the people. Generally, the governments are taking massive tests of the potential infectees, and the positive cases are being quarantined and given medical treatment. Different diagnostic methods, including imaging techniques, are in practice. With the overwhelming number of tests, the complex nature of overlying tissue structure, and the limited number of radiologists and medical experts, considerable delays are experienced in report generation. These delays in identifying the positive infected cases cause further spread of disease, which can be avoided by generating reports efficiently. Also, in the case of medical images, it is not possible to identify the infection with naked eyes. AI technology, especially ML-based automated diagnosis techniques including Deep Learning (DL), can help resolve the issue of identifying infectees more effectively.

ML-based medical image analysis has proved to be the most reliable technique for rationalizing and expediting the diagnosis of COVID-19 patients. Progress has been made in deep learning techniques particularly, Convolution Neural Networks (CNN), in the image recognition task. CNN has the potency to learn automatically from massive image datasets for identifying COVID-19. Hence, it proves to be the mainstream research trend among researchers (Qin et al., 2018).

Machine Learning Algorithms

In machine learning, a model is formed based on a furnished training dataset, and the model predicts the response for the provided data. Following is a description of various ML algorithms that may be trained using the training datasets.

Decision Tree

The decision tree algorithm is one of the supervised learning algorithms. This algorithm can be applied to address the data classification and regression issues, hence known as the classification and regression tree (CART) algorithm (Karim & Rahman, 2013). The decision tree based on the training dataset is developed recursively (LI et al., 2009).

Each tree node represents a collection of traits and has multiple branches, a branch designates the decision rule value, and each leaf node relates to a decision outcome. The algorithm formulates a training model to predict the class or value of the intended variable by discovering fundamental decision rules induced from the training dataset. The tree traversal starts from the root of the tree. Based on comparing the target attribute with the root value, we move to the next desired node.

Random Forest

Random Forest is an aggregate of CARTs, which can discipline an equivalent scope of base datasets termed bootstraps, and conclusively consolidate them for a more precise outcome. The bootstraps comprise anomalous trials from the base dataset (Sarica et al., 2017). Random Forests proved to be more effective than a single tree. This method also produces relatively more accurate results when used with multi-dimensional large datasets.

Gradient Boosting Machine

Gradient Boosting Machine (GBM) is a solitary essential learning algorithm. It is a static-sized decision tree-based algorithm that packages various weak prediction models (Biau et al., 2019). It builds the prototype in a phased mode like variant boosting approaches. It encapsulates them together by authorizing the augmentation of a self-assured discriminable loss function.

Extreme Gradient Boosting

Extreme Gradient Boosting (XGBoost) is a specific application of the Gradient Boosting framework, using further precise estimates to attain the most suitable tree model. It exercises numerous agile skills that make it remarkably victorious, especially with structured data. The most prominent skills employed by XGBoost are measuring second partial derivatives of the loss function and high-level regularization (L1 & L2), which enhances model generalization. XGBoost trains quickly and can address real-world scale matters employing relatively lesser resources (Chen & Guestrin, n.d.).

Support Vector Machine

SVM algorithm creates the decision boundary, a hyperplane, to divide n-dimensional space into classes. The algorithm determines the extreme points/vectors that support composing the hyperplane. These extreme cases are called support vectors, and hence algorithm a Support Vector Machine (SVM). SVM is a solitary trendy, resilient superintended machine training algorithm that addresses data classification and regression problems but is extensively used for classification challenges in machine learning. It is profoundly preferred for producing significant accuracy based on statistical learning frameworks with less computation power. The prime objective of SVM is to find the best decision boundary or absolute hyper-plane, which discriminates the two classes remarkably. The SVM ensures the progression potential of the machine prototype, hence suitable in different fields (Wei & Yuan, 2014).

Preprocessing/Preconditioning of CCT Scan Images

Machine learning algorithms generally require data preprocessing or preconditioning to converge, especially in the case of messy, complicated data. Data preprocessing is a crucial step in building a machine learning model to refine real-world data that generally carries noise or missing values and may not be in such a format that is suitable to use directly. Preprocessing of medical images like CT scans helps normalize the effects of intensity contrasts in consecutive image slices. This process filters out the various artifacts caused by the image reconstruction process. Data preprocessing also helps reduce trivial information and boost the training by recognizing the lungs' boundaries and distinguishing them from encircling respiratory tissue.

In (A. M. Hasan et al., 2020), the authors presented an exemplary preprocessing effort examined in figure 2. They adopted a histogram thresholding routine to segregate lungs by normalizing the sharpness of image slices. Afterward, applying specific morphological processes like dilation and hole-filling eliminate gaps that may arise in the picture segments. Finally, a binary mask helps fragment the respiratory organs (A. M. Hasan et al., 2020).

Figure 2. Preprocessing workflow for CCT scan

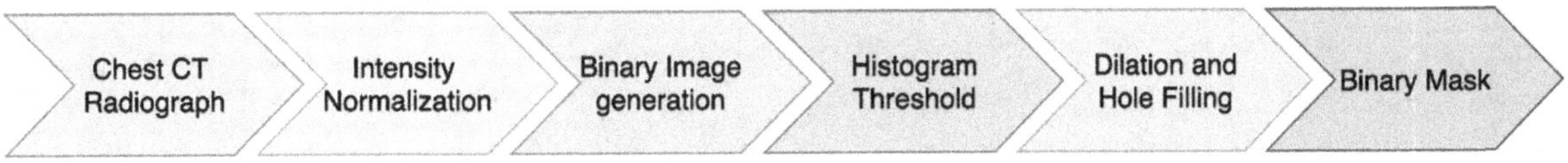

COVID-19 Detection from CCT Scans

Most conventional methods of detecting COVID-19 from CCT scans are not entirely reliable due to false positive or negative results. Still, the development of a trustworthy system is a challenging task. The research community has proposed several state-of-the-art learning algorithms crafted on deep networks detecting COVID-19 from CCT radiographs. In (L. Li et al., 2020), the researchers drafted a deep learning model, COVNet, on Convolutional Neural Networks (CNN) to elicit ocular traits from CCT scans for detecting COVID-19. The hierarchical framework of a robust ML routine CNN and its excellent characteristic derivation feature makes it an ideal dynamic approach to address complex image classification issues. The researchers in (L. Li et al., 2020) suggested a CCN based on multi-objective differential evolution (MODE) for classifying novel coronavirus cases from medical radiographs. The practice and trial measures of the deep convolutional prototype for COVID-19 involve a set of entirely relevant layers that serve as a classifier using deep traits derived from CCT scan images. To avoid overfitting of the proposed model, the researchers employed a twenty-fold cross-validation method. The trial outcomes validated that the CNN approach based on MODE achieved remarkable than the rest fundamental CNN models in responsiveness and certainty.

A method is proposed in (A. M. Hasan et al., 2020), fabricated on an aggregate of quantum deformed entropy (QDE) based characteristics and deep learning traits, blended and improved by applying the analysis of variance (ANOVA) method. Figure 3 portrays the four main steps of the proposed model: image pre-processing, feature extraction, feature selection, and classification.

Figure 3. Block diagram of the system using LSTM classifier (Hasan et al. 2020)

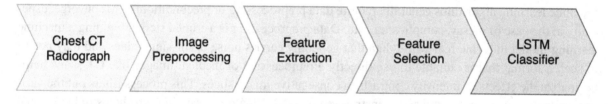

Long short-term memory (LSTM) is an artificial recurrent neural network (RNN) architecture used in deep learning. The authors applied an LSTM classifier; extensive experimental results proved that incorporating an entire or partial set of the elicited deep features (DF) from the DL and QDE elevated the achievement.

Table 3. Classification results (A. M. Hasan et al., 2020)

Method	Accuracy 100%	TP 100%		
		COVID-19	Healthy	Pneumonia
Linear SVM	96.20	94.90	98.10	95.80
KNN	95.30	93.20	97.20	95.80
Logistic Regression	97.20	96.60	98.10	96.80
LSTM (QDE)	97.50	95.70	100	96.80
LSTM (DF)	98	97.40	100	96.80
LSTM (QDE–DF)	99.68	100	100	98.90

Table 3 depicts that the certainty outcomes generated by the combination of DF and QDE features have improved the achievements. Besides, LSTM based classifier has proved more precise as compared to its other competitors.

CONTACT TRACING

The outbreak of the COVID-19 virus has shaken the whole world. Starting from China, there is no country left that has not seen the terrible effects of this virus on human health and lives. The emergence of this virus was so sudden and unexpected that no one was ready to deal with it. Because of its novelty, we do not have any medicine available. Many researchers are working on this virus to find any cure, but they still could not find any real solution. Preventive measures appear to be the most effective strategy until a cure for the COVID-19 disease is developed.

The transmission of this virus among humans is very rapid. It can be easily transferred from one person to another person. So if we have to stop its spread, we have to isolate the infected individuals for the healthy ones. The first step in this process is the identification of infected people. This can be done with the help of contact tracing. According to WHO (World Health Organization) experts, contact tracing is "the process of identifying, assessing, and managing people who have been exposed to a disease to prevent onward transmission." Contact tracing is one of the most effective measures to control the spread of viral infections.

Contact tracing and isolation of infected individuals have been used successfully to control viral diseases like tuberculosis, measles, SARS, and Ebola in the past (Eames & Keeling, 2003; Riley et al., 2020; Saurabh & Prateek, 2017). Before the widespread use of technological solutions, traditional contact tracing relied on the manual process of identifying the source of infection. A contact tracer would interview an infected individual to ascertain any infected close contact the interviewee might have contracted the virus. The interviewer would also try to identify the individuals coming in close contact with the interviewee after virus contraction to identify further infections. This manual process is time-consuming, tedious, inaccurate, and not easily scalable for a highly infectious disease. Figure 4 shows general steps for contact tracing.

Figure 4. Contact tracing

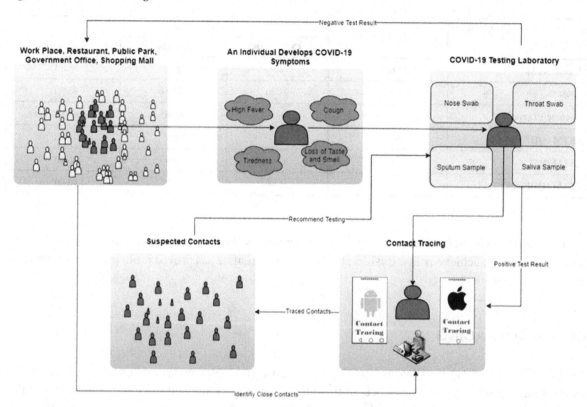

Contact tracing is usually based on a mathematical model that describes the dynamics of infectious diseases. Researchers have proposed four types of models. Individual-based models simulate the behavior of every individual in a community (Zhang et al., 2019). The model can be viewed as a graph comprising individuals represented by nodes and arcs representing interactions between individuals. Various configurations of the graph are possible such as completely connected, randomly connected, or selectively connected nodes. The pair approximation models extend individual-based models by considering pairs of individuals in addition to individual behaviors (Eames & Keeling, 2002). Detailed pairwise equations based on SIR or SIS-based epidemiological models are used to study contact tracing. Stochastic models are another class of models for contact tracing that perform well even with missing data (Kiss et al., 2005). Finally, deterministic models perform discrete-time simulations of contact tracing (Lajmanovich & Yorke, 1976).

Recent advancements in information and communication technologies, coupled with ubiquitous mobile gadgets, call for exploitation of these tools to develop frameworks for swift, reliable, and privacy-preserving contact tracing. The contact tracing technologies can be broadly categorized into three classes, namely Internet of Things (IoT)-based approaches features-based approaches, computer vision-based approaches, and Bluetooth low energy (BLE)-based approaches.

Bluetooth Low Energy-Based Contact Tracing

As the concept of contact tracing involves two individuals coming in close proximity at a given time, real-time location identification of an individual is a mandatory requirement for contact tracing. Several technologies are available for real-time location identification. Buchanan et al.provide an excellent survey of these technologies and summarize the pros and cons of these technologies as given in Table 4 (Buchanan et al., 2020):

Table 4. Contact tracing technologies pros and cons

Technology	Pros	Cons
GPS	Real-time tracking High accuracy Easy identification of clusters of infections	Requires more resources Limited indoor capabilities User reluctance in sharing the exact location
Bluetooth	Ubiquitous technology in mobile devices Low energy requirement Requires fewer computational power Requires less storage	Relatively inaccurate proximity identification Lack of real-time location tracking
Bluetooth with Radiofrequency identification	High accuracy Can also track personal belongings	High cost
WiFi routers	Ubiquitous technology No specialized hardware required	Relatively lower accuracy
Mobile networks	No application required on the user device Works of feature phone too	Needs collaboration of all network operators in an area Proximity identification is not possible
Ultra-wideband 5G	High accuracy	Limited availability
Wearable devices	Highly scalable High indoor accuracy Can also be used for geofencing of infected and quarantined patients Cost-effective No smartphone or other gadget required	Requires a wearable device with the application installed Relatively higher rate of false positives because of a higher range of interaction

Bluetooth Low Energy (BLE)-based proximity is arguably the most popular technology for digital proximity estimation. Google and Apple joined hands to develop a BLE framework for contact tracing (Google & Apple, 2020) based on Decentralized Privacy-Preserving Proximity Tracing (DP-3T) (Troncoso et al., 2020) and Temporary Contact Numbers (TCN) (TCN Coalition, 2020) protocols. Given the proliferation of smartphones having operating systems of these two companies, governments around the globe quickly adopted the proposed framework for digital contact tracing. BLE-based digital contact tracing remains one of the most popular strategies for containment of the spread of COVID-19 to date. The proximity estimation using BLE depends on observing an attenuated signal sent by a nearby device. The distance between a sender and a receiver is measured by the Received Signal Strength Indicator (RSSI), a commonly used measure of signal attenuation. However, RSSI value is affected by various noise sources such as obstacles, phone orientation, owner orientation, and multi-path environments (Lam et al., 2018; Mackey et al., 2020). Madoery et al. improved the RSSI-based distance estimation by deriving other parameters from observed BLE signals (Madoery et al., 2021). They extracted the following features from a series of normalized RSSI values:

- Mean
- Median
- First quartile
- Third quartile
- Minimum
- Maximum
- Standard deviation
- Range
- Interquartile range
- L1 distance to the mean
- L1 distance to the median
- Kurtosis skewness
- Count of different values describing the series

In the next step, Pearson correlation was measured between these features, and three groups of highly correlated features were formed. Various combinations of features from these groups were fed into classifications models. The result analysis showed significant improvement in proximity estimation because of an increased set of features.

NIST TC4TL Challenge

As mentioned above, the RSSI-based distance measurement suffers from several limitations that adversely impact distance measurement between two devices. To further study the limitations of RSSI and propose improved solutions, the "Too Close For Too Long" (TC4TL) challenge was announced by the National Institute of Standards and Technology in collaboration with the MIT Private Automated Contact Tracing (PACT) project in May 2020 (NIST, 2021). The challenge participants used several parameters and machine learning techniques to improve the accuracy of proximity estimation.

He & Printz, the challenge winners, proposed a novel technique for proximity estimation (He & Printz, 2020). Contrary to the commonly used techniques for reducing the noise in RSSI values, He & Printz propose exploiting this noise to improve proximity estimation accuracy. The technique hinges on the fact that Bluetooth signals emanating from a device reach a target device via multiple paths with varying intensities. A histogram is created to represent these RSSI signal values. In addition to RSSI values, the technique also uses data from other smartphone sensors such as magnetic, gyroscope, gravity, linear acceleration, and accelerometer sensors. Bluetooth TxPower is also exploited in proximity measurements.

The authors have employed a deep learning model for classification. As the incubation period of the novel Coronavirus is two weeks, a user device needs to store these parameters for two weeks. The orientation and motion sensors produce vast amounts of data, and keeping this data for two weeks on a user device is practically infeasible for many users. To overcome this limitation, the authors propose a two-step process. In the first step, the user device converts a large amount of data into a fixed-length vector and stores these fixed-length vectors in the local storage. If a user is suspected of coming in close contact with an infected individual, the more intensive second phase is activated for proximity event classification.

Shankar et al. propose to accept the RSSI noise as an inherent limitation and exploit temporal characteristics of the data to accurately measure the proximity of two devices (Shankar et al., 2020). All features extracted from various smartphone sensors are concatenated into one fixed-length vector. The authors converted every four seconds of data to 150 timestamps. Every timestamp captures current sensor readings and converts them into a feature vector of fixed length. After experimenting with several classification models based on deep learning, support vector machines, and decision trees, the authors conclude that the Temporal Conv1D network produces the best results.

Houchens et al. target the non-uniform reporting frequency of various sensors (Houchens et al., 2020). The authors propose to use the frequency of the sensor reporting at the highest rate for all records. While new data records are being generated, the most recent value is used for a sensor reporting at a lower frequency in all records until a new value becomes available. Hence, the length of all records remains the same. The authors used random forests and neural network models for proximity estimation and concluded that neural networks outperformed the random forests model.

Figure 5 presents a generalized machine learning-based framework for contact tracing.

Figure 5. A generalized machine learning framework for contact tracing

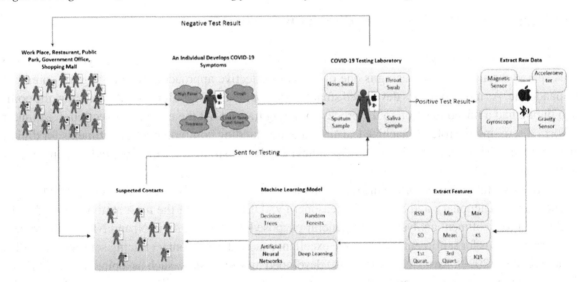

IoT-Based Contact Tracing

Internet of Things (IoT) thrives on its characteristics of uniquity, scalability, connectivity, and smart technologies. These qualities make IoT an excellent vehicle for developing efficient and reliable contact tracing mechanisms. Several proposals have been made to harness the IoT framework for contact tracing.

One of the most straightforward applications, such as those proposed by Yorio et al. (Yorio et al., 2021) and Rajasekar (Rajasekar, 2021), may assist in contact tracing by tracking individuals inside a building through WiFi routers acting as beacons. The data collected from this IoT infrastructure is sent to a random forest algorithm for proximity event detection. However, this simple technique is only feasible in a controlled environment such as an enterprise office, and it cannot be employed in a public setting because of obvious privacy and scalability concerns.

More advanced applications of IoT for contact tracing rely on wearable devices, Radio Frequency Identification (RFID) tags, blockchain technology, and smart infrastructure. Assuming a user owns a wearable device, a centralized or distributed system can easily track the device's real-time location. As soon as a particular individual is confirmed to be COVID positive, all individuals suspected of contracting the virus are alerted and recommended for testing. In case of confirmed infection, these individuals may be isolated to stop the spread of the virus. Researchers have proposed various enhancements to this basic scheme of IoT-based contact tracing. Garg et al. have exploited the presence of RFID tags on mobile objects such as vehicles to receive and send notifications when they come close to suspected or infected individuals or objects (Garg et al., 2020). The authors also propose to use smart contracts to store such encounters in a distributed ledger for later reference and further analysis. Roy et al. consider every individual as a node and ascertain proximity through device-to-device communication among wearable devices (Roy et al., 2020). The authors also propose to exploit the social networks of individuals to link various devices belonging to the same individual. It can help identify close contacts of infected individuals that can help control the spread of the virus.

Jahmunah et al. have presented an excellent exposé of IoT-based contact tracing methods and future directions (Jahmunah et al., 2021).

Sentiment Analysis of User Feedback on Contact Tracing

As stated earlier, in the absence of an effective treatment for the SARS-Cov-2 virus, trying to limit the virus spared through contact tracing is one of the most effective approaches for limiting the spread of the pandemic. As most countries developed smartphone applications for digital contact tracing, the common public had a mixed reaction to the process of being tracked. At the end of this section, we think it is worth discussing the role of machine learning in understanding common public sentiments for digital contact tracing. This understanding can help in devising better strategies with more public support and satisfaction.

Various general-purpose sentiment analysis models are candidates for the domain of contact tracing. Valence Aware Dictionary and Sentiment Reasoner (VADER) is one of the most straightforward techniques used for sentiment analysis. Three general-purpose lexica Bing, AFINN, and NRC, widely used in several domains for sentiment analysis, are excellent candidates for contact tracing sentiments studies (Crable & Sena, 2020). Researchers have also developed hybrid models, such as a pipeline of VADER, TextBlob, and Bidirectional Encoder Representations from Transformers (BERT) (Ahmad et al., 2021).

Acquisition of data is an essential step in all machine learning techniques. Facebook and Twitter are the most popular sources of data for sentiment analysis. Researchers have also used Reddit for this purpose. As almost all digital contact tracing applications are implemented as smartphone apps, some researchers have also analyzed user comments on app stores for sentiment analysis.

The actual value of a sentiment analysis study lies in developing insights into the sentiments to help develop better solutions in the future. In this regard, the researchers have (STM) and Latent Dirichlet Allocation (LDA) techniques to explain user sentiments about contact tracing.

Some of the salient studies on the topic include (Ahmad et al., 2021; Crable & Sena, 2020; Cresswell et al., 2021; Lohar et al., 2021). A general framework for the application of sentiment analysis in contact tracing is shown in figure 6.

Figure 6. Sentiment analysis for contact tracing

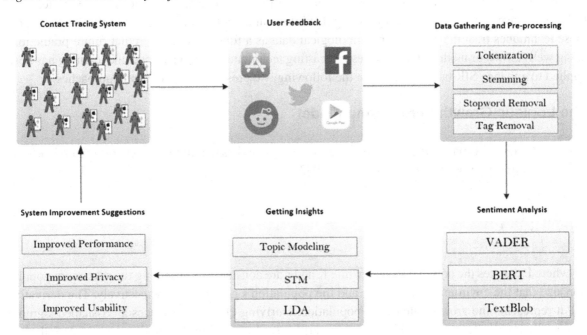

MORTALITY RATE PREDICTION

The ability to correctly predict the mortality rate of COVID-19 patients can significantly assist the decision-making process. The governments can assess the effectiveness of control measures such as social distancing and lockdowns. This prediction ability can also enable service providers to utilize valuable health resources more efficiently. Medical practitioners can analyze the effectiveness of treatments given to critical patients to adjust for future patients. Finally, this forecasting ability can assist various stakeholders in better future planning.

Susceptible, infected, recovered/removed (SIR)-based model is a classic epidemiological model for estimating the spread of a pandemic (Koike & Morimoto, 2018). The model can be used to predict future infections in a population. Researchers have proposed various solutions to solve SIR models (Miller, 2012, 2017). Various modifications have also been suggested to improve the basic SIR-based model. Some researchers have enhanced the model with new states. Others have tweaked model parameters or included external factors such as non-pharmaceutical interventions to achieve higher prediction accuracy. As the response of various nations to the COVID-19 outbreak has not been the same, local conditions need to be considered while applying the SIR model to predict future infections (Maier & Brockmann, 2020). SIS (Susceptible-Infected-Susceptible) (Abouelkheir et al., 2017), SIRD (Susceptible-Infected-Recovered-Dead) (Lalwani et al., 2020), MSIR (Immunity-Susceptible-Infected-Recovered) (Mohamed et al., 2021), SEIR (Suspected-Exposed-Infected-Recovered) (Annas et al., 2020), and MSEIR (meta-population SEIR) (Almeida et al., 2019) are some of the SIR-based models that have been proposed for predicting infections.

Epidemiological model development is a complex task that needs careful consideration of the characteristics of the target population, the virus, and the measures taken by the authorities. Machine learning techniques have been successfully used in such problems before, and these techniques may be used to learn complex relationships among these features to make accurate predictions. Machine learning techniques have been used in two ways for epidemiological model development and refinement. Firstly, these techniques treat the existing epidemiological data as a time series and predict future predictions using well-known or modified time-series forecasting techniques. Secondly, machine learning has been applied to enhance SIR-based models. In the following, we present details of both approaches.

The Logistic Growth Forecasting Model

The logistic model is one of the oldest population growth models (Malthus, 1872). Pierre published the following modified Logistic equation (Miner, 1933):

$$\frac{dP}{dt} = rP\left(1 - \frac{P}{K}\right) \tag{1}$$

where P denotes the population size, while K and r are constants representing the maximum population size that the environment can carry and the population's growth rate, respectively. The derivative dQ/dt represents the growth rate of the population. Deriving Q with t produces an S-shaped sigmoid curve. The logistic function can be applied in various scenarios where an increasing growth follows a slow early growth until the maximum point is achieved. At this point, the growth stabilized. This phenomenon can be easily observed in the COVID-19 pandemic. In the beginning, there was a slow growth in the number of infections. However, soon afterward, the rate of infections started increasing due to a lack of awareness in the common public and no strict containment measures. As the health systems were overwhelmed by the increasing number of patients, the governments started taking pandemic containment measures such as social distancing and lockdowns. These measures resulted in a reduced number of infections, thus straightening the logistic curve. Figure 7 shows the cumulative number of COVID-19 cases in China during the first quarter of 2020. The data points follow the logistic growth model to arrange themselves in a sigmoid curve.

Figure 7. Cumulative number of COVID-19 deaths in China during the first quarter of 2020 (data source: https://www.ecdc.europa.eu)

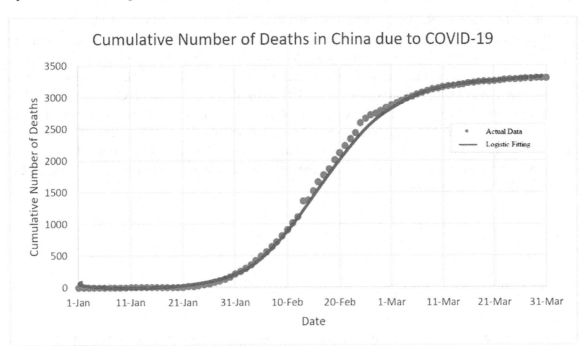

The Prophet Forecasting Model

The logistic growth model can be used to model the spread of COVID-19. However, this model is not very robust as it relies on a point where the population growth rate starts decreasing. Also, when applied to data with missing values, inconsistencies, outliers, pattern changes, or other anomalies, it suffers from low accuracy. Facebook Inc. introduced the Prophet model in 2017 to overcome these shortcomings (Taylor & Letham, 2018). Additionally, the Prophet model is designed to handle seasonal effects in data. This feature becomes very handy in COVID-19 prediction tasks as it is strongly affected by seasonal changes like other respiratory syndromes.

The Prophet model extends the basic logistic growth model by considering parameters K and r as variables instead of constants. An interested reader is referred to (Taylor & Letham, 2018) for further model details.

Many research studies have used the Prophet model for COVID-19 mortality prediction (Devaraj et al., 2021; Gupta et al., 2021; M. N. Hasan et al., 2021; Satpathy et al., 2021; Tulshyan et al., 2020; Wang et al., 2020). However, other models outperformed the Prophet model in all these studies. The reason behind this poor performance is that the Prophet model is a business-oriented model that might not be well-suited to the unique requirements of epidemiology.

Arima Model and its Variations

In its simplest form, an averaging model can predict the mortality rate of COVID-19 by averaging the historical mortality rates. More advanced versions of the averaging models, such as the weighted moving average model or exponential moving average model, assign different weights to values based on relative importance or recency. The Autoregressive Integrated Moving Average (ARIMA) supplements the moving average model with an autoregressive model and the difference method (Campbell & Mankiw, 1987). Using these three components, the ARIMA model calculates a stationary, moving average for non-stationary event data. ARIMA (p,q,d) represents three components AR(p), MA(q), I(d). The parameter p refers to the order of past data to predict the next value of the target variable, q is the order of moving average, and d refers to the order of difference. The following equations can be used to represent AR(p) and MA(q) (Ho & Xie, 1998):

$$Y_t = \alpha_1 Y_{t-1} + \alpha_2 Y_{t-2} + \alpha_3 Y_{t-3} + \ldots \alpha_p Y_{t-p} + \varepsilon_t \tag{2}$$

$$Y_t = \beta_1 \varepsilon_{t-1} + \beta_2 \varepsilon_{t-2} + \beta_3 \varepsilon_{t-3} + \ldots \beta_p \varepsilon_{t-p} + \varepsilon_t \tag{3}$$

where α is the parameter for autoregression and β for moving average, while Yt and εt represent value and error of the series at time t, respectively. Akaike Information Criteria (AIC) or Bayesian Information Criterion (BIC) can be used to select the most appropriate values for parameters p, q, and d.

The ARIMA model has been used for time-series forecasting in various application domains such as stocks exchange, supply chain, weather, and healthcare. For predicting the COVID-19 mortality rate, the first step is to pre-process raw data. In this step, some tasks include filtering the data based on the target country, removing unnecessary features, and handling outliers through a suitable technique. The second step is to check the data for stationarity. Several transforms can be employed to make the data stationary such as log, square root, or Box-Cox transform. In the next step, the confidence interval in the autocorrelation graph can be used to define a range of values for p and q. We may also use hyperparameter optimization to set the most suitable parameter values for the model. After applying the ARIMA model, the model is validated to ensure zero mean value and a constant variance. The model can be assessed through several evaluation measures such as Mean Squared Logarithmic Error (MSLE), Root Mean Squared Logarithmic Error (RMSLE), or Mean Absolute Percentage Error (MAPE). Figure 8 presents the general steps of the ARIMA model for mortality prediction.

Figure 8. Steps of ARIMA model

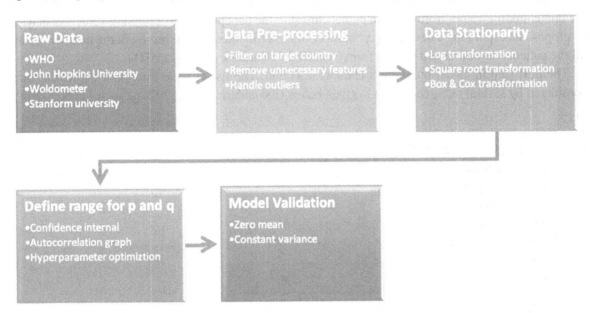

Some of the prominent research works for applying ARIMA for the COVID-19 mortality prediction include (Devaraj et al., 2021; Istaiteh et al., 2020; Satpathy et al., 2021; Zhou et al., 2020). Researchers have also proposed many variations of the basic ARIMA model. Some of these variations include Seasonal Autoregressive Integrated Moving Average (SARIMA) (Meng et al., 2021), SARIMA model with eXogenous variables (SARIMAX) (Toutiaee et al., 2021), and Genetic Algorithm and Autoregressive Integrated Moving Average (GA-ARIMA) (Deif et al., 2021).

Deep Learning Models

Convolutional neural networks (CNN), a particular type of deep learning method, have been widely used for computer vision applications in the past. However, recently, they have also gained popularity for time-series forecasting problems. Recurrent neural network (RNN) and long short term memory (LSTM), a type of RNN, are other deep learning techniques commonly used for such problems.

While the models discussed above can predict the total number of mortalities in a community, the focus of deep learning models is to analyze various features of an individual patient to assess the relative importance of these features. This knowledge can significantly help in optimizing treatment offered to other patients.

Data pre-processing is the first step in building a machine learning model. We assume that one has collected enough data of COVID-19 patients with known illness outcomes, i.e., discharge or death. It is common to have multiple test readings for a single patient in a day. These values should be combined to form one data point for every patient in a day. Exploratory data analytics must be applied to understand the data. Any inconsistencies in the data must be removed. Data normalization might be required if the difference in range of values for some variables is higher than a certain threshold. Missing values are a source of misfit models. Multiple techniques can be used to impute missing values, such as mean, mode, hot-deck imputation, cold-deck imputation, regression imputation, stochastic regression imputation, interpolation, and extrapolation. K-nearest neighborhood algorithm is another commonly used technique to impute missing values from the data points closest to the data point with a missing value. Akshaya et al. propose incorporating a new feature, "number of days till outcome," to represent the number of days till the patient was discharged or died from COVID-19 (Karthikeyan et al., 2021).

The next step in model building is features selection. The list of most important clinical features considered by several research studies is given below (Eskandarian et al., 2021; Yadaw et al., 2020):

- Age
- Demographic characteristics
- Body mass index
- History of hypertension
- History of diabetes
- Cough
- Diarrhea
- Fever
- Dyspnea
- Fatigue
- Myalgia
- Vomiting
- Sputum
- Coronary artery disease
- C-reactive protein (CRP)
- D-dimer
- Oxygen saturation
- Glomerular filtration rate (GFR)
- Neutrophil count
- Lymphocyte count
- Aspartate aminotransferase (AST)
- Lactate dehydrogenase (LDH)
- Chest x-rays
- CT scans
- Respiratory rate
- Serum lactate

Feature selection is an essential step in building machine learning models. While more features may help develop a richer model in some instances, it may result in an overfit model when the number of data points is relatively few. Hence, careful consideration is required for selecting the most appropriate set of parameters for building a model. The logistical issues also influence this step. For example, more expensive features such as CT scans might not be readily available in a low-resource setting. Gradient boosting is another technique that can help select the most prominent features without compromising the prediction accuracy. The popular gradient boosting algorithms, such as XGBoost or LightGBM, provide the added benefit of making the model more interpretable.

After feature selection, we are ready to build the forecast model. A neural network may comprise an input layer, a hidden layer, and an output layer in the simplest form. However, deep learning models usually contain several hidden layers between input and output layers. The network structure, i.e., the number of hidden layers and their inputs, is selected based on the type of neural network. Hyperparameter tuning is the next step where we choose various parameters such as learning rate, activation functions, summation function, and the scheme for updating neuron weights. Finally, the evaluation measures are selected to evaluate the performance of the neural network. Figure 9 shows the application of deep learning in mortality prediction graphically.

Figure 9. Application of deep learning in mortality prediction

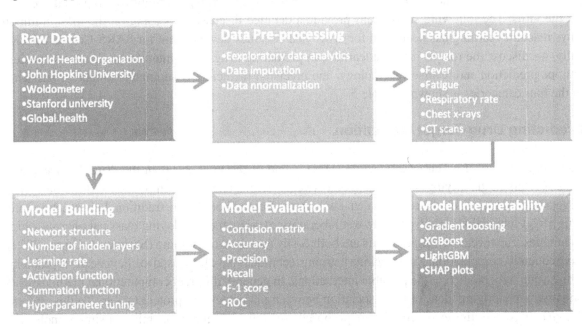

Some studies that used deep learning models for COVID-19 mortality prediction include (Jo et al., 2020; Laponogov et al., 2021; X. Li et al., 2020; Rawal et al., 2021; Verma & Aggarwal, 2021; Zhu et al., 2020).

Hybrid Models

The main drawback of the primitive SIR-based model is a lack of consideration of changes in the behavior of model parameters with time. On the other hand, although machine learning models can provide better forecasts, they are not easily interpretable. Interpretability of predictions is vital because governments and other policymakers use these results for decision-making. Hybrid models of COVID-19 mortality prediction combine the traditional SIR-based epidemiology models with machine learning techniques to make better predictions without compromising forecast interpretability. Hybrid models are also sensitive to government actions such as lockdown, social distancing, and contact tracing to control virus spread. Further, the effect of vaccines, treatments, and medicines can also be modeled.

MACHINE LEARNING TECHNIQUES IN DRUGS DISCOVERY AND VACCINE DEVELOPMENT

Ever since WHO declared the COVID-19 pandemic, besides other strategic measures taken by relevant authorities to mitigate the spread, the medicine, and healthcare industries have been striving towards finding a cure for SARS-CoV-2. Due to exceedingly high infection rates, the contemporary drug development approaches appeared insufficient to achieve the task at sufficient speed. Machine learning methods have played an essential role in improving the situation in several related domains. Researchers have proposed many approaches based on machine learning to help develop and improve drugs and vaccines for the virus. The following subsections provide a detailed discussion focusing on the most common ways machine learning has been used for drugs and vaccines development for SARS-CoV-2. Specifically, we discuss the role of machine learning in predicting drug target interaction, drug repurposing, epitope prediction and reverse vaccinology, and precision medicine. A summary of major approaches in the four categories is shown in Table 5.

Predicting Drug-Target Interaction

The recognition of drug target interaction (DTI) is one of the vital stages in the drug discovery process. It refers to identifying the interactions between various chemical compounds and the biological targets (e.g., proteins, enzymes, or ion channels in the human body). The traditional approach to infer drug target interactions involved using wet laboratory experiments employing classical pharmacology techniques. However, such experiments are both costly and time-consuming (Sachdev & Gupta, 2019). Consequently, computational and machine learning techniques have been used to alleviate these problems while accurately predicting the possible interactions. In general, different computational methods work on varying underlying ideas of the association between molecules and protein targets. To this end, the *ligand methods* (Butina et al., 2002) were proposed, assuming a higher probability of similar molecules being associated with similar protein targets.

Similarly, *docking methods* (Cheng et al., 2007) were offered that employ 3D structures of drugs and proteins to predict their reactions. Another class known as *chemogenomic methods* (Yamanishi et al., 2008) makes predictions based on drug and protein omics data. While studies (e.g., (El-Behery et al., 2021)) have reported some disadvantages related to the former two categories, the chemogenomic techniques are considered effective in overcoming those weaknesses. So far as the prediction technique in the chemogenomic approach is concerned, machine learning-based techniques (opposed to others such as graph-based or network-based) have provided the most reliable predictive results (Sachdev & Gupta, 2019).

Figure 10 shows an overall workflow of machine learning approaches for DTI prediction. These methods work either by discovering the most discriminative features or by calculating the similarity between the drug compounds and target proteins. With the former approach, the drug target pairs and the proteins are encoded as vectors of feature descriptors. The feature vectors containing the properties of drugs and targets are fed into machine learning models that find the most discriminative features to predict the interactions between the drug target pairs.

Figure 10. Overall workflow for machine learning-based DTI prediction

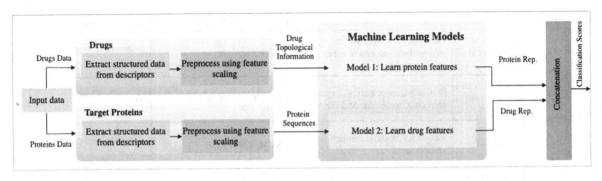

Table 5. Summary of machine learning approaches used in drugs discovery and vaccine development for SARS-CoV-2

Category	Reference	Features	Model	Datasets
DTI	Abdel-Basset et al. (Abdel-Basset et al., 2020)	Binding affinity scores, SMILES	Two-stream bidirectional ConvLSTM	DAVIS, KIBA
	El-Behery et al. (El-Behery et al., 2021)	Protein & drug descriptor, SMILES, reference protein substitution table	Multiple (Best performance achieved with LightBoost and ExtraTree)	DrugBank, Benchmark
	Liu et al. (Liu et al., 2020)	HLA haplotype frequencies, Allele frequencies	Beam search	Self-collected peptides
	Beck et al. (Beck et al., 2020)	SMILES, amino acid sequences	Self-attention-based MT-DTI	Drug Target Common, BindingDB
	Kadioglu et al. (Kadioglu et al., 2021)	Homology models of target proteins	Neural network, naïve bayes	FDA approved drugs, ZINC
Drug repurposing	Bung et al. (Bung et al., 2021)	SMILES, molecule grammar, rule-based filtering	Recurrent neural network	ChEMBL
	Ke et al. (Ke et al., 2020)	ECFP, FCFPs, ALogP_count	Neural network	Self-collected
	Hooshmand et al. (Hooshmand et al., 2021)	Multimodal feature fusion	Stochastic neural network (MM-RBM)	Harmonizome, LINCS
	Rajput et al. (Rajput et al., 2021)	Chemical/structural descriptors and fingerprint	SVMs, KNN, random forest, neural network	DrugRepV
	Verma et al. (Verma & Aggarwal, 2021)	BSAAs, Pharmacophore features	Extended-Connectivity Fingerprints (ECFP6), Bayesian algorithm	PubChem, DrugBank, ZINC
	Laponogov et al. (Laponogov et al., 2021)	GSEA	Random walk propagation	DrugBank, DrugCentral, FooDB, USDA

Continued on following page

Table 5. Continued

Category	Reference	Features	Model	Datasets
Epitope prediction & Reverse vaccinology	Rawal et al. (Rawal et al., 2021)	Pathogen proteomes screening, subcellular localization, CTL epitope prediction	Neural network	Self-curated from open sources
	Yazdani et al. (Yazdani et al., 2020)	Protein sequence retrieval, epitopes screening, molecular docking	Neural network, SVM	NCBI
	Ong et al. (Ong et al., 2020)	Vaxign RV analysis, antigen identification	SVM, KNN, XGB	Self-annotated proteins dataset
Precision medicine	Burdick et al. (Burdick et al., 2020)	Bivariate frequencies calculation	Fine and Gray models, logistic regression	Self-collected from patients in the US hospitals
	Simon et al. (Simon et al., 2021)	Harmonized EHR data	Neural network	Self-collected from patients in the US hospitals
	Dhayne et al. (Dhayne et al., 2021)	BoMT, SNOMED-CT	Neural network	MIMIC-III, Clinical Trials

Abdel-Basset et al. (Abdel-Basset et al., 2020) have addressed the limitations of existing approaches that primarily model DTIs as a binary classification problem, thus ignoring characteristics of protein-ligand interactions. Their proposed framework introduces a two-stream bidirectional ConvLSTM to make a two-fold prediction involving chemical interactions between protein sequences and the homogeneity of drug candidates. They captured spatio-sequential features (encoded in SMILES format) to obtain positional and long-term dependency representation of drug sequences. Later, they trained the proposed framework on two public datasets and applied it to model interactions between several existing drugs and SARS-CoV-2 proteins. Besides providing a detailed record of interactions, the authors further validated the prediction performance of the study by pointing out that some of the predicted output had been previously approved for clinical trials. El-Behery et al. (El-Behery et al., 2021) proposed a DTIs prediction model using a heterogeneous dataset. Specifically, their model combined features and sequence information from amino-acid protein sequences and simplified molecular input line (SMILES) (Weininger, 1988) series. The trained model was used to identify the existing drugs (in DrugBank) that could impact proteins known to be affected by SARS-CoV-2. Liu et al. (Liu et al., 2020) proposed two methods to select amino-acid chains (peptides) for vaccination evaluation. The first method considers the human leukocyte antigens (HLA) allele frequencies, whereas the second considers haplotype frequencies. They then used these methods as the objective functions to formulate a vaccine based on combinatorial optimizations—the vaccine design comprised three key steps. In the first step, peptides with undesired properties were filtered out. Next, the peptides to be presented were scored. Finally, an optimized collection of candidate peptides was selected, a factor in HLA haplotypes' frequency. In this way, the method allows optimized vaccine designs based on the HLA display of multiple peptides. Beck et al. (Beck et al., 2020) used a pre-trained deep learning-based DTI model called MT-DTI to reveal drugs that potentially impact SARS-CoV-2 viral proteins. A model pre-trained on SMILES was fine-tuned on two datasets to get raw prediction results. These results were then screened to identify FDA-approved drugs. Kadioglu et al. (Kadioglu et al., 2021) proposed an approach to identify drugs for SARS-CoV-2 by combining existing methods of virtual drug screening and molecular docking with machine learning. Their workflow essentially consisted of selecting target proteins, constructing compound databases, screening virtual drugs using AutoDock VINA, molecular docking using the Lamarckian algorithm, applying supervised machine learning to learn drug similarities, and identifying candidate compounds with the lowest binding energies.

Drug Repurposing

Drug repurposing (sometimes called drug reprofiling) refers to the practices related to identifying new uses for approved existing drugs (Mohanty et al., 2020). The strategy is specifically an attractive proposition for SARS-CoV-2, given the high infection rates and difficulties involved in speedy drug discovery for the disease.

In this way, for example, clinical trials have been conducted to repurpose existing HIV protease inhibitors, to assess fragment-based drug design methods, and to design more potent new chemical entities (NCEs) to target the known 3C-like (3CL) protease of the virus (Bung et al., 2021), effectively. With the recent advancement of machine learning techniques, several studies have focused on mining existing knowledge and exploiting it to develop novel compounds comprising the desired properties. As shown in figure 11, the existing research has applied machine learning in different ways at various stages of drug repurposing. In particular, the most fundamental use of machine learning is to identify existing drugs that have the potential for anti-SARS-CoV-2 activities. For this purpose, existing datasets of compounds reported being active against the SARS-CoV-2 virus. Specifically, those proven to be active against other related viruses, such as those resulting in human immune deficiency and influenza, are utilized for training (Mohanty et al., 2020). After identifying drugs, another common use case of machine learning techniques is their application for testing the predicted drugs for activities against a mimicked disease state in an in vitro assay (Ke et al., 2020). Furthermore, machine learning models are often used to screen drugs at various stages of the repurposing process.

Figure 11. Overall workflow of machine learning-based approaches for drug repurposing

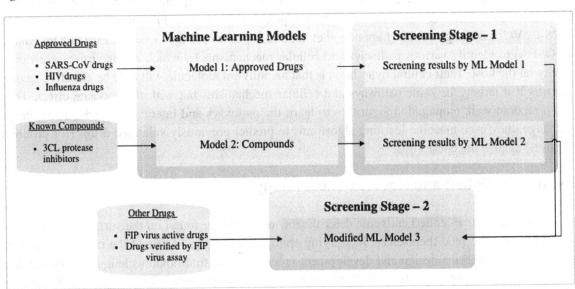

To support the drugs development for SARS-CoV-2 by repurposing existing drugs, Bung et al. (Bung et al., 2021) studied the genome and structural and nonstructural proteins of the virus in a broader context of other coronaviruses. Specifically, they exploited the knowledge of viral protease inhibitors for training a deep neural network model and designed NCEs against the 3CL protease of SARS-CoV-2. For this purpose, their method captured the intrinsic grammar of molecules under study to design protease-specific molecules. The obtained molecules were further optimized by using several property filters and screening the protease's active site. Finally, a ranked list of 31 prospective drug-like NCEs was obtained to be used for drug discovery. Ke et al. (Ke et al., 2020) generated two datasets in which one contained the compounds reported to be active against related viruses (e.g., SARS-CoV, HIV, influenza) and the other comprising the common 3CL protease inhibitors. While generating the datasets, they recorded three types of molecular descriptors: extended connectivity fingerprint (ECFP), functional-class fingerprints (FCFPs), and octanolewater partition coefficient (ALogP_count). They trained a machine learning model in the datasets to make predictions of drugs potentially active against the SARS-CoV-2.

Hooshmand et al. (Hooshmand et al., 2021) have applied the Multimodal Restricted Boltzmann Machine (MM-RBM) approach to the problem of drug repurposing. The RBMs are a class of unsupervised stochastic neural networks containing both visible and hidden layers. As far as the features to be learned are concerned, they use a multimodal feature fusion approach that merges various modalities. In order to find the correlations between different modalities to be learned, they placed the layers of hidden units between the modalities. After feature fusion, they identified 12 potential drug clusters based on the information learned by selected modalities. Rajput et al. (Rajput et al., 2021) employed a recursive feature selection algorithm to select the 50 best-performing features of SARS-CoV-2 and overall coronaviruses from a large set of descriptors (17,968). They used various prediction models to predict the relationship between the structure of drugs and their antiviral activity. The study used the best performing SVM predictors to identify several drug candidates with high efficacies against the virus.

Verma et al. (Verma & Aggarwal, 2021) started by collecting the information of broad-spectrum antiviral agents (BSAAs) that refer to compounds that target viruses belonging to multiple viral families. Next, they identified 45 drug candidates who reported vigorous activity against these BSAAs and got 3D structures of all these compounds. After optimizing the molecular arrangement, they applied computational tools to simulate the molecular interaction between drugs and the target proteins. Finally, Bayesian machine learning models were adopted to predict the compound that may be effective against SARS-CoV-2. Laponogov et al. (Laponogov et al., 2021) performed the Gene Set Enrichment Analysis (GSEA) to identify various pathways and cellular mechanisms by which coronaviruses exert their activity on the host. Their central hypothesis is that an antiviral molecule will only be effective against the virus if it targets the same pathways and cellular mechanisms but with the opposite effect. They used a random walk propagation algorithm to learn the pathways and interaction mechanisms. In the next stage, they used machine learning algorithms to predict previously validated drugs (for different purposes) against the latest virus.

Epitope Prediction and Reverse Vaccinology

An epitope (sometimes called antigenic determinant) is the cluster found on the surface of an antigen to which antibodies bind themselves. Identifying epitopes or antigens is one of the most critical steps in epitope-based vaccine design and development processes. The traditional techniques of biomedical and immunological experiments are both expensive and time-consuming. On the other hand, machine learning methods have been recently adopted, and several studies have reported their success in epitope prediction (Kamalov et al., 2021; Rajput et al., 2021). Reverse vaccinology (RV), as the name implies, uses bioinformatics techniques to scan entire clusters of pathogen genomes to determine genes that could lead to suitable epitopes (Nandy & Basak, 2019). In this way, it is often considered related to the general problem of epitope prediction and a candidate for the adoption of computational and machine learning approaches. Furthermore, after the primary task of epitope prediction and finding vaccine candidates, computational frameworks also adopt machine learning to integrate various algorithms and databases to enable meaningful insight for research teams working on vaccine development. Figure 12 shows the most common application of machine learning in the context of epitope prediction and reverse vaccinology.

Figure 12. Overall view of machine learning-based epitope prediction and reverse vaccinology

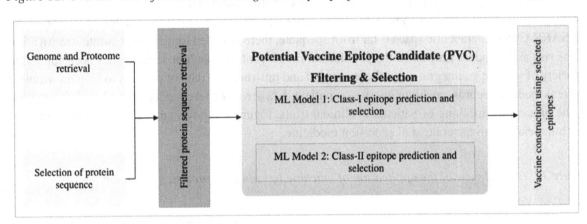

Rawal et al. (Rawal et al., 2021) proposed an integrated framework that integrates various dimensions of vaccine discovery, including immuno-informatics and bioinformatics tools with supervised machine learning tools. To this end, the first step uses specialized computational tools to screen, evaluate, and shortlist the proteins sequences showing relevant features and recognizes them as potential vaccine candidates. In order to guarantee the optimality of thresholds during the selection process, a formal comparison is made between the distributions of the positive dataset (antigenic protein sequences obtained from different organisms) and the negative dataset (non-immunogenic proteins). In the next step, a deep learning model was constructed and trained on the compiled datasets and benchmarked against known prediction tools.

Another vaccine design approach exploiting immune-informatics has been proposed by Yazdani et al. (Yazdani et al., 2020) that is based on six epitopes from four different proteins. They obtained the protein sequences that followed a multi-stage screening of potential epitopes. A consensus method based on neural networks was used during the first stage of screening, whereas in the second stage screening, they carried out prediction of CTL epitopes using a combination of a neural network and a support vector machine. After screening, the selected epitopes between SARS-CoV-2 and SARS were aligned to identify the epitopes that are conserved in two viruses. Next, they evaluated the antigenic properties of the identified epitopes, which finally resulted in selecting three epitopes to be incorporated in the vaccine construction.

A reverse vaccinology approach has been proposed by Ong et al. (Ong et al., 2020). As mentioned previously, the underlying idea of RV techniques is to perform bioinformatics analysis of the pathogen genome to identify promising vaccine candidates. To this end, their work applied Vaxign and Vaxign-ML RV methods to predict SARS-CoV-2 proteins that can be considered as likely candidates for vaccine development. Like Yazdani et al. (Yazdani et al., 2020), they performed a multiple sequence alignment that found that a particular protein (i.e., nsp3) in SARS-CoV-2 was closely related to other existing human coronaviruses. The nsp3 protein was also predicted to contain the most promising epitopes.

Precision Medicine

Precision medicine (Collins & Varmus, 2015) approaches primarily focus on identifying subpopulations of patients that may or may not respond well to a specific treatment. The underlying fact here is that the efficacy of any given drug is inconsistent across patients. Since the existing knowledge base concerning the SARS-CoV-2 therapeutic space is far from adequate, there is a need for a more accurate stratification of the risk and response profiles of the population. In this context, machine learning has been applied in a variety of ways, e.g., integration of the structured and unstructured data of patients to help reformulate patients medical records, enabling queries on patient data across different data sources, and automation of the process of matching of patients to clinical trials. Figure 13 shows an overall view of the use of machine learning in the context of precision medicine.

Figure 13. Overall view of the application of machine learning for precision medicine

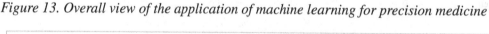
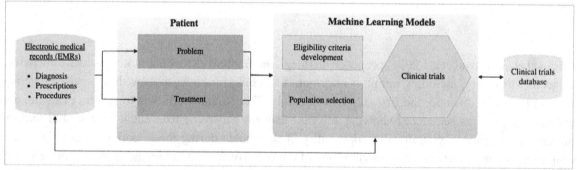

Inspired by observational studies that reported a relationship between hydroxychloroquine treatment and lower mortality, Burdick et al. (Burdick et al., 2020) pursued a controlled study to explore the positive effect of the same on SARS-CoV-2 patients. To this end, they proposed a machine learning-based algorithm to identify groups of patients among those infected with the virus for whom hydroxychloroquine was expected to improve survival rates. The study included patients based on detailed predefined criteria that ensured that the prediction scores (based on various metrics) for all patients were calculated closest to the time of diagnosis. After obtaining the required data, they adopted various machine learning models to investigate the association between hydroxychloroquine treatment and time to hospital discharge and other confounding events. As far as the outcome is concerned, the proposed model could predict over 60% of the receivers of hydroxychloroquine treatment accurately. Simon et al. (Simon et al., 2021) proposed a model to stratify an individual's risk of exposure to drug-induced long QT syndrome (diLQTS) resulting from the use of two medications (i.e., hydroxychloroquine and azithromycin) commonly used for the treatment of SARS-CoV-2 infections. The quantified risk values can then be used to, for example, suggest alternative therapies or, if the use of the same medicine is deemed unavoidable, to guide the monitoring and control. They used a standard data model for harmonized electronic health record (EHR) data forms and identified several variables to select an optimal sampling approach. In the next step, a deep neural network was developed and trained on the available dataset to predict the risk level of developing a diLQTS condition. Dhayne et al. (Dhayne et al., 2021) provided a platform for linking electronic medical records (EMRs) data with clinical trials based on big data vector space. The platform essentially combines machine learning and semantic web techniques to identify individuals eligible for a clinical trial. The proposed system comprises three main modules: derivation of the bag of medical terms (BoMT) based on Systematized Nomenclature of Medicine - Clinical Terms (SNOMED-CT), conversion of patient data into a customized vector, and presentation of the clinical trial as a vector. Different machine learning classifiers (SVM, CNN, LSTM, and BioBERT) were then trained and evaluated to compare their performance regarding the prediction of eligibility of individuals for trials.

REFERENCES

Abdel-Basset, M., Hawash, H., Elhoseny, M., Chakrabortty, R. K., & Ryan, M. (2020). Deeph-DTA: Deep learning for predicting drug-target interactions: A case study of covid-19 drug repurposing. *IEEE Access: Practical Innovations, Open Solutions, 8*, 170433–170451. doi:10.1109/ACCESS.2020.3024238 PMID:34786289

Abouelkheir, I., Rachik, M., Zakary, O., & Elmouki, I. (2017). A multi-regions SIS discrete influenza pandemic model with a travel-blocking vicinity optimal control approach on cells. *Am. J. Comput. Appl. Math, 7*(2), 37–45.

Ahmad, K., Alam, F., Qadir, J., Qolomany, B., Khan, I., Khan, T., Suleman, M., Said, N., Hassan, S. Z., Gul, A., & Al-Fuqaha, A. (2021). Sentiment Analysis of Users' Reviews on COVID-19 Contact Tracing Apps with a Benchmark Dataset. 1–11. https://arxiv.org/abs/2103.01196

Almeida, R., da Cruz, A. M. C. B., Martins, N., & Monteiro, M. T. T. (2019). An epidemiological MSEIR model described by the Caputo fractional derivative. *International Journal of Dynamics and Control, 7*(2), 776–784. doi:10.100740435-018-0492-1

Annas, S., Pratama, M. I., Rifandi, M., Sanusi, W., & Side, S. (2020). Stability analysis and numerical simulation of SEIR model for pandemic COVID-19 spread in Indonesia. *Chaos, Solitons & Fractals, 139*, 110072.

Beck, B. R., Shin, B., Choi, Y., Park, S., & Kang, K. (2020). Predicting commercially available antiviral drugs that may act on the novel coronavirus (SARS-CoV-2) through a drug-target interaction deep learning model. *Computational and Structural Biotechnology Journal, 18*, 784–790. doi:10.1016/j.csbj.2020.03.025 PMID:32280433

Biau, G., Cadre, B., & Rouvière, L. (2019). Accelerated gradient boosting. *Machine Learning, 108*(6), 971–992. doi:10.1007/s10994-019-05787-1 doi:10.1007/s10994-019-05787-1

Broughton, J. P., Deng, X., Yu, G., Fasching, C. L., Singh, J., Streithorst, J., Granados, A., Sotomayor-Gonzalez, A., Zorn, K., Gopez, A., Hsu, E., Gu, W., Miller, S., Pan, C.-Y., Guevara, H., Wadford, D. A., Chen, J. S., & Chiu, C. Y. (2020). *Rapid Detection of 2019 Novel Coronavirus SARS-CoV-2 Using a CRISPR-based DETECTR Lateral Flow Assay.* MedRxiv., doi:10.1101/2020.03.06.20032334

Buchanan, W. J., Imran, M. A., Ur-Rehman, M., Zhang, L., Abbasi, Q. H., Chrysoulas, C., Haynes, D., Pitropakis, N., & Papadopoulos, P. (2020). Review and Critical Analysis of Privacy-Preserving Infection Tracking and Contact Tracing. *Frontiers in Communications and Networks*, *1*(December), 583376. Advance online publication. doi:10.3389/frcmn.2020.583376

Bung, N., Krishnan, S. R., Bulusu, G., & Roy, A. (2021). De novo design of new chemical entities for SARS-CoV-2 using artificial intelligence. *Future Medicinal Chemistry*, *13*(6), 575–585. doi:10.4155/fmc-2020-0262 PMID:33590764

Burdick, H., Lam, C., Mataraso, S., Siefkas, A., Braden, G., Dellinger, R. P., McCoy, A., Vincent, J.-L., Green-Saxena, A., Barnes, G., Hoffman, J., Calvert, J., Pellegrini, E., & Das, R. (2020). Is Machine Learning a Better Way to Identify COVID-19 Patients Who Might Benefit from Hydroxychloroquine Treatment?—The IDENTIFY Trial. *Journal of Clinical Medicine*, *9*(12), 3834. doi:10.3390/jcm9123834 PMID:33256141

Burhan, E., Prasenohadi, P., Rogayah, R., Isbaniyah, F., Reisa, T., & Dharmawan, I. (2020). Clinical Progression of COVID-19 Patient with Extended Incubation Period, Delayed RT-PCR Time-to-positivity, and Potential Role of Chest CT-scan. *Acta Medica Indonesiana*, *52*(1), 80–83. PMID:32291376

Butina, D., Segall, M. D., & Frankcombe, K. (2002). Predicting ADME properties in silico: Methods and models. *Drug Discovery Today*, *7*(11), S83–S88. doi:10.1016/S1359-6446(02)02288-2 PMID:12047885

Campbell, J. Y., & Mankiw, N. G. (1987). Are output fluctuations transitory? *The Quarterly Journal of Economics*, *102*(4), 857–880. doi:10.2307/1884285

Scan, C. A. T. (n.d.). (CT) -. *Chest.*

Chen, T., & Guestrin, C. (n.d.). XGBoost: A Scalable Tree Boosting System. *Proceedings of the 22nd ACM SIGKDD International Conference on Knowledge Discovery and Data Mining.* 10.1145/2939672

Cheng, A. C., Coleman, R. G., Smyth, K. T., Cao, Q., Soulard, P., Caffrey, D. R., Salzberg, A. C., & Huang, E. S. (2007). Structure-based maximal affinity model predicts small-molecule druggability. *Nature Biotechnology*, *25*(1), 71–75. doi:10.1038/nbt1273 PMID:17211405

Chetouani, A., Beghdadi, A., & Deriche, M. (2012). A hybrid system for distortion classification and image quality evaluation. *Signal Processing Image Communication*, *27*(9), 948–960. doi:10.1016/j.image.2012.06.001

Clevert, D.-A. (2020). *Lung Ultrasound in Patients with Coronavirus COVID-19 Disease.* White Paper Published Online by Siemens Medical Solutions USA, Inc.

Collins, F. S., & Varmus, H. (2015). A New Initiative on Precision Medicine. *The New England Journal of Medicine*, *372*(9), 793–795. doi:10.1056/NEJMp1500523 PMID:25635347

Crable, E., & Sena, M. (2020). Exploring Sentiment Towards Contact Tracing. *Proceedings of the Conference on Information*, 1–9. http://proc.conisar.org/2020/pdf/5325.pdf

Cresswell, K., Tahir, A., Sheikh, Z., Hussain, Z., Hernández, A. D., Harrison, E., Williams, R., Sheikh, A., & Hussain, A. (2021). Understanding public perceptions of COVID-19 contact tracing apps: Artificial intelligence-enabled social media analysis. *Journal of Medical Internet Research, 23*(5), 1–8. doi:10.2196/26618 PMID:33939622

Cui, F., & Zhou, H. S. (2020). Diagnostic methods and potential portable biosensors for coronavirus disease 2019. *Biosensors & Bioelectronics, 165*, 112349. doi:10.1016/j.bios.2020.112349 PMID:32510340

Deif, M. A., Solyman, A. A. A., & Hammam, R. E. (2021). ARIMA Model Estimation Based on Genetic Algorithm for COVID-19 Mortality Rates. *International Journal of Information Technology & Decision Making*, 1–24.

Devaraj, J., Madurai Elavarasan, R., Pugazhendhi, R., Shafiullah, G. M., Ganesan, S., Jeysree, A. K., Khan, I. A., & Hossain, E. (2021). Forecasting of COVID-19 cases using deep learning models: Is it reliable and practically significant? *Results in Physics, 21*(January), 103817. doi:10.1016/j.rinp.2021.103817 PMID:33462560

Dhayne, H., Kilany, R., Haque, R., & Taher, Y. (2021). EMR2vec: Bridging the gap between patient data and clinical trial. *Computers and Industrial Engineering, 156*(June), 107236. doi:10.1016/j.cie.2021.107236

Dong, L., Zhou, J., Niu, C., Wang, Q., Pan, Y., Sheng, S., Wang, X., Zhang, Y., Yang, J., Liu, M., Zhao, Y., Zhang, X., Zhu, T., Peng, T., Xie, J., Gao, Y., Wang, D., Zhao, Y., Dai, X., & Fang, X. (2020). Highly accurate and sensitive diagnostic detection of SARS-CoV-2 by digital PCR. MedRxiv, 2020.03.14.20036129. doi:10.1101/2020.03.14.20036129

E, P., A, D., D, B., V, T., E, D., P, M. R., T, M., M, S., A, V., & A, M. (2020). Can Lung US Help Critical Care Clinicians in the Early Diagnosis of Novel Coronavirus (COVID-19) Pneumonia? *Radiology, 295*(3). doi:10.1148/radiol.2020200847 doi:10.1148/radiol.2020200847

Eames, K. T. D., & Keeling, M. J. (2002). Modeling dynamic and network heterogeneities in the spread of sexually transmitted diseases. *Proceedings of the National Academy of Sciences of the United States of America, 99*(20), 13330–13335. doi:10.1073/pnas.202244299 PMID:12271127

Eames, K. T. D., & Keeling, M. J. (2003). Contact tracing and disease control. *Proceedings. Biological Sciences, 270*(1533), 2565–2571. doi:10.1098/rspb.2003.2554 PMID:14728778

El-Behery, H., Attia, A. F., El-Feshawy, N., & Torkey, H. (2021). Efficient machine learning model for predicting drug-target interactions with case study for Covid-19. *Computational Biology and Chemistry, 93*(October), 107536. doi:10.1016/j.compbiolchem.2021.107536 doi:10.1016/j.compbiolchem.2021.107536

Eskandarian, R., Sani, Z. A., Behjati, M., Zahmatkesh, M., Haddadi, A., Kakhi, K., Roshanzamir, M., Shoeibi, A., Alizadehsani, R., Hussain, S., Khozeimeh, F., Keyvani, V., Khosravi, A., Nahavandi, S., & Islam, M. S. (2021). Identification of clinical features associated with mortality in COVID-19 patients. MedRxiv, 2021.04.19.21255715. doi:10.1101/2021.04.19.21255715

Garg, L., Chukwu, E., Nasser, N., Chakraborty, C., & Garg, G. (2020). Anonymity Preserving IoT-Based COVID-19 and Other Infectious Disease Contact Tracing Model. *IEEE Access: Practical Innovations, Open Solutions, 8*, 159402–159414. doi:10.1109/ACCESS.2020.3020513 PMID:34786286

Google & Apple. (2020). *Exposure Notification Bluetooth Specification.* https://covid19-static.cdn-apple.com/applications/covid19/current/static/contact-tracing/pdf/ExposureNotification-BluetoothSpecificationv1.2.pdf?1

Gupta, A. K., Singh, V., Mathur, P., & Travieso-Gonzalez, C. M. (2021). Prediction of COVID-19 pandemic measuring criteria using support vector machine, prophet and linear regression models in Indian scenario. *Journal of Interdisciplinary Mathematics, 24*(1), 89–108. doi:10.1080/09720502.2020.1833458

Hasan, A. M., Al-Jawad, M. M., Jalab, H. A., Shaiba, H., Ibrahim, R. W., & Al-Shamasneh, A. R. (2020). Classification of Covid-19 Coronavirus, Pneumonia and Healthy Lungs in CT Scans Using Q-Deformed Entropy and Deep Learning Features. *Entropy 2020, 22*(5), 517. doi:10.3390/e22050517

Hasan, M. N., Haider, N., Stigler, F. L., Khan, R. A., McCoy, D., Zumla, A., Kock, R. A., & Uddin, M. J. (2021). The global case-fatality rate of COVID-19 has been declining since may 2020. *The American Journal of Tropical Medicine and Hygiene, 104*(6), 2176–2184. doi:10.4269/ajtmh.20-1496 PMID:33882025

He, T., & Printz, M. (2020). *A 2-stage Classifier for Contact Detection with BluetoothLE And INS Signals.* ArXiv.

Ho, S. L., & Xie, M. (1998). The use of ARIMA models for reliability forecasting and analysis. *Computers & Industrial Engineering (American Institute of Industrial Engineers), 35*(1–2), 213–216.

Hooshmand, S. A., Zarei Ghobadi, M., Hooshmand, S. E., Azimzadeh Jamalkandi, S., Alavi, S. M., & Masoudi-Nejad, A. (2021). A multimodal deep learning-based drug repurposing approach for treatment of COVID-19. *Molecular Diversity, 25*(3), 1717–1730. doi:10.100711030-020-10144-9 PMID:32997257

Houchens, J., Gold, J., Maynard, N., Krangle, M., Kikkisetti, S., & Used, A. D. (2020). MITRE TC4TL Challenge System Description. *ArXiv, 20*, 2–4.

Istaiteh, O., Owais, T., Al-Madi, N., & Abu-Soud, S. (2020). Machine Learning Approaches for COVID-19 Forecasting. *2020 International Conference on Intelligent Data Science Technologies and Applications, IDSTA 2020*, 50–57. doi:10.1109/IDSTA50958.2020.926410110.1109/IDSTA50958.2020.9264101

J, W., S, L., M, P., TY, K., MG, P., BY, C., D, K., H, C., VN, K., & CJ, L. (2020). Development of a Laboratory-safe and Low-cost Detection Protocol for SARS-CoV-2 of the Coronavirus Disease 2019 (COVID-19). *Experimental Neurobiology, 29*(2), 107–119. doi:10.5607/en20009

Jahmunah, V., Sudarshan, V. K., Oh, S. L., Gururajan, R., Gururajan, R., Zhou, X., Tao, X., Faust, O., Ciaccio, E. J., Ng, K. H., & Acharya, U. R. (2021). Future IoT tools for COVID-19 contact tracing and prediction: A review of the state-of-the-science. *International Journal of Imaging Systems and Technology, 31*(2), 455–471. doi:10.1002/ima.22552 PMID:33821093

Jaiswal, A. K., Tiwari, P., Kumar, S., Gupta, D., Khanna, A., & Rodrigues, J. J. P. C. (2019). Identifying pneumonia in chest X-rays: A deep learning approach. *Measurement, 145*, 511–518. doi:10.1016/j.measurement.2019.05.076

Jiang, H., Li, Y., Zhang, H., Wang, W., Men, D., Yang, X., Qi, H., Zhou, J., & Tao, S. (2020). *Global profiling of SARS-CoV-2 specific IgG/ IgM responses of convalescents using a proteome microarray.* MedRxiv. doi:10.1101/2020.03.20.20039495

Jo, H., Kim, J., Huang, T.-C., & Ni, Y.-L. (2020). condLSTM-Q: A novel deep learning model for predicting Covid-19 mortality in fine geographical Scale. https://arxiv.org/abs/2011.11507

Kadioglu, O., Saeed, M., Greten, H. J., & Efferth, T. (2021). Identification of novel compounds against three targets of SARS CoV-2 coronavirus by combined virtual screening and supervised machine learning. *Computers in Biology and Medicine*, *133*(March), 104359. doi:10.1016/j.compbiomed.2021.104359 PMID:33845270

Kamalov, F., Cherukuri, A., Sulieman, H., Thabtah, F., & Hossain, A. (2021). *Machine learning applications for COVID-19: A state-of-the-art review.* Academic Press.

Kanne, J. P., Little, B. P., Chung, J. H., Elicker, B. M., & Ketai, L. H. (2020). *Essentials for Radiologists on COVID-19: An Update—Radiology Scientific Expert Panel.* doi:10.1148/radiol.2020200527

Karim, M., & Rahman, R. M. (2013). Decision Tree and Naïve Bayes Algorithm for Classification and Generation of Actionable Knowledge for Direct Marketing. *Journal of Software Engineering and Applications*, *6*(4), 196–206. doi:10.4236/jsea.2013.64025

Karthikeyan, A., Garg, A., Vinod, P. K., & Priyakumar, U. D. (2021). Machine Learning Based Clinical Decision Support System for Early COVID-19 Mortality Prediction. *Frontiers in Public Health*, *9*, 626697. Advance online publication. doi:10.3389/fpubh.2021.626697 PMID:34055710

Ke, Y. Y., Peng, T. T., Yeh, T. K., Huang, W. Z., Chang, S. E., Wu, S. H., Hung, H. C., Hsu, T. A., Lee, S. J., Song, J. S., Lin, W. H., Chiang, T. J., Lin, J. H., Sytwu, H. K., & Chen, C. T. (2020). Artificial intelligence approach fighting COVID-19 with repurposing drugs. *Biomedical Journal*, *43*(4), 355–362. doi:10.1016/j.bj.2020.05.001 PMID:32426387

Kiss, I. Z., Green, D. M., & Kao, R. R. (2005). Disease contact tracing in random and clustered networks. *Proceedings. Biological Sciences*, *272*(1570), 1407–1414. doi:10.1098/rspb.2005.3092 PMID:16006334

Koike, F., & Morimoto, N. (2018). Supervised forecasting of the range expansion of novel non-indigenous organisms: Alien pest organisms and the 2009 H1N1 flu pandemic. *Global Ecology and Biogeography*, *27*(8), 991–1000. doi:10.1111/geb.12754

Lajmanovich, A., & Yorke, J. A. (1976). A deterministic model for gonorrhea in a nonhomogeneous population. *Mathematical Biosciences*, *28*(3–4), 221–236. doi:10.1016/0025-5564(76)90125-5

Lalwani, S., Sahni, G., Mewara, B., & Kumar, R. (2020). Predicting optimal lockdown period with parametric approach using three-phase maturation SIRD model for COVID-19 pandemic. *Chaos, Solitons & Fractals, 138*, 109939.

Lam, C. H., Ng, P. C., & She, J. (2018). Improved Distance Estimation with BLE Beacon Using Kalman Filter and SVM. *2018 IEEE International Conference on Communications (ICC)*, 1–6. 10.1109/ICC.2018.8423010

Laponogov, I., Gonzalez, G., Shepherd, M., Qureshi, A., Veselkov, D., Charkoftaki, G., Vasiliou, V., Youssef, J., Mirnezami, R., Bronstein, M., & Veselkov, K. (2021). Network machine learning maps phytochemically rich "Hyperfoods" to fight COVID-19. *Human Genomics, 15*(1), 1–11. doi:10.118640246-020-00297-x PMID:33386081

Li, F., Li, Y., & Wang, C.LI. (2009). Uncertain data decision tree classification algorithm. *Jisuanji Yingyong, 29*(11), 3092–3095. doi:10.3724/SP.J.1087.2009.03092

Li, L., Qin, L., Xu, Z., Yin, Y., Wang, X., Kong, B., Bai, J., Lu, Y., Fang, Z., Song, Q., Cao, K., Liu, D., Wang, G., Xu, Q., Fang, X., Zhang, S., Xia, J., & Xia, J. (2020). *Using Artificial Intelligence to Detect COVID-19 and Community-acquired Pneumonia Based on Pulmonary CT: Evaluation of the Diagnostic Accuracy.* doi:10.1148/radiol.2020200905 doi:10.1148/radiol.2020200905

Li, X., Ge, P., Zhu, J., Li, H., Graham, J., Singer, A., Richman, P. S., & Duong, T. Q. (2020). Deep learning prediction of likelihood of ICU admission and mortality in COVID-19 patients using clinical variables. *PeerJ, 8*(December), 1–19. doi:10.7717/peerj.10337 doi:10.7717/peerj.10337

Liu, G., Carter, B., Bricken, T., Jain, S., Viard, M., Carrington, M., & Gifford, D. K. (2020). Computationally Optimized SARS-CoV-2 MHC Class I and II Vaccine Formulations Predicted to Target Human Haplotype Distributions. *Cell Systems, 11*(2), 131–144.e6. doi:10.1016/j.cels.2020.06.009 PMID:32721383

Lohar, P., Xie, G., Bendechache, M., Brennan, R., Celeste, E., Trestian, R., & Tal, I. (2021). Irish Attitudes Toward COVID Tracker App & Privacy: Sentiment Analysis on Twitter and Survey Data. *ACM International Conference Proceeding Series.* doi:10.1145/3465481.346919310.1145/3465481.3469193

Mackey, A., Spachos, P., Song, L., & Plataniotis, K. N. (2020). Improving BLE Beacon Proximity Estimation Accuracy Through Bayesian Filtering. *IEEE Internet of Things Journal, 7*(4), 3160–3169. doi:10.1109/JIOT.2020.2965583

Madoery, P. G., Detke, R., Blanco, L., Comerci, S., Fraire, J., Gonzalez Montoro, A., Bellassai, J. C., Britos, G., Ojeda, S., & Finochietto, J. M. (2021). Feature selection for proximity estimation in COVID-19 contact tracing apps based on Bluetooth Low Energy (BLE). *Pervasive and Mobile Computing, 77*, 101474. doi:10.1016/j.pmcj.2021.101474 PMID:34602920

Maier, B. F., & Brockmann, D. (2020). Effective containment explains subexponential growth in recent confirmed COVID-19 cases in China. *Science, 368*(6492), 742–746. doi:10.1126cience.abb4557 PMID:32269067

Malthus, T. R. (1872). *An Essay on the Principle of Population.* Academic Press.

Manych, M. (2020). *X-ray imaging for COVID-19 patients.* Academic Press.

Meng, Y., Wong, M. S., Xing, H., Kwan, M. P., & Zhu, R. (2021). Assessing the country-level excess all-cause mortality and the impacts of air pollution and human activity during the covid-19 epidemic. *International Journal of Environmental Research and Public Health, 18*(13), 1–16. doi:10.3390/ijerph18136883 PMID:34206915

Miller, J. C. (2012). A note on the derivation of epidemic final sizes. *Bulletin of Mathematical Biology, 74*(9), 2125–2141. doi:10.100711538-012-9749-6 PMID:22829179

Miller, J. C. (2017). Mathematical models of SIR disease spread with combined non-sexual and sexual transmission routes. *Infectious Disease Modelling, 2*(1), 35–55. doi:10.1016/j.idm.2016.12.003 PMID:29928728

Miner, J. R. (1933). Pierre-François Verhulst, the discoverer of the logistic curve. *Human Biology, 5*(4), 673.

Mohamed, I. A., Ben Aissa, A., Hussein, L. F., Taloba, A. I., & ... (2021). A new model for epidemic prediction: COVID-19 in kingdom saudi arabia case study. *Materials Today: Proceedings*. Advance online publication. doi:10.1016/j.matpr.2021.01.088 PMID:33520671

Mohanty, S., Harun, A. I., Rashid, M., Mridul, M., Mohanty, C., & Swayamsiddha, S. (2020). Application of Artificial Intelligence in COVID-19 drug repurposing. *Diabetes & Metabolic Syndrome, 14*(5), 1027–1031. doi:10.1016/j.dsx.2020.06.068 PMID:32634717

N, K., M, F., L, F., B, K., & A, I. (2021). Detection of COVID-19 from Chest X-ray Images Using Deep Convolutional Neural Networks. *Sensors (Basel, Switzerland), 21*(17). doi:10.3390/s21175940

Nandy, A., & Basak, S. C. (2019). Bioinformatics in Design of Antiviral Vaccines. Encyclopedia of Biomedical Engineering, 1–3, 280–290. doi:10.1016/B978-0-12-801238-3.10878-5 doi:10.1016/B978-0-12-801238-3.10878-5

NIST. (2021). *TC4TL Challenge.* https://tc4tlchallenge.nist.gov/

Ong, E., Wong, M. U., Huffman, A., & He, Y. (2020). COVID-19 Coronavirus Vaccine Design Using Reverse Vaccinology and Machine Learning. *Frontiers in Immunology, 11*(July), 1581. doi:10.3389/fimmu.2020.01581 PMID:32719684

Woo, P. C. Y., Lau, S. K. P., Wong, B. H. L., Tsoi, H., Fung, A. M. Y., Kao, R. Y. T., Chan, K., Peiris, J. S. M., & Yuen, K.PC. (2005). Differential sensitivities of severe acute respiratory syndrome (SARS) coronavirus spike polypeptide enzyme-linked immunosorbent assay (ELISA) and SARS coronavirus nucleocapsid protein ELISA for serodiagnosis of SARS coronavirus pneumonia. *Journal of Clinical Microbiology, 43*(7), 3054–3058. doi:10.1128/JCM.43.7.3054-3058.2005 PMID:16000415

Qin, C., Yao, D., Shi, Y., & Song, Z. (2018). Computer-aided detection in chest radiography based on artificial intelligence: a survey. *BioMedical Engineering OnLine, 17*(1), 1–23. doi:10.1186/s12938-018-0544-y

Qureshi, M. A., Deriche, M., Beghdadi, A., & Amin, A. (2017). A critical survey of state-of-the-art image inpainting quality assessment metrics. *Journal of Visual Communication and Image Representation, 49*, 177–191. doi:10.1016/j.jvcir.2017.09.006

Peng, Q.-Y., Wang, X.-T., & Zhang, L.-N. (2020). Findings of lung ultrasonography of novel corona virus pneumonia during the 2019-2020 epidemic. *Intensive Care Medicine, 46*(5), 849–850. doi:10.100700134-020-05996-6 PMID:32166346

Rajasekar, S. J. S. (2021). An Enhanced IoT Based Tracing and Tracking Model for COVID -19 Cases. *SN Computer Science, 2*(1), 1–4. doi:10.100742979-020-00400-y PMID:33490971

Rajput, A., Thakur, A., Mukhopadhyay, A., Kamboj, S., Rastogi, A., Gautam, S., Jassal, H., & Kumar, M. (2021). Prediction of repurposed drugs for Coronaviruses using artificial intelligence and machine learning. *Computational and Structural Biotechnology Journal, 19*, 3133–3148. doi:10.1016/j.csbj.2021.05.037 PMID:34055238

Rawal, K., Sinha, R., Abbasi, B. A., Chaudhary, A., Nath, S. K., Kumari, P., Preeti, P., Saraf, D., Singh, S., Mishra, K., Gupta, P., Mishra, A., Sharma, T., Gupta, S., Singh, P., Sood, S., Subramani, P., Dubey, A. K., Strych, U., ... Bottazzi, M. E. (2021). Identification of vaccine targets in pathogens and design of a vaccine using computational approaches. *Scientific Reports, 11*(1), 1–25. doi:10.103841598-021-96863-x PMID:34475453

Riley, S., Fraser, C., Donnelly, C. A., Ghani, A. C., Laith, J., Hedley, A. J., Leung, G. M., Ho, L., Lam, T., Thuan, Q., Chau, P., Chan, K., Lo, S., Leung, P., Tsang, T., Ho, W., Lee, K., Lau, E. M. C., Ferguson, N. M., & Anderson, R. M. (2020). *Transmission Dynamics of the Etiological Agent of SARS in Hong Kong : Impact of Public Health Interventions.* American Association for the Advancement of Science. https://www.jstor.org/stable/3834535

Roy, A., Kumbhar, F. H., Dhillon, H. S., Saxena, N., Shin, S. Y., & Singh, S. (2020). *Efficient Monitoring and Contact Tracing for COVID-19: A Smart IoT-Based Framework.* Academic Press.

Sachdev, K., & Gupta, M. K. (2019). A comprehensive review of feature based methods for drug target interaction prediction. In *Journal of Biomedical Informatics* (Vol. 93, p. 103159). Academic Press. doi:10.1016/j.jbi.2019.103159

Sarica, A., Cerasa, A., & Quattrone, A. (2017). Random Forest Algorithm for the Classification of Neuroimaging Data in Alzheimer's Disease: A Systematic Review. *Frontiers in Aging Neuroscience, 0*(OCT), 329. doi:10.3389/fnagi.2017.00329 PMID:29056906

Satpathy, S., Mangla, M., Sharma, N., Deshmukh, H., & Mohanty, S. (2021). Predicting mortality rate and associated risks in COVID-19 patients. *Spatial Information Research, 29*(4), 455–464. doi:10.100741324-021-00379-5

Saurabh, S., & Prateek, S. (2017). Role of contact tracing in containing the 2014 Ebola outbreak: A review. *African Health Sciences, 17*(1), 225–236. doi:10.4314/ahs.v17i1.28 PMID:29026397

Shankar, S., Chopra, A., Kanaparti, R., Kang, M., Singh, A., & Raskar, R. (2020). *Proximity sensing for contact tracing.* ArXiv.

Simon, S. T., Mandair, D., Tiwari, P., & Rosenberg, M. A. (2021). Prediction of Drug-Induced Long QT Syndrome Using Machine Learning Applied to Harmonized Electronic Health Record Data. *Journal of Cardiovascular Pharmacology and Therapeutics, 26*(4), 335–340. doi:10.1177/1074248421995348 PMID:33682475

Taylor, S. J., & Letham, B. (2018). Forecasting at scale. *The American Statistician, 72*(1), 37–45. doi:10.1080/00031305.2017.1380080

CoalitionT. C. N. (2020). *TCN Protocol.* https://github.com/TCNCoalition/TCN

Toutiaee, M., Li, X., Chaudhari, Y., Sivaraja, S., Venkataraj, A., Javeri, I., Ke, Y., Arpinar, I., Lazar, N., & Miller, J. (2021). Improving COVID-19 Forecasting using eXogenous Variables. ArXiv Preprint ArXiv:2107.10397.

Troncoso, C., Payer, M., Hubaux, J.-P., Salathé, M., Larus, J., Bugnion, E., Lueks, W., Stadler, T., Pyrgelis, A., Antonioli, D., Barman, L., Chatel, S., Paterson, K., Čapkun, S., Basin, D., Beutel, J., Jackson, D., Roeschlin, M., Leu, P., . . . Pereira, J. (2020). Decentralized Privacy-Preserving Proximity Tracing. https://arxiv.org/abs/2005.12273

Tulshyan, V., Sharma, D., & Mittal, M. (2020). An Eye on the Future of COVID'19: Prediction of Likely Positive Cases and Fatality in India over A 30 Days Horizon using Prophet Model. *Disaster Medicine and Public Health Preparedness*, 1–7. Advance online publication. doi:10.1017/dmp.2020.444 PMID:33203489

Verma, A. K., & Aggarwal, R. (2021). Repurposing potential of FDA-approved and investigational drugs for COVID-19 targeting SARS-CoV-2 spike and main protease and validation by machine learning algorithm. *Chemical Biology & Drug Design*, 97(4), 836–853. doi:10.1111/cbdd.13812 PMID:33289334

Wang, P., Zheng, X., Li, J., & Zhu, B. (2020). Prediction of epidemic trends in COVID-19 with logistic model and machine learning technics. *Chaos, Solitons, and Fractals*, 139, 110058. Advance online publication. doi:10.1016/j.chaos.2020.110058 PMID:32834611

Wei, C., & Yuan, H. M. (2014). An improved GA-SVM algorithm. *Proceedings of the 2014 9th IEEE Conference on Industrial Electronics and Applications, ICIEA 2014*, 2137–2141. 10.1109/ICIEA.2014.6931525

Weininger, D. (1988). SMILES, a chemical language and information system. 1. Introduction to methodology and encoding rules. *Journal of Chemical Information and Computer Sciences*, 28(1), 31–36. doi:10.1021/ci00057a005

Wong, H. Y. F., Lam, H. Y. S., Fong, A. H.-T., Leung, S. T., Chin, T. W.-Y., Lo, C. S. Y., Lui, M. M.-S., Lee, J. C. Y., Chiu, K. W.-H., Chung, T. W.-H., Lee, E. Y. P., Wan, E. Y. F., Hung, I. F. N., Lam, T. P. W., Kuo, M. D., & Ng, M.-Y. (2020). *Frequency and Distribution of Chest Radiographic Findings in Patients Positive for COVID-19*. doi:10.1148/radiol.2020201160

Wu, A., Peng, Y., Huang, B., Ding, X., Wang, X., Niu, P., Meng, J., Zhu, Z., Zhang, Z., Wang, J., Sheng, J., Quan, L., Xia, Z., Tan, W., Cheng, G., & Jiang, T. (2020). Genome Composition and Divergence of the Novel Coronavirus (2019-nCoV) Originating in China. *Cell Host & Microbe*, 27(3), 325–328. doi:10.1016/j.chom.2020.02.001 PMID:32035028

Che, X., Qiu, L., Pan, Y., Wen, K., Hao, W., Zhang, L., Wang, Y., Liao, Z., Hua, X., Cheng, V. C. C., & Yuen, K.XY. (2004). Sensitive and specific monoclonal antibody-based capture enzyme immunoassay for detection of nucleocapsid antigen in sera from patients with severe acute respiratory syndrome. *Journal of Clinical Microbiology*, 42(6), 2629–2635. doi:10.1128/JCM.42.6.2629-2635.2004 PMID:15184444

Yadaw, A. S., Li, Y., Bose, S., Iyengar, R., Bunyavanich, S., & Pandey, G. (2020). Clinical features of COVID-19 mortality: Development and validation of a clinical prediction model. *The Lancet. Digital Health*, 2(10), e516–e525. doi:10.1016/S2589-7500(20)30217-X PMID:32984797

Yamanishi, Y., Araki, M., Gutteridge, A., Honda, W., & Kanehisa, M. (2008). Prediction of drug–target interaction networks from the integration of chemical and genomic spaces. *Bioinformatics (Oxford, England), 24*(13), i232–i240. doi:10.1093/bioinformatics/btn162 PMID:18586719

Yazdani, Z., Rafiei, A., Yazdani, M., & Valadan, R. (2020). Design an efficient multi-epitope peptide vaccine candidate against SARS-CoV-2: An in silico analysis. *Infection and Drug Resistance, 13*, 3007–3022. doi:10.2147/IDR.S264573 PMID:32943888

Yorio, Z., El-Tawab, S., & Heydari, M. H. (2021). Room-Level Localization and Automated Contact Tracing via Internet of Things (IoT) Nodes and Machine Learning Algorithm. *2021 IEEE Systems and Information Engineering Design Symposium, SIEDS 2021*, 46–51. 10.1109/SIEDS52267.2021.9483667

Yu, L., Wu, S., Hao, X., Li, X., Liu, X., Ye, S., Han, H., Dong, X., Li, X., Li, J., Liu, N., Liu, J., Zhang, W., Pelechano, V., Chen, W.-H., & Yin, X. (2020). *Rapid colorimetric detection of COVID-19 coronavirus using a reverse transcriptional loop-mediated isothermal amplification (RT-LAMP) diagnostic platform: iLACO.* MedRxiv., doi:10.1101/2020.02.20.20025874

Z, L., Y, Y., X, L., N, X., Y, L., S, L., R, S., Y, W., B, H., W, C., Y, Z., J, W., B, H., Y, L., J, Y., W, C., X, W., J, C., Z, C., … F, Y. (2020). Development and clinical application of a rapid IgM-IgG combined antibody test for SARS-CoV-2 infection diagnosis. *Journal of Medical Virology, 92*(9), 1518–1524. doi:10.1002/jmv.25727 doi:10.1002/jmv.25727

Zhang, J., Tan, L., & Tao, X. (2019). On relational learning and discovery in social networks: A survey. *International Journal of Machine Learning and Cybernetics, 10*(8), 2085–2102. doi:10.100713042-018-0823-8

Zhou, Q., Tao, W., Jiang, Y., & Cui, B. (2020). *A comparative study on the prediction model of COVID-19.* doi:10.1109/ITAIC49862.2020.9338466

Zhu, J. S., Ge, P., Jiang, C., Zhang, Y., Li, X., Zhao, Z., Zhang, L., & Duong, T. Q. (2020). Deep-learning artificial intelligence analysis of clinical variables predicts mortality in COVID-19 patients. *Journal of the American College of Emergency Physicians Open, 1*(6), 1364–1373. doi:10.1002/emp2.12205 PMID:32838390

Zhuoyue, W., Xin, Z., & Xinge, Y. (2003). IFA in testing specific antibody of SARS coronavirus. *South China Journal of Preventive Medicine, 29*(3), 36–37.

Chapter 12
A Rural Healthcare Mobile App:
Urdu Voice–Enabled Mobile App for Disease Diagnosis

Afaq Ahmed
National University of Sciences and Technology, Pakistan

Ahmad Ali Khan
National University of Sciences and Technology, Pakistan

Ismail Shah
National University of Sciences and Technology, Pakistan

Muhammad Ali Tahir
National University of Sciences and Technology, Pakistan

Rafia Mumtaz
National University of Sciences and Technology, Pakistan

ABSTRACT

This chapter presents a voice-enabled mobile application that provides preliminary medical diagnosis of non-fatal diseases. It operates in Urdu language, asking questions from the user and responding in the form of a voice-based dialogue. It addresses the problem of lack of hospitals and medical facilities, especially in rural areas. Furthermore, due to the ongoing pandemic of COVID-19, people are wary of visiting the over-crowded government hospitals. The mobile application developed by the authors provides a quick and inexpensive preliminary diagnosis and medical advice. In case of emergency, it recommends a nearby hospital/doctor. As the system is in Urdu language and voice based, it is well-suited to the Pakistani population's low literacy rate. The application user interface is engineered to be intuitive and simple to use.

DOI: 10.4018/978-1-7998-9201-4.ch012

INTRODUCTION

The purpose of the project described in this chapter is to create a virtual healthcare assistant mobile application for an average Pakistani (literate/illiterate) who converses in Urdu. The healthcare assistant mimics a doctor and talks to its user just like a doctor. It asks certain questions from the user regarding his health and records the user's responses. Based on these responses, the assistant predicts the disease the user might be suffering from. The aim is to create a user-friendly application which will use an audio-visual interface for easy communication with the user. Urdu voice-based conversation and natural language understanding technologies are used in its implementation.

There are inadequate healthcare facilities in the rural areas of Pakistan. So, in case of a medical emergency, the patients often have to travel very long distances to seek medical attention, even for minor diseases. Early diagnosis of a disease can greatly increase the chances of survival of a critically ill person. Also, lack of medical awareness further worsens the patient's condition while travelling. Furthermore, hospitals are already very crowded, given the ongoing COVID-19 pandemic. People have to wait in long queues to see a doctor. This also poses the risk of spreading the pandemic and increases the workload on the doctors. The purpose of creating this application is provide basic healthcare to the average Pakistani using the internet and mobile phone, otherwise free of cost. There is no such system currently available in the market that operates in Urdu language, and that considers the conditions of rural areas of Pakistan.

The mobile application has a chat type interface with access to the microphone and speaker. The user is queried by the chatbot in Urdu about the symptoms which he/she might be facing. The user may either reply verbally through speech using the microphone, or interact with the application through on-screen buttons with possible answers being provided on the user screen. Every next question that the assistant would ask from user depends on the user's answer to the previous question. As a result, the system provides the predicted disease of the user with the corresponding prescription and treatment for that disease in Urdu. By using this application, a person can avoid the following difficulties:

- See a doctor for common, non-fatal diseases.
- Pay medical care fees for mild diseases.
- Travel long distances to a hospital.
- Wait for one's turn to see a doctor.

The system offers the following advantages to its users:

- Receive free healthcare attention.
- Save a lot of time.
- Save money.
- Get instant and preliminary diagnosis.
- Verbal conversation in Urdu for natural interaction

This chapter provides a detailed description of the development of an Urdu voice-based disease diagnosis mobile app. The *background* section provides a review of the previous work in this domain. Details of technology and algorithms used in the proposed system are discussed in the *methodology* section. The *system design and architecture* section describes the overall system and modules developed by the authors. Evaluation of this system is presented in the *experiments and results* section. Finally, a *conclusion* is provided at the end of this chapter.

LITERATURE REVIEW

Speech technology has recently found use in the healthcare industry, and applications for its usage are increasing day by day. There are several speech-based systems already out there. In some systems the personal data and medical history of the patient are transcribed for retrieval (Rosenthal et al, 1998) (David et al, 2009); and in other systems this data is used for disease diagnosis (Toth et al, 2018) (Baldas et al, 2011). Speech recognition is a complex task due to different accents and languages; therefore, there is continuous need of further work to improve its accuracy. The work done in this field is mainly in languages like German, Russian, Chinese and English etc. However, there is little work being done in Urdu language which is Pakistan's national language. The development of this system is aimed at the betterment of Pakistan's society, and that of other developing countries with similar conditions.

Automatic Speech Recognition

Automatic speech recognition is one of the major machine learning areas, allowing the computers to recognize human speech and convert it into a sequence of words. This technology enables a natural mode of interaction between the device and the user. Speech (or sound in general) can be described as representations of frequencies emitted as a characteristic of time. Different speech processing techniques such as speech synthesis, speaker identification, speaker verification come together to create voice-enabled human-machine interfaces.

Traditionally, speech recognition has been done using signal processing-based feature extraction (Han et al, 2006) coupled with statistical models (Juang et al, 1991). Deep neural networks are the most recent strategies that have contributed considerably to the task of speech recognition (Yu and Deng, 2014). Elloumi et al (2018) deployed a multitasking system for performance prediction. This technique is predicated on the convolutional neural network. This came when a comparison between this approach, that is predicated on learned features, and an approach based on predefined traits (engineered features). The data that are employed in this experiment was a set of French-language programs: a set of the Quaero corpus, the information from the ETAPE project, the data from the REPERE evaluation campaign (Galibert and Kahn, 2013). The results obtained during this experiment, the prediction by CNN is healthier than the comparative approach in terms of MAE (Mean Absolute Error) and Edward Calvin Kendall scores. The joint inputs of texts and signals provide positive results and higher performance.

Conversational Systems (Chatbots)

A Conversational User Interface for Stock Analysis

The work of Lauren and Watta (2019) has been presented here because of the technologies used and the flow being similar to the task of this chapter. This system uses NLU (Natural Language Understanding) for the purpose of extraction and comprehending the specifics of conversation. Early work of Cognitive

Psychology influenced the NLU which basically enables the interaction of human and computer which in short is called human-computer interaction.

Extraction of structured semantic information in an unstructured language has been the focus in the current era and for the past few years a lot of research has been going in this field.

In this system, Slack was leveraged as the messaging application with the provision of the Conversation Interface to the user. There were certain APIs like Financial Modelling, Alpha Advantage, News and Twitter APIs used for extracting different kinds of information retrieval based on what the user has asked from the system. Rasa-NLU was specifically used for entity extraction and intent classification. For certain prediction and forecasting tasks, this system used a type of Recurrent Neural Networks, which is Long Short-Term Memory (Hochreiter and Schmidhuber, 1997) implemented in Tensor-Flow.

This system showed an overall average accuracy of 86% utilizing the test records. The system model is shown below in Figure 1. It illustrates how all these different parts worked together.

Figure 1. Stock analysis conversational system (Lauren and Watta, 2019)

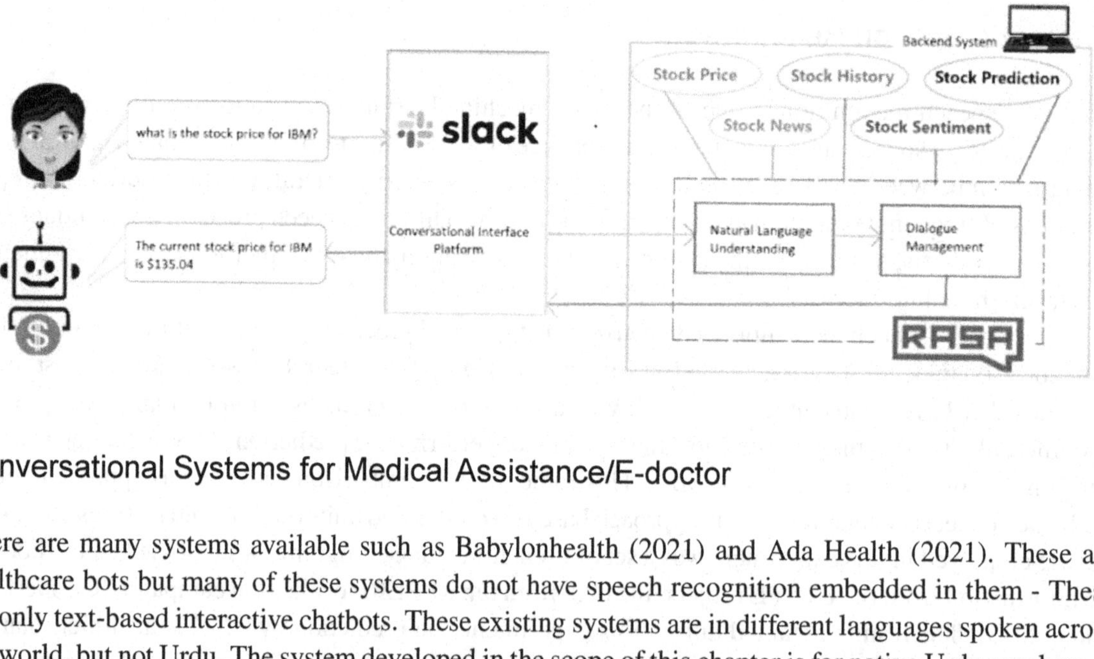

Conversational Systems for Medical Assistance/E-doctor

There are many systems available such as Babylonhealth (2021) and Ada Health (2021). These are healthcare bots but many of these systems do not have speech recognition embedded in them - These are only text-based interactive chatbots. These existing systems are in different languages spoken across the world, but not Urdu. The system developed in the scope of this chapter is for native Urdu speakers. It gives them the service of disease prediction, sitting at home just by having a conversation with the app.

Machine Learning Based Disease Diagnosis

Experimental Disease Prediction Research on Combining Natural Language Processing and Machine Learning

In this era, various Artificial Intelligent (AI) technologies are applied in many different areas to assist knowledge gaining and for various decision-making tasks and in this different kind of fields healthcare information systems could take the most advantage from AI benefits. The attention had been towards the symptom-based disease prediction as well. Various research works have turned their interest in using modern computation techniques to analyse and develop new approaches that are able to predict diseases with quite good accuracy.

The approach used by Yu (2019) is to emply a set of machine learning and natural language processing techniques for disease prediction. This process can be broken down into the following steps.

- Data Extraction and NLP Processing
- Data Evaluation and Clustering
- Prediction

For the first step which is data extraction and processing, a web crawling engine is developed whose purpose is to extract certain diseases information and provision of symptoms description from the website of UK NHS. However, the raw data extracted from the NHS site was composed of repeated conditions and needed cleaning in order to delete the duplicated data. Next tasks were NLP processing which were basic processing methods like capitalization, stop-word removal, Tokenization and Stemming. This pipeline is shown in Figure 2.

Figure 2. Data preparation for ML based disease prediction (Yu, 2019)

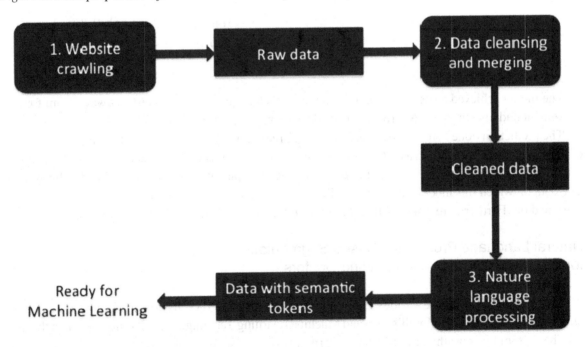

Second step of evaluation was divided into two steps, Finding the frequent symptoms for conditions and then clustering those conditions based on the semantic similarities. For clustering two different clustering algorithms were used which are K-Means clustering and LDA (Latent Dirichlet Allocation). However, the results with K-Means were more suitable and 15 clusters were made in this regard.

Next step was the disease prediction which was carried out by Multinomial Naive Bayes (MNB) probability classification method and this was done in two steps which were calculating the prior probability of each class in the training dataset and then to find the likelihood of a symptom occurring in a particular class. And then both of the probabilities are multiplied and the results will be the probability of the new input belonging to a certain class. These steps are shown in Figure 3.

Figure 3. ML based disease diagnosis (Yu, 2019)

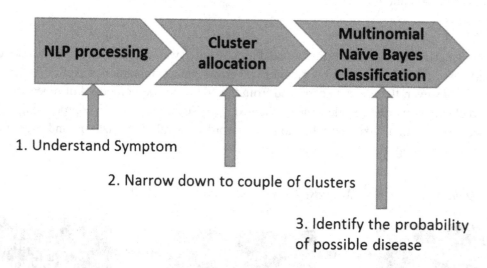

The model achieved an accuracy of around 93% which is quite fair but there is always room for improvement and also there were certain drawbacks which are mentioned below briefly.

The prediction done here is known as Black Box Prediction which means that there is no traceability of why the model is predicting a certain disease given the symptoms which is necessary to know because it is a medical related application and there needs to be an explanation for the step taken by the system. The data on which this model was trained is limited and it can be extended on the number of diseases it is trained on. Furthermore, for updating any knowledge the whole task needs to be redone.

Natural Language Processing Based Segmentation Protocol for Predicting Diseases and Finding Doctors

A lot of current systems use various Data Mining methods to predict diseases. It is considered as the building stone for Artificial Intelligence and Machine Learning Techniques. NLP following up in the line.

The system by Aswathy et al (2019) comprises two main sections, The first one being the disease prediction which used the NLP tools to extract the symptoms from the user and based on those collected symptoms predict the disease accordingly. The second section is doctor suggestion which uses Google API in suggesting the suitable doctor near to the user's circle. Here in this system, NLP purpose in this system can aid with completion and more accurate extraction of medical data by transforming the data from unstructured data to more standardized data.

Figure 4. Symptom extraction (Aswathy et al, 2019)

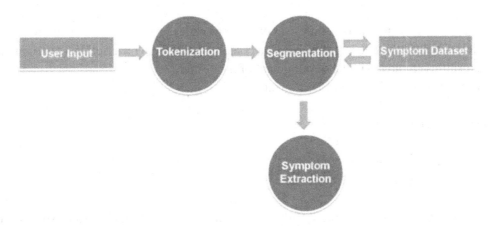

. The first step in disease prediction here is the extraction of the symptoms related to medicine from the unstructured data which is given by the user using certain NLP techniques. This combination of collectively NLP tools is called by a new method NLP Based Unstructured Data Processing Method (NUDPM). The symptoms which are extracted as such are stored in a database known as symptoms database and there is already a database called disease database prepared by doctors which contains the symptoms and its diseases. A cross check is made between the two databases and the disease is predicted in such a manner. Python famous library NLTK is used for the purpose of most type of NLP tasks such as tokenization and segmentation. The flow of information during the symptom extraction phase can be seen in Figure 4.

Doing the disease prediction only is useless unless there is a feature of suggesting a nearby doctor. Based on the predicted disease the Doctor-Type would be recommended. This step basically requires the coordination between the users and doctors in that circle registered with Google Map AI. Basically, the information collected from them will be used in the Doctor-Type selection and provides the details along with the flow being shown in Figure 5.

Figure 5. Doctor suggestion (Aswathy et al, 2019)

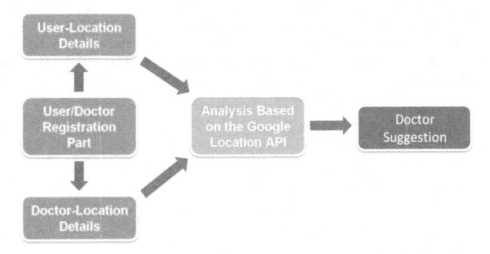

METHODOLOGY

The system developed by the authors comprises of many modules which interact with each other to make up the disease diagnosis mobile app. This section contains a technical description of various technologies and algorithms employed in the development of the system.

Overall, the system comprises of four logical parts:

- Convert Speech to Text
- Collect symptoms from the text
- Disease prediction using the symptoms
- Convert results from Text to Speech and playback.

Out of the above, this research work mainly focuses on the speech recognition and ML-based disease prediction technologies. Therefore, in this section the technical details of these modules are presented. The other parts such as text-to-speech and mobile app development are done using off-the-shelf API and standard methods respectively; therefore these are not discussed at the technological level.

Automatic Speech Recognition

Speech is produced by waves of changing air pressure. Speech is realized through excitation from vocal cords and is modulated by the vocal tract. Vowels, like A, E, I, O, U are simply produced by an open vocal tract. Difference between the sounds of vowels depends upon the width of vocal tract while they are being uttered. Consonants are produced when the vocal cord is constricted with the help of modulators, which are tongue, teeth and lips. Speech is composed of sentences, sentences are composed of words, and words are composed of phonemes. Phonemes are the basic unit of pronunciation, for example phoneme 'Kh' represents the sound of K.

The first step, before doing any kind of processing on speech, is to convert the speech into an electrical signal (voltage). Human hearing has a range of 20Hz to 20kHz, while human speech can have frequency between 85Hz to 8kHz. For this project, all the audio files used for training the speech recognition model or in the actual application, have been sampled at 16000Hz. This means that 16000 data samples are being collected every second. A sampling rate higher than this could have resulted in too much data which can slow down the process of speech recognition, while a rate lower than this means high probability of losing valuable data. The amplitude resolution of each sample is 16 bits.

A speech recognition system is composed of several modules and steps, namely feature extraction, acoustic model, language model, and decoding (Yu and Deng, 2014) (Jelinek, 1997). These are shown below in the Figure 6.

Figure 6. Block diagram of an automatic speech recognition system (RF wireless world, 2012)

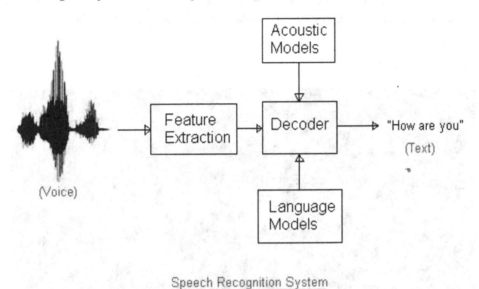

Feature Extraction

The initial analogue speech signal is a simple waveform which provides very little information in its raw form. However, analyzing a very small time-frame shows that waveforms become periodic. The periodic part of the waveform shows one phoneme, for example, the "a" sound in word "Pakistan". Figure 7 shows an audio recording.

Figure 7. Sound wave

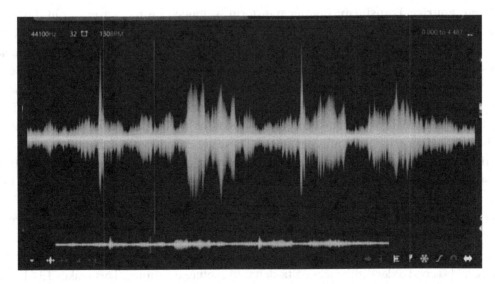

Figure 8 shows a part of the recorded file in Figure 7 on an expanded time scale, where it is seen that wave almost becomes sinusoidal, and each sinusoidal part represents a single phone.

Figure 8. Zoomed-in sound wave

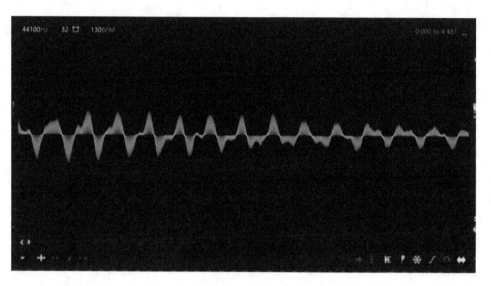

- **Pre-Emphasis:** the sound signal is collected at a fixed sampling rate of 16kHz. Transformations are applied to reduce noise. In our project, we only collect data that has more than 2% sound, which means that low amplitude noise is not recorded.

- **Segmenting and Windowing:** The spectrogram is divided into windows of 25 milliseconds each. The distance between two corresponding windows is 10 ms, that is the windows are overlapping. In this small window, the wave almost becomes sinusoidal.

- **Discrete Fourier Transform:** Fourier transform is applied on the waveform in this step. This provides us with the frequency spectrogram. From this, magnitude information is used only.

- **Mel Frequency Warping:** Human ear has a logarithmic response to the range of frequencies. In this step, all frequencies in the waveform are mapped to specific frequencies using Mel Scale. This means that now we only have linearly spaced, specific frequencies within a given bandwidth. We want a low dimensionality representation, not dependent on speaker, noise, rate etc. FFT is still too high dimensional. We then take local averages of these triangles and downscale it. As a result, we have 40+ dimensional features per frame. These are called log-Mel features.

- **Critical Band Integration and Cepstral Decorrelation:** For traditional systems, 40 features are still too high. Thus, only the first 13 features are taken. These features are obtained by taking discrete cosine transform.

- **Spectral Dynamic Features:** At this stage, our signal is divided into frames. Each frame has 13 dimensions. We now take local averages of the frames. As we have 13 dimensions, we have 13 averages per frame. GMM (Gaussian mixture model) uses local weighted average, local differences deltas, and second order differences delta-deltas to capture dynamics. This makes $13+13+13 = 39$ dimensions. [s9]. Delta features are obtained by: delta = frame(t) − frame(t-1) where t represents number of frames. We now have MFCC + deltas + delta-deltas per frame.

- **Cepstral Mean and Energy Normalization:** We subtract mean to eliminate unknown transfer function. Furthermore, the 0th cepstral coefficient is normalized so that variation due to amplitude is removed.
- **Linear Discriminant Analysis:** The features are subjected to a linear transformation that maximizes the inter-class data variance, while at the same time minimizing the intra-class variance. LDA is calculated using an eigenvalue-based algorithm. Ultimately, a 39-dimensional output vector is used.

Acoustic Model

Acoustic modelling enables the matching of pronunciation units called phonemes with speech feature vectors. For training the acoustic model, speech audio files and their corresponding transcriptions are required. However, vital information missing from the transcriptions is the time alignment of the transcription words with the speech feature vectors. We only have words, which are converted to phonemes, but we do not have any information regarding the length of a word in the transcriptions. This makes training very challenging. In order to model the relationship between acoustic feature vectors and their corresponding transcription, Hidden Markov Models (HMMs) are used to estimate the time alignments.

Hidden Markov Models are statistical models based on the dynamic Bayesian network (Young and Woodland, 1994). In HMM, it is assumed that there are some hidden states. As noted above, while training our acoustic model, we face the challenge of not knowing the time alignments of our transcriptions. HMMs work in a similar fashion with hidden states and the transition probabilities between these states. It can help us overcome the discrepancies regarding the length of phonemes. This makes HMM a very good choice for training our model.

As mentioned above, we have a list of words in our vocabulary and their corresponding pronunciations, written in phonemes. Note that it is possible to have multiple pronunciations for a single word, based on the fact that people with different accents, ethnic and educational background may have different pronunciations.

An HMM phone model has the following features:

- There are self-loops
- One state can be skipped

On entering a state, a feature vector is generated using the distribution associated with the state being entered. Now we have conditional independence needed for HMM; the states are conditionally independent of all other states given the previous state. Observations are also conditionally independent of all other observations given the state that generated it.

Time Alignment and Dynamic Programming

A critical issue in training and decoding is time alignment. In both cases, phones have to be aligned with words in such a way that maximum accuracy is achieved. This issue is discussed in this document under HMM section, where we are not sure in matching our sound signals to our transcription. This problem is catered by using dynamic programming algorithm. Dynamic programming can be used to find minimum distance between two vectors. Normal Euclidean distance is not very accurate, specially because we do not know if there is a substitution, elimination or replacement that makes two vectors different from each other. Dynamic programming allows us to find this combination in minimum time. We use this while calculating probabilities of words.

N-gram language models

A language model has probabilities of the occurrence of each word. The n-gram language model is trained using the word ordered pair counts on a given text corpus (Zitouni, 2007). These word counts are normalized to yield probabilities of occurrence of given n-grams of words. Ideally, the text should contain the kind of sentences and words that are expected to be found in the testing data. Various types of n-gram language models are given below.

- **Unigram**: This model has probabilities of occurrence of each single word. The probabilities of the occurrences of the words are independent from the previous states. The probability of the sequence to be recognized as $w_1, w_2, w_3 \ldots w_N$ is the product of probability of w_1 times w_2 till w_N.
- **Bigram**: In this model, probability of the occurrence of a word depends on the previous word. For example the word "کامیاب" has a higher probability to come after the word "ہیپلونولوج" as compared to the word "ملانا". In bigram language model. This means that probability of the word to be w is dependent on the previously recognized word.
- **Trigram**: In trigram model, probability of a word to be w depends on the two previously recognized words. This means that given the last two recognized words were "گردوں" and "اک", the probability of the next word to be "اوڑڈ" is more than the probability of it to be "راستہ". This is because the phrase "گردوں اک اوڑڈ" makes more sense than the phrase "گردوں اک راستہ".

The n-gram model is trained on a large text corpus. This text corpus contains the text of all the transcriptions that were used for the acoustic model training. Additionally, it also contains other text scraped from internet, books, newspaper websites etc. This large amount of text allows the language model to estimate the probabilities of even those words and phrases which are not present in the audio transcriptions. More recently, recurrent neural networks have been used in conjunction with n-grams for language modelling (Mikolov et al 2010).

Decoding (Recognition)

Decoding is the process of converting voice signals to text, using a pre-trained acoustic model, language model and graphs. Kaldi facilitates multiple ways of decoding, depending on the way training was done, that is using Gaussian mixture model, deep neural networks etc. Two main categories of decoding are online decoding and offline decoding.

- **Online Decoding**: An online decoder collects sound signals in real time and decodes them on the fly. Such decoders are used in real time systems, where the input audio has to be transcribed instantly. Systems like Google Assistant, Siri and Alexa use online decoding. Furthermore, it can also be used to generate the subtitles for a live television broadcast, e.g., the president's press conference. However, because sound is expected at all times, and there is no boundary to describe when a person stopped speaking, online decoders are more prone to errors and have less accuracy.

It is also hard to compute cepstral mean-variance normalization features in online decoding. This is because CMVN features are calculated per speaker. These are mean and averages calculated on all the data so that features can be normalized. Since we do not have all the data at once, on which we can compute the normalization, a "windowing" approach is used, where a window of six seconds by default is used, and moving averages are calculated. However, by this method, the calculated features are only an estimation of a more accurate answer, which could have been achieved by offline decoding. Also, online decoding needs very fast processors, so that a sentence can be decoded before the speaker speaks the next sentence. Otherwise, it is possible that the system either decodes half a sentence, completely miss a sentence or merge parts of two different sentences, which will then make no sense to the NLP system.

- **Offline Decoding** - In offline decoding, the acoustic signal to be decoded are already available in form of files like a .wav or .flac file. As opposed to online decoders, clear boundaries for each sound signal are available. Offline decoding is free from the problems of online decoders (as described above). This makes offline decoders highly accurate and suitable to run on slow processors. Applications of offline decoder include transcribing speech that is not time-critical.

The Viterbi algorithm is a dynamic programming algorithm that finds the most likely hidden states, which, in our case, are the words. As described above, acoustic vectors to be decoded are matched with every word in the vocabulary. This means that if there are M words to be recognized and N words in our vocabulary, we have to match MN words, which can take a lot of computational time. Viterbi algorithm reduces the computational time significantly. It is a recursive algorithm that builds up a matrix called lattice recursively, calculates probabilities of all possible paths, stores the pointers to paths, and keeps the most probable path in last column. A lattice is a matrix with one dimension as states and the other dimension as time, containing probabilities of multiple paths. Viterbi exploits first-order Markov property to finds the most probable path using a lattice. The most probable path is found through the equation below

$$W^* = \text{argmax } W \; P(X|W) \; P(W)$$

While decoding, we have to determine the hypothesis word sequence W^* using the above equation, from the acoustic model $P(X|W)$ and language model $P(W)$. However, Viterbi performs exact search, which is not possible when vocabulary is extensive. N-gram models like trigram increase the size of search space.

Disease Prediction using Machine Learning

There are several applications of machine learning algorithms in medical field. Due to the sensitivity and criticality of the applications, it is necessary that the decisions being made are reliable and effective. Thus, simple decision-making models are most suitable for medical applications. The process of disease classification involves two phases – learning and prediction. During the learning phase, a model is trained using available data. In the prediction phase, the model is used to classify the input data. Voting Classifiers provide very high classification accuracy and are widely used throughout the medical field for decision-making. Our system uses a Voting Classifier to predict the disease from a set of collected symptoms.

Voting Classifier

A Voting Classifier is a classifier which is trained on an ensemble of multiple models (Ruta and Gabrys, 2005). While making a decision, it classifies a certain sample based on highest probability. In simple words, it aggregates the predictions of all its constituent classifiers and makes a decision based on voting. So, it is a combination of different classifiers, as shown in Figure 9. A voting classifier can have two types of voting:

- **Hard Voting:** The class having the greatest number of votes is chosen as the predicted output class.
- **Soft Voting:** The class having the greatest average of the probabilities given by the classifiers is chosen as the predicted output class.

Figure 9. Voting classifier (Sanjay, 2020)

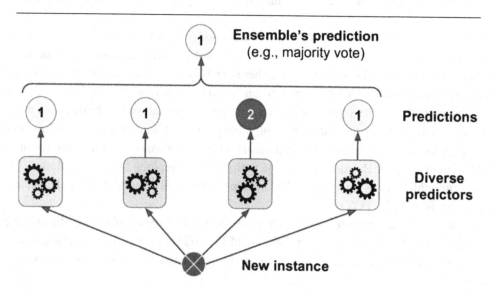

We have used a Voting classifier with soft voting for our system. We are using three classifiers, which are used to calculate the class probabilities of our sample. The class with greatest cumulative probability for all the classifiers is chosen as the final prediction. Our voting classifier is made by aggregating the decisions from the following constituent classifiers.

- Random forests
- AdaBoost
- Bagging classifier

Random Forests

Random Forests is an ensemble classifier which is created using numerous decision trees (Breiman, 2001). It uses bootstrapping, random feature sub-setting and voting between the decision trees to make predictions, as illustrated in Figure 10. It is very good at avoiding overfitting.

Figure 10. Random Forests (Sharma, 2020)

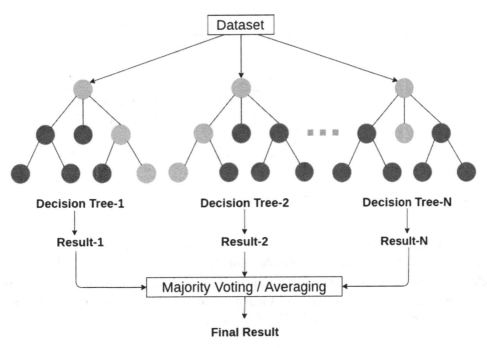

AdaBoost

AdaBoost is short for Adaptive Boosting (Schapire, 2013). It is also an ensemble classifier which can be used to combine multiple weak classifiers to create a strong classifier. AdaBoost is commonly used with decision trees. It works by increasing the weight of samples which are hard to classify and decreases the weight of samples which are classified easily.

Figure 11. Boosting technique (Dangeti, 2018)

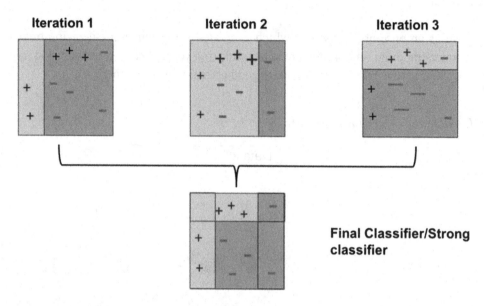

Bagging Classifier

A bagging classifier (Skurichina and Duin, 1998) is also an ensemble classifier which trains each of the constituent classifiers on a random set of samples from the original dataset and aggregates the predictions of all these classifiers for final prediction. Its working is depicted in Figure 12. It can be used to tackle the variance in a data for a simple model like decision tree. It decreases overfitting but gives rise to bias which is compensated by decrement in the variance.

Figure 12. Bagging classifier (Kumar, 2020)

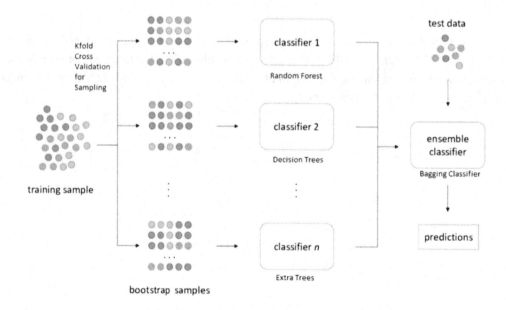

SYSTEM ARCHITECTURE AND IMPLEMENTATION

System Architecture

The architecture of the whole system is divided into four parts which are the mobile app, Urdu speech recognition, text-to-speech, and machine learning disease diagnosis. The mobile app is the user-facing module, while the other three modules make up the functionality of the system. There are also other auxiliary functions like showing related images/videos/animations based on the symptom/disease but that those are mainly to enhance the user experience and to make it more user-friendly.

The system asks the user a question using text-to-speech. The user in return presses a button or utters a verbal answer. The speech recognition module accepts raw audio as input. It extracts features from this audio and creates feature vectors. These vectors are passed to the acoustic model which provides input for the language model. The language model translates the output to individual phonemes which results in conversion of speech to text. This text response resulting from this answer is sent for further use by the machine learning model for the disease prediction. The response is converted to speech and is sent back to the user interface. In this way, the user interacts with the application user interface in the form of a dialogue. This whole procedure takes place in certain different steps and is shown in the high-level architecture diagram given in Figure 13.

Figure 13. System architecture diagram of disease diagnosis mobile app

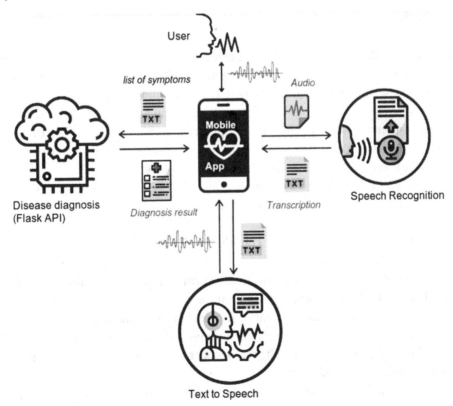

Speech Recognition

Speech recognition is an important module in the system because the follow-up response of the system depends on speech recognition module's accuracy. The application prompts it when the user clicks on the microphone button on the user interface. This module converts the spoken sound signals of the user response/query from speech to text by recognizing what the words have been spoken. It has to capture and normalize the speech variability resulting from different dialects and accents. As this is an environment sensitive module, the noise in the surroundings should be kept to a minimum. Our module caters the noise of the room/surrounding to some extent. However, if it is too loud, even human ears cannot distinguish between the actual noise and the speech, let alone a computer model. The module will only be able to recognize those words/responses from the user which are available in the dictionary. As the training data is in Urdu, it could be also considered as one of the constraints because the training data does not cover every single word in the Urdu language. The module uses the tool Kaldi and it is composed of many sub parts. Those sub parts are explained in detail in the methodology part of this chapter.

In this work, trigram language model has been used, as it is context dependent and so is more accurate. It it is a reduced vocabulary speech recognizer; therefore the number of n-grams is small, as given below.

- Number of unigrams =169
- Number of bigrams =718
- Number of trigrams =843

Kaldi ASR toolkit

We have used Kaldi for developing Urdu speech recognition model for transcription purposes in our application. Kaldi is a speech recognition toolkit created in C++ programming language (Povey et al, 2011). It has been licensed under the Apache License v2.0. Developed by a team led by Daniel Povey – a John Hopkins university associate research professor, Kaldi is intended to be used by speech recognition researchers. It runs on Linux platforms and has multiple scripts which can be used in different processes related to speech recognition. Kaldi also comes with some pre-trained models like Voxforge (recognizer for English language) and YesNo (recognizer that recognizes only "yes" and "no"). Kaldi is similar in aims and scope to HTK. The goal is to have modern and flexible code, written in C++, which is easy to modify and extend. Features of Kaldi are as follows:

- Code-level integration with Finite State Transducers (FSTs). The finite state transduces have been compiled with the OpenFST toolkit.
- Kaldi extensively supports linear algebra. It has a matrix manipulation library which includes the standard BLAS and LAPACK functions and routines.
- Kaldi has a very extensible design because it provides the algorithms in most generic form as much as possible. For example, the decoders have been made on the template of an object which gives tuple-indexed scores. As a result, the decoder can be made to work from any source of scores, for example a neural network.
- Kaldi code has been licensed under the Apache 2.0, which imposes the least restrictions compared to other licenses. The code is available for research as well as commercial purposes, and accessible to everyone from the associated Github repository.

The goal of releasing complete recipes is an important aspect of Kaldi. Since the code is publicly available under a license that permits modifications and re-release, the creators of Kaldi encourage people to release their code, along with their script directories, in a similar format to Kaldi's own example script. Kaldi comes with sufficient documentation online, as well as help forums. However, Kaldi documentation is not as thorough as it should be. This is what Kaldi developers have to say about the documentation:

"We have tried to make Kaldi's documentation as complete as possible given time constraints, but in the short term we cannot hope to generate documentation that is as thorough as HTK's. Much of Kaldi's documentation is written in such a way that it will only be accessible to an expert. In the future we hope to make it somewhat more accessible, bearing in mind that our intended audience is speech recognition researchers or researchers-in-training. In general, Kaldi is not a speech recognition toolkit "for dummies." It will allow you to do many kinds of operations that don't make sense."

Our system requires an online decoder due to its conversational nature. However, as the accuracy of online decoders is less than offline decoders, and they do not run well on slow systems, we have implemented offline decoding embedded in a voice-chat style. The user presses a button to initiate recording a message. After the audio has been recorded, the user presses the button again to stop recording. This audio is saved as a mono wave file sampled at 16 KHz with 16 bits. This file is sent to the backend speech recognition server where the MFCC and CMVN features are calculated. These features are then sent for decoding. After we have extracted feature vectors from our speech sample using the methods described, we have to find the most likely word sequence. From our transition model, named as final.mdl, we have transition probabilities of phonemes. Bayes theorem is used to find out the most likely word sequence. In our system, given an observation Y that is a sequence of feature vectors x_1, x_2 ..., x_T. We want to find the probability that the word sequence is $W = w_1$, w_2 ..., w_N.

Kaldi Gstreamer Server

The Kaldi G-Streamer Server is a real-time full-duplex speech recognition server which is based on the Kaldi toolkit and utilizes the G-Streamer framework. It has been implemented in the Python programming language. It has been used in our system to set-up our Kaldi Urdu model in a client-server architecture. It accepts speech through an HTTP request and returns transcription from the Kaldi model. It is very scalable and can be used to handle unlimited speech recognition sessions in parallel, if the resources permit. Thus, whenever a user inputs speech into our application chat-screen, the cloud server makes an HTTP request to the G-Streamer server, which gets it transcribed using the trained Kaldi model and returns the transcription in JSON format using the same connection.

Disease Prediction/Classification

Machine learning based disease prediction/classification is the second module and one of the most important modules of this whole system. The application prompts it when the symptoms are passed in the text form in Urdu from the questionnaire on which the processing is done. There could be many options available for the specified task but as it is the Medical Application, we will use Voting Classifier in this module for the prediction of the disease and the prescription of that particular disease accordingly. It has to predict different diseases based on the different combination of symptoms it receives in questionnaires which might involve images/animations in questions and from users with good accuracy as being a medical assistance model, there should be no compromise on the accuracy.

The main constraint in this module is that the user can only give an answer in no or yes in Urdu to every question being asked by the bot or something similar. It can be extended to the complexity of taking the response in the form of a sentence but according to our dataset that wasn't needed so we tried to keep it to this level. Also, there are a limited number of 51 diseases which can be predicted by this Model and it can be extended to the diseases which can be predicted by mere questionnaire only. The interaction of the user with the Machine Learning Model is through the Kaldi model which basically passes the response of the user in text form and that response along with the symptom is passed to Machine Learning Model for the prediction task. The specific images/animations are shown for assistance in the questionnaire.

Diseases Dataset

The text generated by the speech-to-text module is used to collect symptoms from the user. Since the symptoms would map to one or more diseases, an authentic diseases dataset is required to help pick the disease based on the symptoms. A diseases dataset available on Kaggle was acquired for our system, which has been created by Patil (2020) in the English language. The dataset contains the following information:

- Disease description
- Symptom precaution
- Symptom severity
- All possible combinations of symptoms for all the diseases.

Cumulatively, the dataset contains 132 different symptoms and 41 common diseases.

We require an Urdu dataset, so we converted the dataset to Urdu using Google Translate API. However, Google Translate was not very accurate in for our case, so we checked the translated diseases and symptoms one by one, skipping the correct translations and altering the incorrect translation in all the files. In order to make the dataset consistent, we cleaned and pre-processed the dataset by dropping null values and stripping all the entries in the dataset of trailing white spaces. For each of the symptoms, our system has to ask a separate question. Since, the Urdu grammar is quite diverse, and the same sentence structure can't be used for all the symptoms, we had to create a separate question in Urdu for each symptom. As a result, we had a complete diseases dataset in the Urdu language along with all the symptoms and precautions in Urdu as well.

Text to Speech

Once the prediction is done from the Machine Learning model, the conversion of the text back to speech needs to be done (Taylor, 2009). In this way, the questionnaire is presented in speech form along with text. The results of the prediction and prescription are provided to the user in both speech and text form, whatever the user sees as convenient. This module is simply for the conversion of each question asked and in the end results and prescription to speech. This module has the responsibility to set every action like a series of questions asked and its results to speech to make it more convenient for users who should have the feel that they are just having a normal kind of conversation with a doctor. Also, this module is to make it more accessible for the disabled users.

The module uses the Google Text-To-Speech API for the conversion of the text to speech. The main constraints for this module are that there should be an internet connection, also that the text should be less than 5000 characters at a time in the free version. Apart from that as its the medical application, it consists of medical terms so there could be problems in accurate conversion to speech if the terms are not correctly specified to it.

Google Cloud Text-to-Speech API enables developers to generate synthetic yet realistic human speech from text and play it as audio in their applications. This API supports the Urdu language as well, which can generate clear Urdu speech from Urdu text. We have used Google's text-to-speech API to generate Urdu speech from text. This API works as an HTTP request, which is made using a URL. The Urdu text, that needs to be converted to speech, is plugged into the body of that URL. In return, the API instantly returns an audio file which can be played in our application for the user. We have used this API in our application to generate Urdu speech for all the questions being asked during the diagnosis process. It has also been used to play the welcome audio message whenever the application is launched. Furthermore, it has been used to read all the on-screen text for the any user who might be illiterate. For example, it reads all the precautions on the diagnosis screen.

We are just converting small Urdu sentences to audio, which include questions, precautions and other conversational sentences. We have used this API in our application to provide Urdu audio feedback wherever necessary, such as question asking, diagnosis result, precautions etc.

Mobile Application

The above-mentioned components have been integrated into a user-friendly android mobile app. This app has been developed using React native. React native is an open-source mobile application development framework, released by Facebook. It is used to develop Android & iOS applications. Using text-to-speech, the app asks the user verbal questions. The user's responses are understood by the system using speech recognition. The disease diagnosis module then analyzes the symptoms and then tells the user about the predicted disease.

The application refers to the nearest hospital, when the application is not able to gather enough symptoms to accurately predict the user's illness or if it gathers the symptoms and the user's condition is critical. It refers the user to the nearest hospital using Google Maps API if there is an ambiguity in the predicted disease or the situation is critical. It will display the hospital on the screen and ask the user to get himself/herself examined as soon as possible. The user should have a working GPS module in this phone. The user must allow the application to access his location so that it can get the location details of the users and depending on that referred to the nearby hospital. The working of the mobile application is illustrated in Figure 14.

Figure 14. Two screenshots depicting the usage of mobile app

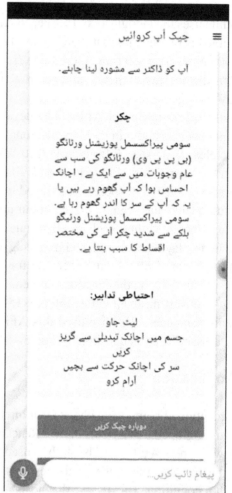

EXPERIMENTS AND RESULTS

In this section, we have discussed the results obtained by letting the potential users interact with the developed mobile application. Statistics have been calculated based on users' responses. These numbers have been used to improve the quality of the system. The details of different experiments are provided below in the form of graphs and numbers.

An Example Use Case

When the application starts, the medical assistant asks user for initial symptoms as a first input. From this input, the machine learning model predicts the possible diseases and collects all their symptoms. Then, the assistant asks the users whether he/she is having these symptoms. Based on user's responses to these questions, the machine learning model again predicts the disease using all the collected symptoms.

And in the end, it generates a diagnosis report and advises precautions. And if need be, it displays the nearest hospital on the map near the user's location.

Figure 15. Depiction of a general use-case for the mobile app

Figure 15 shows a general use-case of the developed application. Its general conversational flow would start with the user providing a symptom to the application. From there, the system will take over and lead the user by asking different questions related to his symptoms. And in the end, it will predict the disease and advise the patient on what to do next.

Testing Strategy

Since our application is speech-based, we gathered users belonging to different areas across Pakistan and having different ethnic backgrounds. They were asked to test our application multiple times and try to input symptoms if they are suffering from any disease. Or if they aren't sick, they may input symptoms from the last time they were sick. After testing the application multiple times, the users were asked to fill a feedback form which collected information regarding their experience while using our application. Their responses are as follows.

Figure 16. User responses - 1

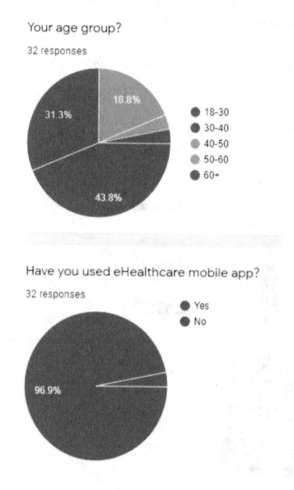

As shown in Figure 16; there are users from different age groups with most of them between 18 to 30 years old, and only 1 or 2 users in 50 to 60 and the 60+ age group.

Figure 17. User responses - 2

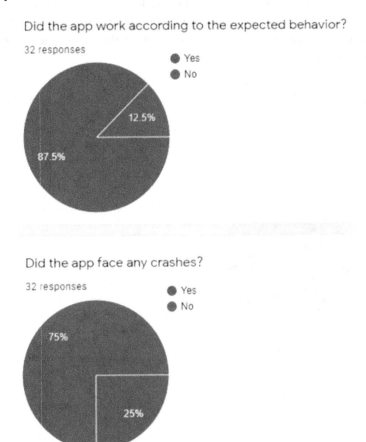

Regarding the results in Figure 17, the application was crashing for some users which was happening because the speech recognition server crashed due to overloading.

Figure 18. User responses - 3

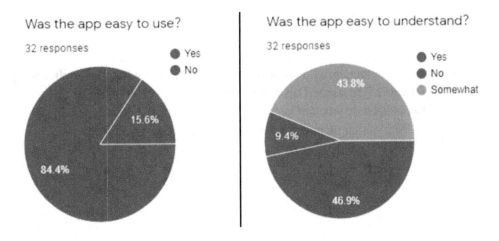

It can be seen from Figure 18 that most of the people found our application quite easy to use and understandable, while the users in the older age group found it hard. Furthermore, speech recognition model performed quite well, transcribing wrong only 15% of the times it was used. 75% users were satisfied with the diagnosis provided by our application.

Figure 19. User responses - 4

If yes, did the actual diagnosis correspond with the app's diagnosis result?

32 responses

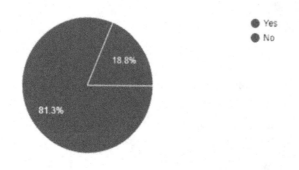

Rate your experience with the app out of 5.

32 responses

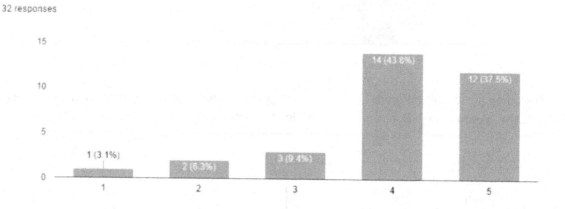

81% of the users found the diagnosis our application to be correct while the application predicted incorrect disease for the rest of the users. This can be seen in Figure 19. It might be due to the fact that sometimes even the user is not sure whether he/she is having a certain symptom or not. Around 80% of the users were satisfied with our application and rated the application 4+ stars.

CONCLUSION AND FUTURE WORK

This chapter discusses the implementation of a voice enabled disease diagnosis mobile app. This app has been designed for an average Pakistani to help them get free medical advice in Urdu from the comfort of their home. It converses with its user just like a doctor and accurately predicts a patient's disease based on symptoms collected during conversation. It eliminates long waiting times in hospitals and diagnoses common and non-fatal diseases free of cost.

For now, the system accepts answers only in YES/NO form. However, it can be extended further to accept full Urdu sentences as answers and extract relevant symptoms from these answers. Furthermore, it can support all the Pakistani regional languages as well, such as Pashto, Punjabi, Saraiki, Balochi, Sindhi etc. A prescription module can be added, which would prescribe relevant medicines after disease prediction, thus cutting down the patient's effort to a bare minimum of just buying medicine from the nearest medical store. Our system can be adopted by doctors and hospitals across Pakistan and as a result, they would accept medical reports generated by our application in order to speed up the diagnosis and treatment process at hospitals.

REFERENCES

Aswathy, K. P., Rathi, R., & Shyam Shankar, E. P. (2019). NLP based segmentation protocol for predicting diseases and finding doctors. *International Research Journal of Engineering and Technology, 6*(2).

Babylon Health. (2021, January 28). *Babylon Health UK - The Online Doctor and Prescription Services App*. https://www.babylonhealth.com/

Baldas, V., Lampiris, C., Capsalis, C., & Koutsouris, D. (2011). Early Diagnosis of Alzheimer's Type Dementia Using Continuous Speech Recognition. *Lecture Notes of the Institute for Computer Sciences, Social Informatics and Telecommunications Engineering*, 105–110.

Breiman, L. (2001). Random Forests. *Machine Learning, 45*(1), 5–32. doi:10.1023/A:1010933404324

Dangeti, P., Yu, A., & Chung, C. (2018). *Numerical Computing with Python*. Packt publishing.

David, G. C., Garcia, A. C., Rawls, A. W., & Chand, D. (2009). Listening to what is said - transcribing what is heard: The impact of speech recognition technology (SRT) on the practice of medical transcription (MT). *Sociology of Health & Illness, 31*(6), 924–938. doi:10.1111/j.1467-9566.2009.01186.x PMID:19843274

Elloumi, Z., Besacier, L., Galibert, O., Kahn, J., & Lecouteux, B. (2018). ASR Performance Prediction on Unseen Broadcast Programs Using Convolutional Neural Networks. *IEEE International Conference on Acoustics, Speech and Signal Processing (ICASSP)*, Calgary, Canada. 10.1109/ICASSP.2018.8461751

Galibert, O., & Kahn, J. (2013). The first official REPERE evaluation. *First Workshop on Speech, Language and Audio in Multimedia*, Marseille, France.

Han, W., Chan, C., Choy, C., & Pun, C. (2006). *An efficient MFCC extraction method in speech recognition. IEEE International Symposium on Circuits and Systems*, Kos, Greece.

Health. (2021). *Ada*. https://ada.com/

Hochreiter, S., & Schmidhuber, J. (1997). Long Short-Term Memory. *Neural Computation, 9*(8), 1735–1780. doi:10.1162/neco.1997.9.8.1735 PMID:9377276

Jelinek, F. (1997). *Statistical Methods for Speech Recognition*. MIT Press.

Juang, B. H., & Rabiner, L. R. (1991). Hidden Markov Models for Speech Recognition. *Technometrics, 33*(3), 251–272. doi:10.1080/00401706.1991.10484833

Kumar, A. (2020). *Bagging Classifier Python Code Example. Data Analytics*. https://vitalflux.com/bagging-classifier-python-code-example/

Lauren, P., & Watta, P. (2019). *A Conversational User Interface for Stock Analysis. IEEE International Conference on Big Data*, Los Angeles, CA. 10.1109/BigData47090.2019.9005635

Mikolov, T., Karafiat, M., Burget, L., Cernocky, J., & Khudanpur, S. (2010). *Recurrent Neural Network Based Language Model. 11th Annual Conference of the International Speech Communication Association*, Makuhari, Japan.

Patil, P. (2020). *Disease Symptom Prediction*. Kaggle. https://www.kaggle.com/itachi9604/disease-symptom-description-dataset

Povey, D., Ghoshal, A., & Boulianne, G. (2011). The Kaldi Speech Recognition Toolkit. *IEEE Workshop on Automatic Speech Recognition and Understanding*.

RF Wireless World. (2012). *Automatic speech recognition system*. https://www.rfwireless-world.com/Terminology/automatic-speech-recognition-system.html

Rosenthal, D. I., Chew, F. S., Dupuy, D. E., Kattapuram, S. V., Palmer, W. E., Yap, R. M., & Levine, L. A. (1998). Computer-based speech recognition as a replacement for medical transcription. *AJR. American Journal of Roentgenology, 170*(1), 23–25. doi:10.2214/ajr.170.1.9423591 PMID:9423591

Ruta, D., & Gabrys, B. (2005). Classifier selection for majority voting. *Information Fusion, 6*(1), 63–81. doi:10.1016/j.inffus.2004.04.008

Sanjay, M. (2020). Model Performance boosting with Voting-Classifier - Analytics Vidhya. *Medium*. https://medium.com/analytics-vidhya/performance-boosting-with-voting-classifier-ea69313a367c

Schapire, R. E. (2013). Explaining adaboost. In *Empirical inference* (pp. 37–52). Springer. doi:10.1007/978-3-642-41136-6_5

Sharma, A. (2020). *Decision Tree vs. Random Forest – Which Algorithm Should you Use? Analytics Vidhya*. https://www.analyticsvidhya.com/blog/2020/05/decision-tree-vs-random-forest-algorithm/

Skurichina, M., & Duin, R. P. (1998). Bagging for linear classifiers. *Pattern Recognition, 31*(7), 909–930. doi:10.1016/S0031-3203(97)00110-6

Taylor, P. (2009). *Text-to-speech Synthesis*. Cambridge University Press. doi:10.1017/CBO9780511816338

Toth, L., Hoffmann, I., Gosztolya, G., Vincze, V., Szatloczki, G., Banreti, Z., Pakaski, M., & Kalman, J. (2018). A Speech Recognition-based Solution for the Automatic Detection of Mild Cognitive Impairment from Spontaneous Speech. *Current Alzheimer Research*, *15*(2), 130–138. doi:10.2174/156720501 4666171121114930 PMID:29165085

Young, S., & Woodland, P. (1994). State clustering in hidden Markov model-based continuous speech recognition. *Computer Speech & Language*, *8*(4), 369–383. doi:10.1006/csla.1994.1019

Yu, D., & Deng, L. (2014). *Automatic Speech Recognition: A Deep Learning Approach*. Springer.

Yu, H. Q. (2019). *Experimental Disease Prediction Research on Combining Natural Language Processing and Machine Learning*. IEEE 7th International Conference on Computer Science and Network Technology, Dalian, China. 10.1109/ICCSNT47585.2019.8962507

Zitouni, I. (2007). Backoff hierarchical class n-gram language models: Effectiveness to model unseen events in speech recognition. *Computer Speech & Language*, *21*(1), 88–104. doi:10.1016/j.csl.2006.01.001

ADDITIONAL READING

Alim, S. A., & Rashid, N. K. A. (2018). Some Commonly Used Speech Feature Extraction Algorithms. In R. Lopez-Ruiz (Ed.), *From Natural to Artificial Intelligence - Algorithms and Applications*. IntechOpen. doi:10.5772/intechopen.80419

Dhole, G. (2014). NLP based retrieval of medical information for diagnosis of human diseases. *International Journal of Research in Engineering and Technology*, *03*(10), 243–248. doi:10.15623/ijret.2014.0310037

El-Baz, S. A., & Suri, J. S. (2021). *Diabetes and Cardiovascular Disease: Diabetes and Cardiovascular Disease (Computer-Assisted Diagnosis)* (1st ed.). Elsevier.

Giri, A. K., & Rana, D. R. (2020). Charting the challenges behind the testing of COVID-19 in developing countries: Nepal as a case study. *Biosafety and Health*, *2*(2), 53–56. doi:10.1016/j.bsheal.2020.05.002

Johnson, M., Lapkin, S., Long, V., Sanchez, P., Suominen, H., Basilakis, J., & Dawson, L. (2014). A systematic review of speech recognition technology in health care. *BMC Medical Informatics and Decision Making*, *14*(1), 94. Advance online publication. doi:10.1186/1472-6947-14-94 PMID:25351845

Kalva, S. (2021). Semantic NLP Technologies in Information Retrieval Systems for Legal Research. *Advances in Machine Learning & Artificial Intelligence*, *2*(1). Advance online publication. doi:10.33140/AMLAI.02.01.05

Rabiner, L., & Juang, B. (1993). *Fundamentals of Speech Recognition*. Pearson College Div.

Themistocleous, M., & Morabito, V. (2012). How can user-centred design affect the acceptance and adoption of service oriented healthcare information systems? *International Journal of Healthcare Technology and Management*, *13*(5/6), 321. doi:10.1504/IJHTM.2012.052550

KEY TERMS AND DEFINITIONS

Automatic Speech Recognition: A technology that allows a human being to converse with a computer in order to give some commands or generate a written transcript.

COVID-19: A contagious disease which is caused by severe acute respiratory syndrome coronavirus 2. Currently as of 2021, it is a worldwide pandemic.

Machine Learning: It is a branch of artificial intelligence that uses large amounts of data to learn how to classify it or gain some insights from it. In some aspects, it strives to mimic a human brain's learning from data.

Medical Diagnosis: It is the practice of determining which disease or ailment a particular person is affected with.

Natural Language Processing: A field of computer science related to automatic processing of human generated language data. It can be used for information retrieval as well as command and control.

Rural Healthcare: The study of providing healthcare systems and services in a rural environment. There are aspects of policy making as well as technology.

Urdu: It is the national language of Pakistan. It is primarily an Indo-Aryan language, with influence of middle eastern languages.

Compilation of References

Abdel-Basset, M., Hawash, H., Elhoseny, M., Chakrabortty, R. K., & Ryan, M. (2020). Deeph-DTA: Deep learning for predicting drug-target interactions: A case study of covid-19 drug repurposing. *IEEE Access: Practical Innovations, Open Solutions, 8*, 170433–170451. doi:10.1109/ACCESS.2020.3024238 PMID:34786289

Abouelkheir, I., Rachik, M., Zakary, O., & Elmouki, I. (2017). A multi-regions SIS discrete influenza pandemic model with a travel-blocking vicinity optimal control approach on cells. *Am. J. Comput. Appl. Math, 7*(2), 37–45.

Ahmad, K., Alam, F., Qadir, J., Qolomany, B., Khan, I., Khan, T., Suleman, M., Said, N., Hassan, S. Z., Gul, A., & Al-Fuqaha, A. (2021). Sentiment Analysis of Users' Reviews on COVID-19 Contact Tracing Apps with a Benchmark Dataset. 1–11. https://arxiv.org/abs/2103.01196

Ahmad, I., Ali, S., Tariq, M., & Ikram, M. (2001). Water pollution in Rawal lake Islamabad (part-1). *Pakistan Journal of Analytical Chemistry, 2*(1), 66–69.

Ahmed, U., Mumtaz, R., Anwar, H., Mumtaz, S., & Qamar, A. M. (2020). Water quality monitoring: from conventional to emerging technologies. *Water Supply, 20*(1), 28-45.

Akhmedov, R. (2017). *Implementation of CRM Strategies to Increase Customer Loyalty, Case of Kazakhstan Companies.* Academic Press.

Alex, S. A., & Nayahi, J. J. V. (2018). Deep Incremental Learning for Big Data Stream Analytics. In *International conference on Computer Networks, Big data and IoT*. Springer.

Ali, Z., Ali, H., Badawy, M., Alam, S., Chowdhury, M., Noll, J., Kalmar, A., Vida, R., Maliosz, M., Gubbi, J., Buyya, R., Marusic, S., Palaniswami, M., Li, D., Chen, Y., Ning, H., Liu, H., Wang, N., Wu, W., ... García, C. G. (2015). Internet of Things (IoT): Definitions, challenges and recent research directions. *International Journal of Computers and Applications, 128*(1), 37–47. doi:10.5120/ijca2015906430

Aljumah, A. A., Ahamad, M. G., & Siddiqui, M. K. (2013). Application of data mining: Diabetes health care in young and old patients. *Journal of King Saud University - Computer and Information Sciences, 25*(2), 127–136. doi:10.1016/j.jksuci.2012.10.003

Al-Lawati, J. A. (2017). Diabetes mellitus: A local and global public health emergency! *Oman Medical Journal, 32*(3), 177–179. doi:10.5001/omj.2017.34 PMID:28584596

Almeida, R., da Cruz, A. M. C. B., Martins, N., & Monteiro, M. T. T. (2019). An epidemiological MSEIR model described by the Caputo fractional derivative. *International Journal of Dynamics and Control, 7*(2), 776–784. doi:10.100740435-018-0492-1

Al-Turjman, F., & Baali, I. (2019). Machine learning for wearable IoT-based applications: A survey. *Transactions on Emerging Telecommunications Technologies*, e3635. doi:10.1002/ett.3635

Amruthnath, N., & Gupta, T. (2018). A research study on unsupervised machine learning algorithms for early fault detection in Predictive Maintenance. In *2018 5th International Conference on Industrial Engineering and Applications (ICIEA)* (pp. 355-361). IEEE. 10.1109/IEA.2018.8387124

Annas, S., Pratama, M. I., Rifandi, M., Sanusi, W., & Side, S. (2020). Stability analysis and numerical simulation of SEIR model for pandemic COVID-19 spread in Indonesia. *Chaos, Solitons & Fractals, 139*, 110072.

Antunes, J. G., Pinto, A., Nogueira Reis, P., & Henriques, C. (2018). Industry 4.0: A challenge of competition. *Millenium*, (6), 89–97. Advance online publication. doi:10.29352/mill0206.08.00159

Appavu, A., Thangavelu, S., Muthukannan, S., Jesudoss, J. S., & Pandi, B. (2016). Study of water quality parameters of Cauvery River water in erode region. *Journal of Global Biosciences, 5*(9), 4556–4567.

Ariffin, N. H. M. (2013). The Development of a Strategic CRM-i Framework: Case Study in Public Institutions of Higher Learning. *Procedia: Social and Behavioral Sciences, 75*, 36–43. doi:10.1016/j.sbspro.2013.04.005

Aswathy, K. P., Rathi, R., & Shyam Shankar, E. P. (2019). NLP based segmentation protocol for predicting diseases and finding doctors. *International Research Journal of Engineering and Technology, 6*(2).

Aswini, D., Santhya, S., Nandheni, T. S., & Sukirthini, N. (2017). Cattle health and environment monitoring system. *International Research Journal of Engineering and Technology, 4*(3), 1899–1903.

Attallah, O., Karthikesalingam, A., Holt, P. J., Thompson, M. M., Sayers, R., Bown, M. J., Choke, E. C., & Ma, X. (2017). Using multiple classifiers for predicting the risk of endovascular aortic aneurysm repair re-intervention through hybrid feature selection. *Proceedings of the Institution of Mechanical Engineers. Part H, Journal of Engineering in Medicine, 231*(11), 1048–1063. doi:10.1177/0954411917731592 PMID:28925817

Attallah, O., & Ma, X. (2014). Bayesian neural network approach for determining the risk of re-intervention after endovascular aortic aneurysm repair. *Proceedings of the Institution of Mechanical Engineers. Part H, Journal of Engineering in Medicine, 228*(9), 857–866. doi:10.1177/0954411914549980 PMID:25212212

Awan, A. G., & Azhar, S. (2014). Customer Relationship Management System: A case study of Floor Mills Bahawalpur District. *British Journal of Marketing Studies, 2*(7), 1–13.

Ayvaz, S., & Alpay, K. (2021). Predictive Maintenance system for production lines in manufacturing: A machine learning approach using IoT data in real-time. *Expert Systems with Applications, 173*. doi:10.1016/j.eswa.2021.114598

Babylon Health. (2021, January 28). *Babylon Health UK - The Online Doctor and Prescription Services App.* https://www.babylonhealth.com/

Bagherighadikolaei, S., Ghousi, R., & Haeri, A. (2020). A Data Mining approach for forecasting failure root causes: A case study in an Automated Teller Machine (ATM) manufacturing company. *Journal of Optimization in Industrial Engineering, 13*(2), 101–121. doi:10.22094/joie.2020.1863364.1630

Baldas, V., Lampiris, C., Capsalis, C., & Koutsouris, D. (2011). Early Diagnosis of Alzheimer's Type Dementia Using Continuous Speech Recognition. *Lecture Notes of the Institute for Computer Sciences, Social Informatics and Telecommunications Engineering*, 105–110.

Ballinger, N. (2020). *Three transformative trends in predictive maintenance.* https://www.electronicspecifier.com/news/three-transformative-trends-in-predictive-maintenance

Balpande, V. R., & Wajgi, R. D. (2017). Prediction and severity estimation of diabetes using data mining technique. *IEEE International Conference on Innovative Mechanisms for Industry Applications, ICIMIA 2017 - Proceedings, Icimia*, 576–580. 10.1109/ICIMIA.2017.7975526

Beck, B. R., Shin, B., Choi, Y., Park, S., & Kang, K. (2020). Predicting commercially available antiviral drugs that may act on the novel coronavirus (SARS-CoV-2) through a drug-target interaction deep learning model. *Computational and Structural Biotechnology Journal, 18*, 784–790. doi:10.1016/j.csbj.2020.03.025 PMID:32280433

Benjamin, E. J., Muntner, P., Alonso, A., Bittencourt, M. S., Callaway, C. W., Carson, A. P., Chamberlain, A. M., Chang, A. R., Cheng, S., Das, S. R., Delling, F. N., Djousse, L., Elkind, M. S. V., Ferguson, J. F., Fornage, M., Jordan, L. C., Khan, S. S., Kissela, B. M., Knutson, K. L., ... Virani, S. S. (2019). Heart disease and stroke statistics—2019 update: A report from the American Heart Association. *Circulation, 139*(10), e56–e528. doi:10.1161/CIR.0000000000000659 PMID:30700139

Bersano, A., Kraemer, M., Burlina, A., Mancuso, M., & Finsterer, J. (2020). Heritable and non-heritable uncommon causes of stroke. *Journal of Neurology*, 1–28. PMID:32318851

Bezabeh, B. B. (2017). The application of data mining techniques to support customer relationship management: the case of ethiopian revenue and customs authority. *arXiv preprint arXiv:1706.10050*.

Biau, G., Cadre, B., & Rouvière, L. (2019). Accelerated gradient boosting. *Machine Learning, 108*(6), 971–992. doi:10.1007/s10994-019-05787-1 doi:10.1007/s10994-019-05787-1

Bilal, M., Usmani, R. S. A., Tayyab, M., Mahmoud, A. A., & Abdalla, R. M. (2020). Smart cities data: Framework, applications, and challenges. Handbook of Smart Cities, 1-29.

Block, L., El-Merhi, A., Liljencrantz, J., Naredi, S., Staron, M., & Odenstedt Hergès, H. (2020). Cerebral ischemia detection using artificial intelligence (CIDAI)—A study protocol. *Acta Anaesthesiologica Scandinavica, 64*(9), 1335–1342. doi:10.1111/aas.13657 PMID:32533722

Bochkovskiy, A., Wang, C.-Y., & Liao, H.-Y. M. (2020). *YOLOv4: Optimal Speed and Accuracy of Object Detection.* Retrieved from https://arxiv.org/abs/2004.10934

Boukobza, M., Nahmani, S., Deschamps, L., & Laissy, J.-P. (2019). Brain abscess complicating ischemic embolic stroke in a patient with cardiac papillary fibroelastoma–Case report and literature review. *Journal of Clinical Neuroscience, 66*, 277–279. doi:10.1016/j.jocn.2019.03.041 PMID:31097380

Bousdekis, A., Lepenioti, K., Apostolou, D., & Mentzas, G. (2019). Decision making in Predictive Maintenance: Literature review and research agenda for industry 4.0. *IFAC-PapersOnLine, 52*(13), 607-612. doi:10.1016/j.ifacol.2019.11.226

Bousdekis, A., Papageorgiou, N., Magoutas, B., Apostolou, D., & Mentzas, G. (2017). A proactive event-driven decision model for joint equipment Predictive Maintenance and spare parts inventory optimization. *Procedia CIRP, 59*, 184–189. doi:10.1016/j.procir.2016.09.015

Breiman, L. (2001). Random Forests. *Machine Learning, 45*(1), 5–32. doi:10.1023/A:1010933404324

Bresciani, M., Giardino, C., Stroppiana, D., Dessena, M. A., Buscarinu, P., Loretta Cabras, K. S., & Tzimas, A. (2019). Monitoring water quality in two dammed reservoirs from multispectral satellite data. *European Journal of Remote Sensing, 52*(sup4), 113–122. doi:10.1080/22797254.2019.1686956

Broughton, J. P., Deng, X., Yu, G., Fasching, C. L., Singh, J., Streithorst, J., Granados, A., Sotomayor-Gonzalez, A., Zorn, K., Gopez, A., Hsu, E., Gu, W., Miller, S., Pan, C.-Y., Guevara, H., Wadford, D. A., Chen, J. S., & Chiu, C. Y. (2020). *Rapid Detection of 2019 Novel Coronavirus SARS-CoV-2 Using a CRISPR-based DETECTR Lateral Flow Assay.* MedRxiv., doi:10.1101/2020.03.06.20032334

Buchanan, W. J., Imran, M. A., Ur-Rehman, M., Zhang, L., Abbasi, Q. H., Chrysoulas, C., Haynes, D., Pitropakis, N., & Papadopoulos, P. (2020). Review and Critical Analysis of Privacy-Preserving Infection Tracking and Contact Tracing. *Frontiers in Communications and Networks*, *1*(December), 583376. Advance online publication. doi:10.3389/frcmn.2020.583376

Bung, N., Krishnan, S. R., Bulusu, G., & Roy, A. (2021). De novo design of new chemical entities for SARS-CoV-2 using artificial intelligence. *Future Medicinal Chemistry*, *13*(6), 575–585. doi:10.4155/fmc-2020-0262 PMID:33590764

Burdick, H., Lam, C., Mataraso, S., Siefkas, A., Braden, G., Dellinger, R. P., McCoy, A., Vincent, J.-L., Green-Saxena, A., Barnes, G., Hoffman, J., Calvert, J., Pellegrini, E., & Das, R. (2020). Is Machine Learning a Better Way to Identify COVID-19 Patients Who Might Benefit from Hydroxychloroquine Treatment?—The IDENTIFY Trial. *Journal of Clinical Medicine*, *9*(12), 3834. doi:10.3390/jcm9123834 PMID:33256141

Burhan, E., Prasenohadi, P., Rogayah, R., Isbaniyah, F., Reisa, T., & Dharmawan, I. (2020). Clinical Progression of COVID-19 Patient with Extended Incubation Period, Delayed RT-PCR Time-to-positivity, and Potential Role of Chest CT-scan. *Acta Medica Indonesiana*, *52*(1), 80–83. PMID:32291376

Butina, D., Segall, M. D., & Frankcombe, K. (2002). Predicting ADME properties in silico: Methods and models. *Drug Discovery Today*, *7*(11), S83–S88. doi:10.1016/S1359-6446(02)02288-2 PMID:12047885

Campbell, J. Y., & Mankiw, N. G. (1987). Are output fluctuations transitory? *The Quarterly Journal of Economics*, *102*(4), 857–880. doi:10.2307/1884285

Canchi, T., Ng, E. Y. K., Narayanan, S., & Finol, E. A. (2018). On the assessment of abdominal aortic aneurysm rupture risk in the Asian population based on geometric attributes. *Proceedings of the Institution of Mechanical Engineers. Part H, Journal of Engineering in Medicine*, *232*(9), 922–929. doi:10.1177/0954411918794724 PMID:30122103

Caradu, C., Spampinato, B., Vrancianu, A. M., Bérard, X., & Ducasse, E. (2021). Fully automatic volume segmentation of infrarenal abdominal aortic aneurysm computed tomography images with deep learning approaches versus physician controlled manual segmentation. *Journal of Vascular Surgery*, *74*(1), 246–256.e246. doi:10.1016/j.jvs.2020.11.036 PMID:33309556

Cavique, L. (2005). *Next-item discovery in the market basket analysis.* Paper presented at the 2005 portuguese conference on artificial intelligence. 10.1109/EPIA.2005.341294

Chahine, M. T., McCleese, D. J., Rosenkranz, P. W., & Staelin, D. H. (1983). Interaction mechanisms within the atmosphere. *Manual of Remote Sensing, 1*(5).

Chaikof, E. L., Dalman, R. L., Eskandari, M. K., Jackson, B. M., Lee, W. A., Mansour, M. A., Mastracci, T. M., Mell, M., Murad, M. H., & Nguyen, L. L. (2018). The Society for Vascular Surgery practice guidelines on the care of patients with an abdominal aortic aneurysm. *Journal of Vascular Surgery, 67*(1), 2-77.

Chang, N.-B., Ba, K., & Chen, C.-F. (2017). Integrating multisensor satellite data merging and image reconstruction in support of machine learning for better water quality management. *Journal of Environmental Management*, *201*, 227–240. doi:10.1016/j.jenvman.2017.06.045 PMID:28667841

Cheng, A. C., Coleman, R. G., Smyth, K. T., Cao, Q., Soulard, P., Caffrey, D. R., Salzberg, A. C., & Huang, E. S. (2007). Structure-based maximal affinity model predicts small-molecule druggability. *Nature Biotechnology*, *25*(1), 71–75. doi:10.1038/nbt1273 PMID:17211405

Chen, T., & Guestrin, C. (n.d.). XGBoost: A Scalable Tree Boosting System. *Proceedings of the 22nd ACM SIGKDD International Conference on Knowledge Discovery and Data Mining*. 10.1145/2939672

Chen, Y.-H., & Sawan, M. (2021). Trends and Challenges of Wearable Multimodal Technologies for Stroke Risk Prediction. *Sensors (Basel)*, *21*(2), 460. doi:10.339021020460 PMID:33440697

Chetouani, A., Beghdadi, A., & Deriche, M. (2012). A hybrid system for distortion classification and image quality evaluation. *Signal Processing Image Communication*, *27*(9), 948–960. doi:10.1016/j.image.2012.06.001

Che, X., Qiu, L., Pan, Y., Wen, K., Hao, W., Zhang, L., Wang, Y., Liao, Z., Hua, X., Cheng, V. C. C., & Yuen, K.XY. (2004). Sensitive and specific monoclonal antibody-based capture enzyme immunoassay for detection of nucleocapsid antigen in sera from patients with severe acute respiratory syndrome. *Journal of Clinical Microbiology*, *42*(6), 2629–2635. doi:10.1128/JCM.42.6.2629-2635.2004 PMID:15184444

Chiu, Y. C., Gheng, F. T., & Huang, H. C. (2017). Developing a factory-wide intelligent Predictive Maintenance system based on Industry 4.0. *Zhongguo Gongcheng Xuekan*, *40*(7), 562–571. doi:10.1080/02533839.2017.1362357

Choi, Y.-A., Park, S., Jun, J.-A., Ho, C. M. B., Pyo, C.-S., Lee, H., & Yu, J. (2021). Machine-Learning-Based Elderly Stroke Monitoring System Using Electroencephalography Vital Signals. *Applied Sciences (Basel, Switzerland)*, *11*(4), 1761. doi:10.3390/app11041761

Çınar, Z. M., Abdussalam Nuhu, A., Zeeshan, Q., Korhan, O., Asmael, M., & Safaei, B. (2020). Machine Learning in Predictive Maintenance towards sustainable smart manufacturing in industry 4.0. *Sustainability*, *12*(8211). Advance online publication. doi:10.3390u121982

Cleartheair.scottishairquality.scot. (2020). *How do we monitor air pollution? Clear the Air, a learning resource in Scotland*. Available at: http://cleartheair.scottishairquality.scot/about/how-do-we-monitor

Clevert, D.-A. (2020). *Lung Ultrasound in Patients with Coronavirus COVID-19 Disease*. White Paper Published Online by Siemens Medical Solutions USA, Inc.

CoalitionT. C. N. (2020). *TCN Protocol*. https://github.com/TCNCoalition/TCN

Coleman, C., Damofaran, S., & Deuel, E. (2017). *Predictive Maintenance and the Smart Factory*. https://www2.deloitte.com/content/dam/Deloitte/us/Documents/process-and-operations/us-cons-predictive-maintenance.pdf

Collins, F. S., & Varmus, H. (2015). A New Initiative on Precision Medicine. *The New England Journal of Medicine*, *372*(9), 793–795. doi:10.1056/NEJMp1500523 PMID:25635347

Columbo, J. A., Kang, R., Hoel, A. W., Kang, J., Leinweber, K. A., Tauber, K. S., Hila, R., Ramkumar, N., Sedrakyan, A., & Goodney, P. P. (2019). A comparison of reintervention rates after endovascular aneurysm repair between the Vascular Quality Initiative registry, Medicare claims, and chart review. *Journal of Vascular Surgery*, *69*(1), 74–79.e76. doi:10.1016/j.jvs.2018.03.423 PMID:29914838

Consolata, G., & Jeniffer, J. (2019) *A classification model for water quality analysis using decision tree*. Academic Press.

Convertino, V. A., Moulton, S. L., Grudic, G. Z., Rickards, C. A., Hinojosa-Laborde, C., Gerhardt, R. T., Blackbourne, L. H., & Ryan, K. L. (2011). Use of advanced machine-learning techniques for noninvasive monitoring of hemorrhage. *The Journal of Trauma and Acute Care Surgery*, *71*(1), S25–S32. doi:10.1097/TA.0b013e3182211601 PMID:21795890

Cornuz, J., Sidoti Pinto, C., Tevaearai, H., & Egger, M. (2004). Risk factors for asymptomatic abdominal aortic aneurysm: Systematic review and meta-analysis of population-based screening studies. *European Journal of Public Health*, *14*(4), 343–349. doi:10.1093/eurpub/14.4.343 PMID:15542867

Costa, D., Turco, S., Ramos, R., Silva, F., & Freire, M. (2018). Electronic monitoring system for measuring heart rate and skin temperature in small ruminants. *Engenharia Agrícola*, *38*(2), 166–172. doi:10.1590/1809-4430-eng.agric.v38n2p166-172/2018

Cota, A., Omer, A., Jaipersad, A., & Wilson, N. (2005). Elective versus ruptured abdominal aortic aneurysm repair: A 1-year cost-effectiveness analysis. *Annals of Vascular Surgery*, *19*(6), 858–861. doi:10.100710016-005-7457-5 PMID:16177868

Crable, E., & Sena, M. (2020). Exploring Sentiment Towards Contact Tracing. *Proceedings of the Conference on Information*, 1–9. http://proc.conisar.org/2020/pdf/5325.pdf

Cresswell, K., Tahir, A., Sheikh, Z., Hussain, Z., Hernández, A. D., Harrison, E., Williams, R., Sheikh, A., & Hussain, A. (2021). Understanding public perceptions of COVID-19 contact tracing apps: Artificial intelligence-enabled social media analysis. *Journal of Medical Internet Research*, *23*(5), 1–8. doi:10.2196/26618 PMID:33939622

Cronenwett, J. L., Sargent, S. K., Wall, M. H., Hawkes, M. L., Freeman, D. H., Dain, B. J., Curé, J. K., Walsh, D. B., Zwolak, R. M., McDaniel, M. D., & Schneider, J. R. (1990). Variables that affect the expansion rate and outcome of small abdominal aortic aneurysms. *Journal of Vascular Surgery*, *11*(2), 260–269. doi:10.1016/0741-5214(90)90269-G PMID:2405198

Cs229.stanford.edu. (2021). Available at: http://cs229.stanford.edu/proj2019spr/report/22.pdf

Cui, F., & Zhou, H. S. (2020). Diagnostic methods and potential portable biosensors for coronavirus disease 2019. *Biosensors & Bioelectronics*, *165*, 112349. doi:10.1016/j.bios.2020.112349 PMID:32510340

Curtis, A. (2020). *Using MODIS Satellite Imagery to Estimate Particulate Matter*. Academic Press.

Dangeti, P., Yu, A., & Chung, C. (2018). *Numerical Computing with Python*. Packt publishing.

Darmawan, T. B., & Syafei, A. D. (2019, May). Characterizing NO_2 in Indonesia Using Satellite Ozone Monitoring Instruments. *IOP Conference Series. Earth and Environmental Science*, *284*(1), 012011. doi:10.1088/1755-1315/284/1/012011

Das, H., Naik, B., & Behera, H. S. (2018). *Classification of Diabetes Mellitus Disease (DMD): A Data Mining (DM) Approach*. doi:10.1007/978-981-10-7871-2_52

Daud, M. K. (2017). Drinking Water Quality Status and Contamination in Pakistan. Academic Press.

Daud, M. K., Nafees, M., Ali, S., Rizwan, M., Bajwa, R., Shakoor, M., ... Shui, Z. J. (2017). Drinking Water Quality Status and Contamination in Pakistan. *BioMed Research International*, *2017*, 18. doi:10.1155/2017/7908183 PMID:28884130

David, G. C., Garcia, A. C., Rawls, A. W., & Chand, D. (2009). Listening to what is said - transcribing what is heard: The impact of speech recognition technology (SRT) on the practice of medical transcription (MT). *Sociology of Health & Illness*, *31*(6), 924–938. doi:10.1111/j.1467-9566.2009.01186.x PMID:19843274

Deif, M. A., Solyman, A. A. A., & Hammam, R. E. (2021). ARIMA Model Estimation Based on Genetic Algorithm for COVID-19 Mortality Rates. *International Journal of Information Technology & Decision Making*, 1–24.

Devaraj, J., Madurai Elavarasan, R., Pugazhendhi, R., Shafiullah, G. M., Ganesan, S., Jeysree, A. K., Khan, I. A., & Hossain, E. (2021). Forecasting of COVID-19 cases using deep learning models: Is it reliable and practically significant? *Results in Physics*, *21*(January), 103817. doi:10.1016/j.rinp.2021.103817 PMID:33462560

Dhayne, H., Kilany, R., Haque, R., & Taher, Y. (2021). EMR2vec: Bridging the gap between patient data and clinical trial. *Computers and Industrial Engineering*, *156*(June), 107236. doi:10.1016/j.cie.2021.107236

Dilmegani, C. (2021). *Predictive Maintenance (PdM): Why it Matters & How it Works*. https://research.aimultiple.com/predictive-maintenance/

Dixon, A., Lawrence, J., & Mitchell, J. (1984). Age-related changes in the abdominal aorta shown by computed tomography. *Clinical Radiology*, *35*(1), 33–37. doi:10.1016/S0009-9260(84)80228-7 PMID:6690178

Dong, L., Zhou, J., Niu, C., Wang, Q., Pan, Y., Sheng, S., Wang, X., Zhang, Y., Yang, J., Liu, M., Zhao, Y., Zhang, X., Zhu, T., Peng, T., Xie, J., Gao, Y., Wang, D., Zhao, Y., Dai, X., & Fang, X. (2020). Highly accurate and sensitive diagnostic detection of SARS-CoV-2 by digital PCR. MedRxiv, 2020.03.14.20036129. doi:10.1101/2020.03.14.20036129

E, P., A, D., D, B., V, T., E, D., P, M. R., T, M., M, S., A, V., & A, M. (2020). Can Lung US Help Critical Care Clinicians in the Early Diagnosis of Novel Coronavirus (COVID-19) Pneumonia? *Radiology, 295*(3). doi:10.1148/radiol.2020200847 doi:10.1148/radiol.2020200847

Eames, K. T. D., & Keeling, M. J. (2002). Modeling dynamic and network heterogeneities in the spread of sexually transmitted diseases. *Proceedings of the National Academy of Sciences of the United States of America, 99*(20), 13330–13335. doi:10.1073/pnas.202244299 PMID:12271127

Eames, K. T. D., & Keeling, M. J. (2003). Contact tracing and disease control. *Proceedings. Biological Sciences, 270*(1533), 2565–2571. doi:10.1098/rspb.2003.2554 PMID:14728778

El-Behery, H., Attia, A. F., El-Feshawy, N., & Torkey, H. (2021). Efficient machine learning model for predicting drug-target interactions with case study for Covid-19. *Computational Biology and Chemistry, 93*(October), 107536. doi:10.1016/j.compbiolchem.2021.107536 doi:10.1016/j.compbiolchem.2021.107536

El-Hajj, C., & Kyriacou, P. A. (2020). A review of machine learning techniques in photoplethysmography for the non-invasive cuff-less measurement of blood pressure. *Biomedical Signal Processing and Control, 58*, 101870. doi:10.1016/j.bspc.2020.101870

Elloumi, Z., Besacier, L., Galibert, O., Kahn, J., & Lecouteux, B. (2018). ASR Performance Prediction on Unseen Broadcast Programs Using Convolutional Neural Networks. *IEEE International Conference on Acoustics, Speech and Signal Processing (ICASSP)*, Calgary, Canada. 10.1109/ICASSP.2018.8461751

Embong, M. (2019, November 29). *What is prediabetes?* https://www.thestar.com.my/lifestyle/health/2013/11/14/what-is-prediabetes

Eren, L. (2017). Bearing fault detection by one-dimensional convolutional neural networks. *Mathematical Problems in Engineering, 2017*, 1–9. Advance online publication. doi:10.1155/2017/8617315

Eskandarian, R., Sani, Z. A., Behjati, M., Zahmatkesh, M., Haddadi, A., Kakhi, K., Roshanzamir, M., Shoeibi, A., Alizadehsani, R., Hussain, S., Khozeimeh, F., Keyvani, V., Khosravi, A., Nahavandi, S., & Islam, M. S. (2021). Identification of clinical features associated with mortality in COVID-19 patients. MedRxiv, 2021.04.19.21255715. doi:10.1101/2021.04.19.21255715

Everingham, M., Gool, V., Williams, Winn, J., & Zisserman, A. (n.d.a). *The PASCAL Visual Object Classes Challenge 2007 (VOC2007) Results*. Academic Press.

Everingham, M., Gool, V., Williams, Winn, J., & Zisserman, A. (n.d.b). *The PASCAL Visual Object Classes Challenge 2012 (VOC2012) Results*. Academic Press.

Faiyaz Ahmad, M. (2020). Machine Learning Approach for Predicting the Quality of Water. *International Journal of Advanced Science and Technology, 29*(5s), 275–282.

Felzenszwalb, P. F., Girshick, R. B., & McAllester, D. (2010, June 1). *Cascade object detection with deformable part models*. doi:10.1109/CVPR.2010.5539906

Felzenszwalb, P., McAllester, D., & Ramanan, D. (2008, June 1). *A discriminatively trained, multiscale, deformable part model*. doi:10.1109/CVPR.2008.4587597

Felzenszwalb, P. F., Girshick, R. B., McAllester, D., & Ramanan, D. (2010). Object Detection with Discriminatively Trained Part-Based Models. *IEEE Transactions on Pattern Analysis and Machine Intelligence*, *32*(9), 1627–1645. doi:10.1109/TPAMI.2009.167 PMID:20634557

Fugate, J. E., Lyons, J. L., Thakur, K. T., Smith, B. R., Hedley-Whyte, E. T., & Mateen, F. J. (2014). Infectious causes of stroke. *The Lancet. Infectious Diseases*, *14*(9), 869–880. doi:10.1016/S1473-3099(14)70755-8 PMID:24881525

Galibert, O., & Kahn, J. (2013). The first official REPERE evaluation. *First Workshop on Speech, Language and Audio in Multimedia*, Marseille, France.

Garg, L., Chukwu, E., Nasser, N., Chakraborty, C., & Garg, G. (2020). Anonymity Preserving IoT-Based COVID-19 and Other Infectious Disease Contact Tracing Model. *IEEE Access: Practical Innovations, Open Solutions*, *8*, 159402–159414. doi:10.1109/ACCESS.2020.3020513 PMID:34786286

General Complications of EVAR. (n.d.). *Thoracic Key*. https://thoracickey.com/general-complications-of-evar/

Gil-Gomez, H., Guerola-Navarro, V., Oltra-Badenes, R., & Lozano-Quilis, J. A. (2020). Customer relationship management: Digital transformation and sustainable business model innovation. *Economic Research-Ekonomska Istraživanja*, *33*(1), 2733–2750. doi:10.1080/1331677X.2019.1676283

Girshick, R., Donahue, J., Darrell, T., & Malik, J. (2014). Rich Feature Hierarchies for Accurate Object Detection and Semantic Segmentation. *2014 IEEE Conference on Computer Vision and Pattern Recognition*. 10.1109/CVPR.2014.81

Girum, T., Lentiro, K., Geremew, M., Migora, B., Shewamare, S., & Shimbre, M. (2021). Optimal strategies for COVID-19 prevention from global evidence achieved through social distancing, stay at home, travel restriction and lockdown: A systematic review. *Archives of Public Health*, *79*(1), 150. Advance online publication. doi:10.118613690-021-00663-8 PMID:34419145

Goldstein, B. A., Navar, A. M., Pencina, M. J., & Ioannidis, J. (2017). Opportunities and challenges in developing risk prediction models with electronic health records data: A systematic review. *Journal of the American Medical Informatics Association: JAMIA*, *24*(1), 198–208. doi:10.1093/jamia/ocw042 PMID:27189013

Google & Apple. (2020). *Exposure Notification Bluetooth Specification*. https://covid19-static.cdn-apple.com/applications/covid19/current/static/contact-tracing/pdf/ExposureNotification-BluetoothSpecificationv1.2.pdf?1

Graffy, P. M., Liu, J., O'Connor, S., Summers, R. M., & Pickhardt, P. J. (2019). Automated segmentation and quantification of aortic calcification at abdominal CT: Application of a deep learning-based algorithm to a longitudinal screening cohort. *Abdominal Radiology*, *44*(8), 2921–2928. doi:10.100700261-019-02014-2 PMID:30976827

Griggs, M. (1975). Measurements of Atmospheric Aerosol Optical Thickness over Water Using ERTS-1 Data. *Journal of the Air Pollution Control Association*, *25*(6), 622–626. doi:10.1080/00022470.1975.10470118 PMID:1141544

Guo, Y., He, D., & Chai, L. (2020). A Machine Vision-Based Method for Monitoring Scene-Interactive Behaviors of Dairy Calf. *Animals (Basel)*, *10*(2), 190. doi:10.3390/ani10020190 PMID:31978962

Gupta, A. K., Singh, V., Mathur, P., & Travieso-Gonzalez, C. M. (2021). Prediction of COVID-19 pandemic measuring criteria using support vector machine, prophet and linear regression models in Indian scenario. *Journal of Interdisciplinary Mathematics*, *24*(1), 89–108. doi:10.1080/09720502.2020.1833458

Gurudath, S. (2020). *Market Basket Analysis & Recommendation System Using Association Rules*. Academic Press.

Habibi, S., Ahmadi, M., & Alizadeh, S. (2015). Type 2 Diabetes Mellitus Screening and Risk Factors Using Decision Tree: Results of Data Mining. *Global Journal of Health Science*, *7*(5), 304–310. doi:10.5539/gjhs.v7n5p304 PMID:26156928

Hadjianastassiou, V. G., Franco, L., Jerez, J. M., Evangelou, I. E., Goldhill, D. R., Tekkis, P. P., & Hands, L. J. (2006). Informed prognosis after abdominal aortic aneurysm repair using predictive modeling techniques. *Journal of Vascular Surgery, 43*(3), 467–473. doi:10.1016/j.jvs.2005.11.022 PMID:16520157

Hafeez, S., Wong, M. S., Ho, H. C., Nazeer, M., Nichol, J., Abbas, S., Tang, D., Lee, K. H., & Pun, L. (2019). Comparison of machine learning algorithms for retrieval of water quality indicators in case-II waters: A case study of Hong Kong. *Remote Sensing, 11*(6), 617. doi:10.3390/rs11060617

Haghiabi, A. H., Nasrolahi, A. H., & Parsaie, A. (2018). Water quality prediction using machine learning methods. *Water Quality Research Journal, 53*(1), 3–13. doi:10.2166/wqrj.2018.025

Hahn, S., Perry, M., Morris, C. S., Wshah, S., & Bertges, D. J. (2020). Machine deep learning accurately detects endoleak after endovascular abdominal aortic aneurysm repair. *JVS: Vascular Science, 1*, 5-12.

Hamid, Jhanjhi, & Humayun. (2020). Digital Governance for Developing Countries Opportunities. *Employing Recent Technologies for Improved Digital Governance*, 36-58.

Han, W., Chan, C., Choy, C., & Pun, C. (2006). *An efficient MFCC extraction method in speech recognition. IEEE International Symposium on Circuits and Systems*, Kos, Greece.

Harrar, D. B., Salussolia, C. L., Kapur, K., Danehy, A., & Kleinman, M. E. (2020). A stroke alert protocol decreases the time to diagnosis of brain attack symptoms in a pediatric emergency department. The Journal of Pediatrics, 216, 136-141. doi:10.1016/j.jpeds.2019.09.027

Hasan, A. M., Al-Jawad, M. M., Jalab, H. A., Shaiba, H., Ibrahim, R. W., & Al-Shamasneh, A. R. (2020). Classification of Covid-19 Coronavirus, Pneumonia and Healthy Lungs in CT Scans Using Q-Deformed Entropy and Deep Learning Features. *Entropy 2020, 22*(5), 517. doi:10.3390/e22050517

Hasan, M. N., Haider, N., Stigler, F. L., Khan, R. A., McCoy, D., Zumla, A., Kock, R. A., & Uddin, M. J. (2021). The global case-fatality rate of COVID-19 has been declining since may 2020. *The American Journal of Tropical Medicine and Hygiene, 104*(6), 2176–2184. doi:10.4269/ajtmh.20-1496 PMID:33882025

He, K., Zhang, X., Ren, S., & Sun, J. (2015). *Deep residual learning for image recognition.* arXiv preprint arXiv:1512.03385.

Health. (2021). *Ada.* https://ada.com/

Hellwegera, F., Schlossera, P., Lall, U., & Weissel, J. (2004). Use of satellite imagery for water quality studies in New York Harbor. *Coastal and Shelf Science, 61*(3), 437–448. doi:10.1016/j.ecss.2004.06.019

Hernesniemi, J. A., Vänni, V., & Hakala, T. (2015). The prevalence of abdominal aortic aneurysm is consistently high among patients with coronary artery disease. *Journal of Vascular Surgery, 62*(1), 232-240.

He, T., & Printz, M. (2020). *A 2-stage Classifier for Contact Detection with BluetoothLE And INS Signals.* ArXiv.

He, Y., Gu, C., Chen, Z., & Han, X. (2017). Integrated Predictive Maintenance strategy for manufacturing systems by combining quality control and mission reliability analysis. *International Journal of Production Research, 55*(19), 5841–5862. doi:10.1080/00207543.2017.1346843

Hirata, K., Nakaura, T., Nakagawa, M., Kidoh, M., Oda, S., Utsunomiya, D., & Yamashita, Y. (2020). Machine Learning to Predict the Rapid Growth of Small Abdominal Aortic Aneurysm. *Journal of Computer Assisted Tomography, 44*(1), 37–42. doi:10.1097/RCT.0000000000000958 PMID:31939880

Hobbs, S., Claridge, M., Quick, C., Day, N., Bradbury, A., & Wilmink, A. (2003). LDL cholesterol is associated with small abdominal aortic aneurysms. *European Journal of Vascular and Endovascular Surgery*, 26(6), 618–622. doi:10.1016/S1078-5884(03)00412-X PMID:14603421

Hochreiter, S., & Schmidhuber, J. (1997). Long Short-Term Memory. *Neural Computation*, 9(8), 1735–1780. doi:10.1162/neco.1997.9.8.1735 PMID:9377276

Hong, H. A., & Sheikh, U. U. (2016, March 4-6). Automatic detection, segmentation and classification of abdominal aortic aneurysm using deep learning. *2016 IEEE 12th International Colloquium on Signal Processing & Its Applications (CSPA)*.

Hong, K.-S., Bang, O. Y., Kang, D.-W., Yu, K.-H., Bae, H.-J., Lee, J. S., Heo, J. H., Kwon, S. U., Oh, C. W., Lee, B.-C., Kim, J. S., & Yoon, B.-W. (2013). Stroke statistics in Korea: part I. Epidemiology and risk factors: a report from the korean stroke society and clinical research center for stroke. *Journal of Stroke*, 15(1), 2. doi:10.5853/jos.2013.15.1.2 PMID:24324935

HonorHealth. (2020). *Signs, Symptoms and Diagnosis of Diabetes*. https://www.honorhealth.com/medical-services/diabetes/signs-symptoms-diagnosis%0A

Hooshmand, S. A., Zarei Ghobadi, M., Hooshmand, S. E., Azimzadeh Jamalkandi, S., Alavi, S. M., & Masoudi-Nejad, A. (2021). A multimodal deep learning-based drug repurposing approach for treatment of COVID-19. *Molecular Diversity*, 25(3), 1717–1730. doi:10.100711030-020-10144-9 PMID:32997257

Hope, T. M., Seghier, M. L., Leff, A. P., & Price, C. J. (2013). Predicting outcome and recovery after stroke with lesions extracted from MRI images. *NeuroImage. Clinical*, 2, 424–433. doi:10.1016/j.nicl.2013.03.005 PMID:24179796

Ho, S. L., & Xie, M. (1998). The use of ARIMA models for reliability forecasting and analysis. *Computers & Industrial Engineering (American Institute of Industrial Engineers)*, 35(1–2), 213–216.

Hosanee, M., Chan, G., Welykholowa, K., Cooper, R., Kyriacou, P. A., Zheng, D., Allen, J., Abbott, D., Menon, C., Lovell, N. H., Howard, N., Chan, W.-S., Lim, K., Fletcher, R., Ward, R., & Elgendi, M. (2020). Cuffless single-site photoplethysmography for blood pressure monitoring. *Journal of Clinical Medicine*, 9(3), 723. doi:10.3390/jcm9030723 PMID:32155976

Houchens, J., Gold, J., Maynard, N., Krangle, M., Kikkisetti, S., & Used, A. D. (2020). MITRE TC4TL Challenge System Description. *ArXiv*, 20, 2–4.

Howell, A. (2011). *Snail-borne diseases in bovids at high and low altitude in Eastern Uganda: Integrated parasitological and malacological mapping*. Liverpool School of Tropical Medicine.

Huang, G., Sun, Y., Liu, Z., Sedra, D., & Weinberger, K. (2016). Deep Networks with Stochastic Depth. doi:10.1007/978-3-319-46493-0_39

Huang, G., Liu, Z., Van Der Maaten, L., & Weinberger, K. Q. (2017). Densely Connected Convolutional Networks. *2017 IEEE Conference on Computer Vision and Pattern Recognition (CVPR)*. 10.1109/CVPR.2017.243

Huawei. (2018, August 31). *Kirin 980, the World's First 7nm Process Mobile AI*. Retrieved from https://consumer.huawei.com/en/campaign/kirin980/

Humayun, M., & Alsayat, A. (2022). Prediction Model for Coronavirus Pandemic Using Deep Learning. *Computer Systems Science and Engineering*, 40(3), 947–961. doi:10.32604/csse.2022.019288

Hung, C.-Y., Lin, C.-H., Lan, T.-H., Peng, G.-S., & Lee, C.-C. (2019). Development of an intelligent decision support system for ischemic stroke risk assessment in a population-based electronic health record database. *PLoS One*, *14*(3), e0213007. doi:10.1371/journal.pone.0213007 PMID:30865675

Inan, O. T., Baran Pouyan, M., Javaid, A. Q., Dowling, S., Etemadi, M., Dorier, A., Heller, J. A., Bicen, A. O., Roy, S., De Marco, T., & Klein, L. (2018). Novel wearable seismocardiography and machine learning algorithms can assess clinical status of heart failure patients. *Circulation: Heart Failure*, *11*(1), e004313. doi:10.1161/CIRCHEARTFAIL-URE.117.004313 PMID:29330154

Inc. (2020, September 16). *Apple Keynote*. Retrieved from www.apple.com/newsroom/2020/09/apple-unveils-all-new-ipad-air-with-a14-bionic-apples-most-advanced-chip/

Isakadze, N., & Martin, S. S. (2020). How useful is the smartwatch ECG? *Trends in Cardiovascular Medicine*, *30*(7), 442–448. doi:10.1016/j.tcm.2019.10.010 PMID:31706789

Istaiteh, O., Owais, T., Al-Madi, N., & Abu-Soud, S. (2020). Machine Learning Approaches for COVID-19 Forecasting. *2020 International Conference on Intelligent Data Science Technologies and Applications, IDSTA 2020*, 50–57. doi:10.1109/IDSTA50958.2020.926410110.1109/IDSTA50958.2020.9264101

Iyer, A., S, J., & Sumbaly, R. (2015). Diagnosis of Diabetes Using Classification Mining Techniques. *International Journal of Data Mining & Knowledge Management Process*, *5*(1), 1–14. doi:10.5121/ijdkp.2015.5101

J, W., S, L., M, P., TY, K., MG, P., BY, C., D, K., H, C., VN, K., & CJ, L. (2020). Development of a Laboratory-safe and Low-cost Detection Protocol for SARS-CoV-2 of the Coronavirus Disease 2019 (COVID-19). *Experimental Neurobiology*, *29*(2), 107–119. doi:10.5607/en20009

Jahmunah, V., Sudarshan, V. K., Oh, S. L., Gururajan, R., Gururajan, R., Zhou, X., Tao, X., Faust, O., Ciaccio, E. J., Ng, K. H., & Acharya, U. R. (2021). Future IoT tools for COVID-19 contact tracing and prediction: A review of the state-of-the-science. *International Journal of Imaging Systems and Technology*, *31*(2), 455–471. doi:10.1002/ima.22552 PMID:33821093

Jaiswal, A. K., Tiwari, P., Kumar, S., Gupta, D., Khanna, A., & Rodrigues, J. J. P. C. (2019). Identifying pneumonia in chest X-rays: A deep learning approach. *Measurement*, *145*, 511–518. doi:10.1016/j.measurement.2019.05.076

Janssens, O., Slavkovikj, V., Vervisch, B., Stockman, K., Loccufier, M., Verstockt, S., Van de Walle, R., & Van Hoecke, S. (2016). Convolutional neural network based fault detection for rotating machinery. *Journal of Sound and Vibration*, *377*, 331-345. doi:10.1016/j.jsv.2016.05.027

Jardim-Goncalves, R., Romero, D., & Grilo, A. (2017). Factories of the future: Challenges and leading innovations in intelligent manufacturing. *International Journal of Computer Integrated Manufacturing*, *30*(1), 4–14. doi:10.1080/09 51192X.2016.1258120

Jelinek, F. (1997). *Statistical Methods for Speech Recognition*. MIT Press.

Jerman, A., & Dominici, G. (2018). Smart factories from business, management and accounting perspective: A systemic analysis of current research. *Management*, *13*(4), 355–365. doi:10.26493/1854-4231.13.355-365

Jiang, H., Li, Y., Zhang, H., Wang, W., Men, D., Yang, X., Qi, H., Zhou, J., & Tao, S. (2020). *Global profiling of SARS-CoV-2 specific IgG/IgM responses of convalescents using a proteome microarray*. MedRxiv. doi:10.1101/2020.03.20.20039495

Jiang, Z., Do, H. N., Choi, J., Lee, W., & Baek, S. (2020). A deep learning approach to predict abdominal aortic aneurysm expansion using longitudinal data. *Frontiers in Physics*, *7*, 235.

Jo, H., Kim, J., Huang, T.-C., & Ni, Y.-L. (2020). condLSTM-Q: A novel deep learning model for predicting Covid-19 mortality in fine geographical Scale. https://arxiv.org/abs/2011.11507

Johnson, C. O., Nguyen, M., Roth, G. A., Nichols, E., Alam, T., Abate, D., Abd-Allah, F., Abdelalim, A., Abraha, H. N., Abu-Rmeileh, N. M. E., Adebayo, O. M., Adeoye, A. M., Agarwal, G., Agrawal, S., Aichour, A. N., Aichour, I., Aichour, M. T. E., Alahdab, F., Ali, R., ... Murray, C. J. L. (2019). Global, regional, and national burden of stroke, 1990–2016: A systematic analysis for the Global Burden of Disease Study 2016. *Lancet Neurology*, *18*(5), 439–458. doi:10.1016/S1474-4422(19)30034-1 PMID:30871944

Joo, G., Song, Y., Im, H., & Park, J. (2020). H. Im and J. Park, "Clinical implication of machine learning in predicting the occurrence of cardiovascular disease using big data (Nationwide Cohort Data in Korea). *IEEE Access: Practical Innovations, Open Solutions*, *8*, 157643–157653. doi:10.1109/ACCESS.2020.3015757

Jordanski, M., Radovic, M., Milosevic, Z., Filipovic, N., & Obradovic, Z. (2018). Machine Learning Approach for Predicting Wall Shear Distribution for Abdominal Aortic Aneurysm and Carotid Bifurcation Models. *IEEE Journal of Biomedical and Health Informatics*, *22*(2), 537–544. doi:10.1109/JBHI.2016.2639818 PMID:28113333

Joshi, R., Cardona, M., Iyengar, S., Sukumar, A., Raju, C. R., Raju, K. R., Raju, K., Reddy, K. S., Lopez, A., & Neal, B. (2006). Chronic diseases now a leading cause of death in rural India—Mortality data from the Andhra Pradesh Rural Health Initiative. *International Journal of Epidemiology*, *35*(6), 1522–1529. doi:10.1093/ije/dyl168 PMID:16997852

Jouppi, N., Young, C., Patil, N., Patterson, D., Agrawal, G., Bajwa, R., . . . Yoon, D. (2017). *In-Datacenter Performance Analysis of a Tensor Processing Unit TM*. Retrieved from https://arxiv.org/ftp/arxiv/papers/1704/1704.04760.pdf

Juang, B. H., & Rabiner, L. R. (1991). Hidden Markov Models for Speech Recognition. *Technometrics*, *33*(3), 251–272. doi:10.1080/00401706.1991.10484833

Kadioglu, O., Saeed, M., Greten, H. J., & Efferth, T. (2021). Identification of novel compounds against three targets of SARS CoV-2 coronavirus by combined virtual screening and supervised machine learning. *Computers in Biology and Medicine*, *133*(March), 104359. doi:10.1016/j.compbiomed.2021.104359 PMID:33845270

Kamalov, F., Cherukuri, A., Sulieman, H., Thabtah, F., & Hossain, A. (2021). *Machine learning applications for CO-VID-19: A state-of-the-art review*. Academic Press.

Kanaroglou, P. S., Soulakellis, N. A., & Sifakis, N. I. (2002). Improvement of satellite derived pollution maps with the use of a geostatistical interpolation method. *Journal of Geographical Systems*, *4*(2), 193–208. doi:10.1007101090100080

Kang, Y.-N., Shen, H.-N., Lin, C.-Y., Elwyn, G., Huang, S.-C., Wu, T.-F., & Hou, W.-H. (2019). Does a Mobile app improve patients' knowledge of stroke risk factors and health-related quality of life in patients with stroke? A randomized controlled trial. *BMC Medical Informatics and Decision Making*, *19*(1), 1–9. doi:10.118612911-019-1000-z PMID:31864348

Kanne, J. P., Little, B. P., Chung, J. H., Elicker, B. M., & Ketai, L. H. (2020). *Essentials for Radiologists on COVID-19: An Update—Radiology Scientific Expert Panel*. doi:10.1148/radiol.2020200527

Karim, M., & Rahman, R. M. (2013). Decision Tree and Naïve Bayes Algorithm for Classification and Generation of Actionable Knowledge for Direct Marketing. *Journal of Software Engineering and Applications*, *6*(4), 196–206. doi:10.4236/jsea.2013.64025

Karpathy, A. (2019). *PyTorch at Tesla - Andrej Karpathy, Tesla* [YouTube Video]. Retrieved from https://www.youtube.com/watch?v=oBklltKXtDE&t=553s

Karpathy, A. (2020). *Tesla Andrej Karpathy in CVPR 2020: Scalability in Autonomous Driving Workshop*. Retrieved from https://www.youtube.com/watch?v=X2CpuabzRaY

Karthikesalingam, A., Attallah, O., Ma, X., Bahia, S. S., Thompson, L., Vidal-Diez, A., Choke, E. C., Bown, M. J., Sayers, R. D., Thompson, M. M., & Holt, P. J. (2015). An Artificial Neural Network Stratifies the Risks of Reintervention and Mortality after Endovascular Aneurysm Repair; a Retrospective Observational study. *PLoS One*, *10*(7), e0129024. doi:10.1371/journal.pone.0129024 PMID:26176943

Karthikeyan, A., Garg, A., Vinod, P. K., & Priyakumar, U. D. (2021). Machine Learning Based Clinical Decision Support System for Early COVID-19 Mortality Prediction. *Frontiers in Public Health*, *9*, 626697. Advance online publication. doi:10.3389/fpubh.2021.626697 PMID:34055710

Kaufman, Y. J., Fraser, R. S., & Ferrare, R. A. (1990). Satellite measurements of large-scale air pollution: Methods. *Journal of Geophysical Research, D, Atmospheres*, *95*(D7), 9895–9909. doi:10.1029/JD095iD07p09895

Kavakiotis, I., Tsave, O., Salifoglou, A., Maglaveras, N., Vlahavas, I., & Chouvarda, I. (2017). Machine Learning and Data Mining Methods in Diabetes Research. In *Computational and Structural Biotechnology Journal* (Vol. 15, pp. 104–116). Elsevier B.V. doi:10.1016/j.csbj.2016.12.005

Kazerouni, F., Bayani, A., Asadi, F., Saeidi, L., Parvizi, N., & Mansoori, Z. (2020). Type2 diabetes mellitus prediction using data mining algorithms based on the long-noncoding RNAs expression: A comparison of four data mining approaches. *BMC Bioinformatics*, *21*(1), 1–13. doi:10.118612859-020-03719-8 PMID:32854616

Keisler, B., & Carter, C. (2015). Abdominal aortic aneurysm. *American Family Physician*, *91*(8), 538–543. PMID:25884861

Kernan, W. N., Ovbiagele, B., Black, H. R., Bravata, D. M., Chimowitz, M. I., Ezekowitz, M. D., Fang, M. C., Fisher, M., Furie, K. L., Heck, D. V., Johnston, S. C. C., Kasner, S. E., Kittner, S. J., Mitchell, P. H., Rich, M. W., Richardson, D. J., Schwamm, L. H., & Wilson, J. A. (2014). Guidelines for the prevention of stroke in patients with stroke and transient ischemic attack: A guideline for healthcare professionals from the American Heart Association/American Stroke Association. *Stroke*, *45*(7), 2160–2236. doi:10.1161/STR.0000000000000024 PMID:24788967

Kerut, E. K., To, F., Summers, K. L., Sheahan, C., & Sheahan, M. (2019). Statistical and machine learning methodology for abdominal aortic aneurysm prediction from ultrasound screenings. *Echocardiography (Mount Kisco, N.Y.)*, *36*(11), 1989–1996. doi:10.1111/echo.14519 PMID:31682022

Ke, Y. Y., Peng, T. T., Yeh, T. K., Huang, W. Z., Chang, S. E., Wu, S. H., Hung, H. C., Hsu, T. A., Lee, S. J., Song, J. S., Lin, W. H., Chiang, T. J., Lin, J. H., Sytwu, H. K., & Chen, C. T. (2020). Artificial intelligence approach fighting COVID-19 with repurposing drugs. *Biomedical Journal*, *43*(4), 355–362. doi:10.1016/j.bj.2020.05.001 PMID:32426387

Khwaja, H., Fatmi, Z., Malashock, D., Aminov, Z., Kazi, A., Siddique, A., Qureshi, J., & Carpenter, D. (2012). Effect of air pollution on daily morbidity in Karachi, Pakistan. *Journal of Local and Global Health Science*, *2012*(1). Available at: https://www.qscience.com/content/journals/10.5339/jlghs.2012.3

Kiss, I. Z., Green, D. M., & Kao, R. R. (2005). Disease contact tracing in random and clustered networks. *Proceedings. Biological Sciences*, *272*(1570), 1407–1414. doi:10.1098/rspb.2005.3092 PMID:16006334

Kobeissi, E., Hibino, M., Pan, H., & Aune, D. (2019). Blood pressure, hypertension and the risk of abdominal aortic aneurysms: A systematic review and meta-analysis of cohort studies. *European Journal of Epidemiology*, *34*(6), 547–555. doi:10.100710654-019-00510-9 PMID:30903463

Koike, F., & Morimoto, N. (2018). Supervised forecasting of the range expansion of novel non-indigenous organisms: Alien pest organisms and the 2009 H1N1 flu pandemic. *Global Ecology and Biogeography*, *27*(8), 991–1000. doi:10.1111/geb.12754

Krizhevsky, A., Sutskever, I., & Hinton, G. E. (2012). *ImageNet Classification with Deep Convolutional Neural Networks*. Retrieved from https://papers.nips.cc/paper/4824-imagenet-classification-with-deep-convolutional-neural-networks.pdf

Kubota, Y., Folsom, A. R., Ballantyne, C. M., & Tang, W. (2018). Lipoprotein (a) and abdominal aortic aneurysm risk: The Atherosclerosis Risk in Communities study. *Atherosclerosis, 268*, 63–67. doi:10.1016/j.atherosclerosis.2017.10.017 PMID:29182987

Kumar Paul, P., & Ghose, M. K. (2012). Cloud Computing: Possibilities, challenges and opportunities with special reference to its emerging need in the academic and working area of information science. *Procedia Engineering, 38*, 2222-2227. doi:10.1016/j.proeng.2012.06.267

Kumar, A. (2020). *Bagging Classifier Python Code Example. Data Analytics.* https://vitalflux.com/bagging-classifier-python-code-example/

Kumar, D. (2013). A Case Study of Customer Relationship Management using Data Mining Techniques. *International Journal of Technological Exploration and Learning, 2*, 275–280.

Lajmanovich, A., & Yorke, J. A. (1976). A deterministic model for gonorrhea in a nonhomogeneous population. *Mathematical Biosciences, 28*(3–4), 221–236. doi:10.1016/0025-5564(76)90125-5

Lalwani, S., Sahni, G., Mewara, B., & Kumar, R. (2020). Predicting optimal lockdown period with parametric approach using three-phase maturation SIRD model for COVID-19 pandemic. *Chaos, Solitons & Fractals, 138*, 109939.

Lam, C. H., Ng, P. C., & She, J. (2018). Improved Distance Estimation with BLE Beacon Using Kalman Filter and SVM. *2018 IEEE International Conference on Communications (ICC)*, 1–6. 10.1109/ICC.2018.8423010

Laponogov, I., Gonzalez, G., Shepherd, M., Qureshi, A., Veselkov, D., Charkoftaki, G., Vasiliou, V., Youssef, J., Mirnezami, R., Bronstein, M., & Veselkov, K. (2021). Network machine learning maps phytochemically rich "Hyperfoods" to fight COVID-19. *Human Genomics, 15*(1), 1–11. doi:10.118640246-020-00297-x PMID:33386081

Larsson, E., Granath, F., Swedenborg, J., & Hultgren, R. (2009). A population-based case-control study of the familial risk of abdominal aortic aneurysm. *Journal of Vascular Surgery, 49*(1), 47–51. doi:10.1016/j.jvs.2008.08.012 PMID:19028058

Lattanzi, S., & Silvestrini, M. (2016). Blood pressure in acute intra-cerebral hemorrhage. *Annals of Translational Medicine, 4*(16), 320. doi:10.21037/atm.2016.08.04 PMID:27668240

Lauren, P., & Watta, P. (2019). *A Conversational User Interface for Stock Analysis. IEEE International Conference on Big Data*, Los Angeles, CA. 10.1109/BigData47090.2019.9005635

Lee, W. J., Wu, H., Yun, H., Kim, H., Jun, M. B. G., & Sutherland, J. W. (2019). Predictive maintenance of machine tool systems using artificial intelligence techniques applied to machine condition data, *Procedia CIRP, 80*, 506-511. doi:10.1016/j.procir.2018.12.019

Lee, R., Jarchi, D., Perera, R., Jones, A., Cassimjee, I., Handa, A., Clifton, D. A., Bellamkonda, K., Woodgate, F., Killough, N., Maistry, N., Chandrashekar, A., Darby, C. R., Halliday, A., Hands, L. J., Lintott, P., Magee, T. R., Northeast, A., Perkins, J., & Sideso, E. (2018). Applied Machine Learning for the Prediction of Growth of Abdominal Aortic Aneurysm in Humans. *EJVES Short Reports, 39*, 24–28. doi:10.1016/j.ejvssr.2018.03.004 PMID:29988820

Lema, P. C., Kim, J. H., & St James, E. (2017). Overview of common errors and pitfalls to avoid in the acquisition and interpretation of ultrasound imaging of the abdominal aorta. *Journal of Vascular Diagnostics and Interventions, 5*, 41–46. doi:10.2147/JVD.S124327

Li, F.-F., Johnson, J., & Karpathy, A. (2016, March). *CS231n Convolutional Neural Networks for Visual Recognition.* Retrieved from CS231n: Convolutional Neural Networks for Visual Recognition website: http://cs231n.stanford.edu/2016/

Li, L., Qin, L., Xu, Z., Yin, Y., Wang, X., Kong, B., Bai, J., Lu, Y., Fang, Z., Song, Q., Cao, K., Liu, D., Wang, G., Xu, Q., Fang, X., Zhang, S., Xia, J., & Xia, J. (2020). *Using Artificial Intelligence to Detect COVID-19 and Community-acquired Pneumonia Based on Pulmonary CT: Evaluation of the Diagnostic Accuracy*. doi:10.1148/radiol.2020200905 doi:10.1148/radiol.2020200905

Li, X., Ge, P., Zhu, J., Li, H., Graham, J., Singer, A., Richman, P. S., & Duong, T. Q. (2020). Deep learning prediction of likelihood of ICU admission and mortality in COVID-19 patients using clinical variables. *PeerJ, 8*(December), 1–19. doi:10.7717/peerj.10337 doi:10.7717/peerj.10337

Liakos, K., Moustakidis, S. P., Tsiotra, G., Bartzanas, T., Bochtis, D., & Parisses, C. (2017). Machine Learning Based Computational Analysis Method for Cattle Lameness Prediction. *HAICTA, 128*, 139.

Liao, Y., Deschamps, F., Freitas Rocha Loures, E., & Pierin Ramos, L. F. (2017). Past, present and future of Industry 4.0 - a systematic literature review and research agenda proposal. *International Journal of Production Research, 55*(12), 3609–3629. doi:10.1080/00207543.2017.1308576

Li, F., Li, Y., & Wang, C.LI. (2009). Uncertain data decision tree classification algorithm. *Jisuanji Yingyong, 29*(11), 3092–3095. doi:10.3724/SP.J.1087.2009.03092

Lin, T.-Y., Goyal, P., Girshick, R., He, K., & Dollár, P. (2018). *Focal Loss for Dense Object Detection*. Retrieved from https://arxiv.org/abs/1708.02002

Lin, T.-Y., Maire, M., Belongie, S., Hays, J., Perona, P., Ramanan, D., ... Zitnick, C. L. (2014). Microsoft COCO: Common Objects in Context. *Computer Vision – ECCV 2014*, 740–755. doi:10.1007/978-3-319-10602-1_48

Lin, R., Ye, Z., Wang, H., & Wu, B. (2018). Chronic diseases and health monitoring big data: A survey. *IEEE Reviews in Biomedical Engineering, 11*, 275–288. doi:10.1109/RBME.2018.2829704 PMID:29993699

Lin, S.-S., Lan, C.-W., Hsu, H.-Y., & Chen, S.-T. (2018). Data analytics of a wearable device for heat stroke detection. *Sensors (Basel), 18*(12), 4347. doi:10.339018124347 PMID:30544887

Lin, T.-Y., Dollar, P., Girshick, R., He, K., Hariharan, B., & Belongie, S. (2017). Feature Pyramid Networks for Object Detection. *2017 IEEE Conference on Computer Vision and Pattern Recognition (CVPR)*. 10.1109/CVPR.2017.106

Lip, G. Y., Genaidy, A., Tran, G., Marroquin, P., & Estes, C. (2021). Improving Stroke Risk Prediction in the General Population: A Comparative Assessment of Common Clinical Rules, a New Multimorbid Index, and Machine-Learning-Based Algorithms. *Thrombosis and Haemostasis*. PMID:33765685

Liu, W., Anguelov, D., Erhan, D., Szegedy, C., Reed, S., Fu, C.-Y., & Berg, A. C. (2016). SSD: Single Shot MultiBox Detector. *Computer Vision – ECCV 2016*, 21–37. doi:10.1007/978-3-319-46448-0_2

Liu, G., Carter, B., Bricken, T., Jain, S., Viard, M., Carrington, M., & Gifford, D. K. (2020). Computationally Optimized SARS-CoV-2 MHC Class I and II Vaccine Formulations Predicted to Target Human Haplotype Distributions. *Cell Systems, 11*(2), 131–144.e6. doi:10.1016/j.cels.2020.06.009 PMID:32721383

Liu, P., Wang, J., Sangaiah, A. K., Xie, Y., & Yin, X. (2019). Analysis and Prediction of Water Quality Using LSTM Deep Neural Networks in IoT Environment. *Sustainability, 11*(7), 2058. doi:10.3390u11072058

Liu, R.-Q., Lee, Y.-C., & Mu, H. (2018). Customer Classification and Market Basket Analysis Using K-Means Clustering and Association Rules. *Evidence from Distribution Big Data of Korean Retailing Company., 19*, 59–76. doi:10.15813/kmr.2018.19.4.004

Liu, X., Li, B., Jiang, A., Qi, S., Xiang, C., & Xu, N. (2015, June). A bicycle-borne sensor for monitoring air pollution near roadways. In *2015 IEEE International Conference on Consumer Electronics-Taiwan* (pp. 166-167). IEEE. 10.1109/ICCE-TW.2015.7216835

Liu, Y., Yin, B., & Cong, Y. (2020). The Probability of Ischaemic Stroke Prediction with a Multi-Neural-Network Model. *Sensors (Basel)*, *20*(17), 4995. doi:10.339020174995 PMID:32899242

Li, W., Luo, S., Luo, J., Liu, Y., Ning, B., Huang, W., Xue, L., & Chen, J. (2017). Predictors associated with increased prevalence of abdominal aortic aneurysm in Chinese patients with atherosclerotic risk factors. *European Journal of Vascular and Endovascular Surgery*, *54*(1), 43–49. doi:10.1016/j.ejvs.2017.04.004 PMID:28527818

Lohar, P., Xie, G., Bendechache, M., Brennan, R., Celeste, E., Trestian, R., & Tal, I. (2021). Irish Attitudes Toward COVID Tracker App & Privacy: Sentiment Analysis on Twitter and Survey Data. *ACM International Conference Proceeding Series.* doi:10.1145/3465481.346919310.1145/3465481.3469193

López-Linares, K., García, I., García-Familiar, A., Macía, I., & Ballester, M. A. G. (2019). *3D convolutional neural network for abdominal aortic aneurysm segmentation.* arXiv preprint arXiv:1903.00879.

López-Linares, K., Aranjuelo, N., Kabongo, L., Maclair, G., Lete, N., Ceresa, M., García-Familiar, A., Macía, I., & González Ballester, M. A. (2018). Fully automatic detection and segmentation of abdominal aortic thrombus in post-operative CTA images using Deep Convolutional Neural Networks. *Medical Image Analysis*, *46*, 202–214. doi:10.1016/j.media.2018.03.010 PMID:29609054

Loshin, D., & Reifer, A. (2013). Customer Data Analytics. In D. Loshin & A. Reifer (Eds.), *Using Information to Develop a Culture of Customer Centricity* (pp. 68–78). Morgan Kaufmann. doi:10.1016/B978-0-12-410543-0.00009-3

Lu, J.-T., Brooks, R., Hahn, S., Chen, J., Buch, V., Kotecha, G., Andriole, K. P., Ghoshhajra, B., Pinto, J., & Vozila, P. (2019). DeepAAA: clinically applicable and generalizable detection of abdominal aortic aneurysm using deep learning. *International Conference on Medical Image Computing and Computer-Assisted Intervention.*

Ma, N., Zhang, X., Zheng, H.-T., & Sun, J. (2018). ShuffleNet V2: Practical Guidelines for Efficient CNN Architecture Design. *Computer Vision – ECCV 2018*, 122–138. doi:10.1007/978-3-030-01264-9_8

Mackey, A., Spachos, P., Song, L., & Plataniotis, K. N. (2020). Improving BLE Beacon Proximity Estimation Accuracy Through Bayesian Filtering. *IEEE Internet of Things Journal*, *7*(4), 3160–3169. doi:10.1109/JIOT.2020.2965583

Madaric, J., Vulev, I., Bartunek, J., Mistrik, A., Verhamme, K., De Bruyne, B., & Riecansky, I. (2005). Frequency of abdominal aortic aneurysm in patients> 60 years of age with coronary artery disease. *The American Journal of Cardiology*, *96*(9), 1214–1216. doi:10.1007/978-3-030-32245-8_80

Madoery, P. G., Detke, R., Blanco, L., Comerci, S., Fraire, J., Gonzalez Montoro, A., Bellassai, J. C., Britos, G., Ojeda, S., & Finochietto, J. M. (2021). Feature selection for proximity estimation in COVID-19 contact tracing apps based on Bluetooth Low Energy (BLE). *Pervasive and Mobile Computing*, *77*, 101474. doi:10.1016/j.pmcj.2021.101474 PMID:34602920

Mahboob Alam, T., Iqbal, M. A., Ali, Y., Wahab, A., Ijaz, S., Imtiaz Baig, T., Hussain, A., Malik, M. A., Raza, M. M., Ibrar, S., & Abbas, Z. (2019). A model for early prediction of diabetes. *Informatics in Medicine Unlocked*, *16*, 100204. doi:10.1016/j.imu.2019.100204

Maier, B. F., & Brockmann, D. (2020). Effective containment explains subexponential growth in recent confirmed COVID-19 cases in China. *Science*, *368*(6492), 742–746. doi:10.1126cience.abb4557 PMID:32269067

Majeed, A. (2017, Jan. 20). *Multi-sensor board.* https://www.electronicproducts.com/multi-sensor-board-speeds-up-wearable-iot-designs/#

Malthus, T. R. (1872). *An Essay on the Principle of Population.* Academic Press.

Mamatha Bai, B. G., Nalini, B. M., & Majumdar, J. (2019). Analysis and Detection of Diabetes Using Data Mining Techniques: A Big Data Application in Health Care. In Advances in Intelligent Systems and Computing (Vol. 882). Springer Singapore. doi:10.1007/978-981-13-5953-8_37

Manjusree, M., & Sateesh Kumar, K. A. (2019). Diabetes prediction using data mining classification techniques. *International Journal of Recent Technology and Engineering, 8*(3), 5901–5905. doi:10.35940/ijrte.C4735.098319

Mannini, A., & Sabatini, A. M. (2010). Machine learning methods for classifying human physical activity from on-body accelerometers. *Sensors (Basel), 10*(2), 1154–1175. doi:10.3390100201154 PMID:22205862

Manych, M. (2020). *X-ray imaging for COVID-19 patients.* Academic Press.

Maraghi, M., Adibi, M. A., & Mehdizadeh, E. (2020). Using RFM Model and Market Basket Analysis for Segmenting Customers and Assigning Marketing Strategies to Resulted Segments. *Journal of Applied Intelligent Systems and Information Sciences, 1*(1), 35–43. doi:10.22034/jaisis.2020.102488

McGehee, D. V., Mazzae, E. N., & Baldwin, G. H. S. (2000). Driver Reaction Time in Crash Avoidance Research: Validation of a Driving Simulator Study on a Test Track. *Proceedings of the Human Factors and Ergonomics Society Annual Meeting, 44*(20). 10.1177/154193120004402026

Menard, S. (2002). *Applied logistic regression analysis* (Vol. 106). Sage. doi:10.4135/9781412983433

Meng, Y., Wong, M. S., Xing, H., Kwan, M. P., & Zhu, R. (2021). Assessing the country-level excess all-cause mortality and the impacts of air pollution and human activity during the covid-19 epidemic. *International Journal of Environmental Research and Public Health, 18*(13), 1–16. doi:10.3390/ijerph18136883 PMID:34206915

Menzel, C. M., & Reiners, T. (2014). Customer Relationship Management System a Case Study on Small-Medium-Sized Companies in North Germany. In J. Devos, H. van Landeghem, & D. Deschoolmeester (Eds.), *Information Systems for Small and Medium-sized Enterprises: State of Art of IS Research in SMEs* (pp. 169–197). Springer Berlin Heidelberg. doi:10.1007/978-3-642-38244-4_9

Mikolov, T., Karafiat, M., Burget, L., Cernocky, J., & Khudanpur, S. (2010). *Recurrent Neural Network Based Language Model. 11th Annual Conference of the International Speech Communication Association,* Makuhari, Japan.

Miller, J. C. (2012). A note on the derivation of epidemic final sizes. *Bulletin of Mathematical Biology, 74*(9), 2125–2141. doi:10.100711538-012-9749-6 PMID:22829179

Miller, J. C. (2017). Mathematical models of SIR disease spread with combined non-sexual and sexual transmission routes. *Infectious Disease Modelling, 2*(1), 35–55. doi:10.1016/j.idm.2016.12.003 PMID:29928728

Miner, J. R. (1933). Pierre-François Verhulst, the discoverer of the logistic curve. *Human Biology, 5*(4), 673.

Miotto, R., Wang, F., Wang, S., Jiang, X., & Dudley, J. T. (2018). Deep learning for healthcare: Review, opportunities and challenges. *Briefings in Bioinformatics, 19*(6), 1236–1246. doi:10.1093/bib/bbx044 PMID:28481991

Mladenovic, A., Markovic, Z., Grujicic-Sipetic, S., & Hyodoh, H. (2012). Abdominal Aortic Aneurysm in Different Races Epidemiologic Features and Morphologic-Clinical Implications Evaluated by CT Aortography. *Aneurysm,* 109.

Mobley, R. K. (2001). *Plant Engineer's Handbook.* Butterworth-Heinemann.

Mohamed, I. A., Ben Aissa, A., Hussein, L. F., Taloba, A. I., & ... (2021). A new model for epidemic prediction: COVID-19 in kingdom saudi arabia case study. *Materials Today: Proceedings*. Advance online publication. doi:10.1016/j.matpr.2021.01.088 PMID:33520671

Mohanty, S., Harun, A. I., Rashid, M., Mridul, M., Mohanty, C., & Swayamsiddha, S. (2020). Application of Artificial Intelligence in COVID-19 drug repurposing. *Diabetes & Metabolic Syndrome*, *14*(5), 1027–1031. doi:10.1016/j.dsx.2020.06.068 PMID:32634717

Monsalve-Torra, A., Ruiz-Fernandez, D., Marin-Alonso, O., Soriano-Payá, A., Camacho-Mackenzie, J., & Carreño-Jaimes, M. (2016). Using machine learning methods for predicting inhospital mortality in patients undergoing open repair of abdominal aortic aneurysm. *Journal of Biomedical Informatics*, *62*, 195–201. doi:10.1016/j.jbi.2016.07.007 PMID:27395372

Muhammad, S., Makhtar, M., Rozaimee, A., Aziz, A., & Jamal, A. A. (2015). Classification Model for Water Quality using Machine Learning Techniques. *International Journal of Software Engineering and Its Applications*, *9*(6), 45–52. doi:10.14257/ijseia.2015.9.6.05

Munsadwala, Y., Joshi, P., Patel, P., & Rana, K. (2019, April). Identification and visualization of hazardous gases using IoT. In *2019 4th International Conference on Internet of Things: Smart Innovation and Usages (IoT-SIU)* (pp. 1-6). IEEE. 10.1109/IoT-SIU.2019.8777481

Murray, N. L., Holmes, H. A., Liu, Y., & Chang, H. H. (2019). A Bayesian ensemble approach to combine PM2. 5 estimates from statistical models using satellite imagery and numerical model simulation. *Environmental Research*, *178*, 108601. doi:10.1016/j.envres.2019.108601 PMID:31465992

Musk, E., Karpathy, A., & Bannon, P. (2019). *Tesla Autonomy Day* [YouTube Video]. Retrieved from https://www.youtube.com/watch?v=Ucp0TTmvqOE

N, K., M, F., L, F., B, K., & A, I. (2021). Detection of COVID-19 from Chest X-ray Images Using Deep Convolutional Neural Networks. *Sensors (Basel, Switzerland)*, *21*(17). doi:10.3390/s21175940

Nagarajan, S., & Chandrasekaran, R. M. (2015). Design and implementation of expert clinical system for diagnosing diabetes using data mining techniques. *Indian Journal of Science and Technology*, *8*(8), 771. Advance online publication. doi:10.17485/ijst/2015/v8i8/69272

Nandy, A., & Basak, S. C. (2019). Bioinformatics in Design of Antiviral Vaccines. Encyclopedia of Biomedical Engineering, 1–3, 280–290. doi:10.1016/B978-0-12-801238-3.10878-5 doi:10.1016/B978-0-12-801238-3.10878-5

NASA. (2017). Satellite Observations of Water Quality for Sustainable Development Goal 6. *GEO Week 2017*.

Negandhi, V., Sreenivasan, L., Giffen, R., Sewak, M., & Rajasekharan, A. (2015). *IBM Predictive Maintenance and Quality 2.0 Technical Overview*. IBM Redbooks.

Ngai, E. W. T., Xiu, L., & Chau, D. C. K. (2009). Application of data mining techniques in customer relationship management: A literature review and classification. *Expert Systems with Applications*, *36*(2, Part 2), 2592–2602. doi:10.1016/j.eswa.2008.02.021

NIST. (2021). *TC4TL Challenge*. https://tc4tlchallenge.nist.gov/

Noriega, L. (2005). *Multilayer perceptron tutorial. School of Computing*. Staffordshire University.

Nugroho, A., Suharmanto, A., & Masugino. (2018). Customer relationship management implementation in the small and medium enterprise. *AIP Conference Proceedings*, *1941*(1), 020018. doi:10.1063/1.5028076

Ohnsman, A. (2020, October 8). *Waymo Restarts Robotaxi Service Without Human Safety Drivers.* Retrieved December 22, 2020, from Forbes website: https://www.forbes.com/sites/alanohnsman/2020/10/08/waymo-restarts-robotaxi-service-without-human-safety-drivers/?sh=6ab6c85d69d8

Olsen, N. (2020). *The Effects of Diabetes on Your Body.* Academic Press.

Ong, E., Wong, M. U., Huffman, A., & He, Y. (2020). COVID-19 Coronavirus Vaccine Design Using Reverse Vaccinology and Machine Learning. *Frontiers in Immunology, 11*(July), 1581. doi:10.3389/fimmu.2020.01581 PMID:32719684

Osterrieder, P., Budde, L., & Friedli, T. (2020). The smart factory as a key construct of industry 4.0: A systematic literature review. *International Journal of Production Economics, 221.* doi:10.1016/j.ijpe.2019.08.011

Painuli, D., Mishra, D., Bhardwaj, S., & Aggarwal, M. (2021). Forecast and prediction of COVID-19 using machine learning. *Data Science for COVID-19*, 381–397. doi:10.1016/B978-0-12-824536-1.00027-7

Paliouras, K., & Siakas, K. (2017). Social Customer Relationship Management: A Case Study. *International Journal of Entrepreneurial Knowledge, 5*(1), 20–34. Advance online publication. doi:10.1515/ijek-2017-0002

Patil, P. (2020). *Disease Symptom Prediction.* Kaggle. https://www.kaggle.com/itachi9604/disease-symptom-description-dataset

Peng, Q.-Y., Wang, X.-T., & Zhang, L.-N. (2020). Findings of lung ultrasonography of novel corona virus pneumonia during the 2019-2020 epidemic. *Intensive Care Medicine, 46*(5), 849–850. doi:10.100700134-020-05996-6 PMID:32166346

Perveen, S., Shahbaz, M., Guergachi, A., & Keshavjee, K. (2016). Performance Analysis of Data Mining Classification Techniques to Predict Diabetes. *Procedia Computer Science, 82*, 115–121. doi:10.1016/j.procs.2016.04.016

Pisner, D. A., & Schnyer, D. M. (2020). Support vector machine. In A. Mechelli & S. Vieira (Eds.), *Machine Learning* (pp. 101–121). Academic Press. doi:10.1016/B978-0-12-815739-8.00006-7

Poongodi, T., Krishnamurthi, R., Indrakumari, R., Suresh, P., & Balusamy, B. (2020). Wearable devices and IoT. In *A handbook of Internet of Things in biomedical and cyber physical system* (pp. 245–273). Springer. doi:10.1007/978-3-030-23983-1_10

Poór, P., & Basl, J. (2019). Predictive maintenance as an intelligent service in Industry 4.0. *Journal of Systems Integration, 10*, 3–10. doi:10.20470/jsi.v10i1.364

Povey, D., Ghoshal, A., & Boulianne, G. (2011). The Kaldi Speech Recognition Toolkit. *IEEE Workshop on Automatic Speech Recognition and Understanding.*

Powell, J. T., & Wanhainen, A. (2020). Analysis of the differences between the ESVS 2019 and NICE 2020 guidelines for abdominal aortic aneurysm. *European Journal of Vascular and Endovascular Surgery, 60*(1), 7–15. doi:10.1016/j.ejvs.2020.04.038 PMID:32439141

Qin, C., Yao, D., Shi, Y., & Song, Z. (2018). Computer-aided detection in chest radiography based on artificial intelligence: a survey. *BioMedical Engineering OnLine, 17*(1), 1–23. doi:10.1186/s12938-018-0544-y

Qureshi, M. A., Deriche, M., Beghdadi, A., & Amin, A. (2017). A critical survey of state-of-the-art image inpainting quality assessment metrics. *Journal of Visual Communication and Image Representation, 49*, 177–191. doi:10.1016/j.jvcir.2017.09.006

R.M., B., Salif, D., Oscar E., N., Niekerk, H. v., Sherbinin, A. d., Vijselaar, L., . . . Al-Lami, A. A.-Z. (2007). *Global Drinking Water Quality Index Development and Sensitivity Analysis Report.* United Nations Environment Programme Global Environment Monitoring System/Water Programme.

Raffort, J., Adam, C., Carrier, M., Ballaith, A., Coscas, R., Jean-Baptiste, E., Hassen-Khodja, R., Chakfé, N., & Lareyre, F. (2020). Artificial intelligence in abdominal aortic aneurysm. *Journal of Vascular Surgery, 72*(1), 321-333.

Rahmanian, N., Ali, S. H., Homayoonfard, M., Ali, N. J., Rehan, M., Sadef, Y., & Nizami, A. S. (2015). Analysis of Physiochemical Parameters to Evaluate the Drinking Water Quality in the State of Perak, Malaysia. *Journal of Chemistry, 2015*, 10. doi:10.1155/2015/716125

Rajasekar, S. J. S. (2021). An Enhanced IoT Based Tracing and Tracking Model for COVID -19 Cases. *SN Computer Science, 2*(1), 1–4. doi:10.100742979-020-00400-y PMID:33490971

Rajput, A., Thakur, A., Mukhopadhyay, A., Kamboj, S., Rastogi, A., Gautam, S., Jassal, H., & Kumar, M. (2021). Prediction of repurposed drugs for Coronaviruses using artificial intelligence and machine learning. *Computational and Structural Biotechnology Journal, 19*, 3133–3148. doi:10.1016/j.csbj.2021.05.037 PMID:34055238

Rani, S. (2018). mining in Continuous data for Diabetes Prediction. *2018 Second International Conference on Intelligent Computing and Control Systems (ICICCS), Iciccs*, 1209–1214. 10.1109/ICCONS.2018.8662909

Rasmussen, M., Valentin, J. B., & Simonsen, C. Z. (2020). Blood Pressure Thresholds During Endovascular Therapy in Ischemic Stroke—Reply. *JAMA Neurology, 77*(12), 1579–1580. doi:10.1001/jamaneurol.2020.3819 PMID:33044508

Rawal, K., Sinha, R., Abbasi, B. A., Chaudhary, A., Nath, S. K., Kumari, P., Preeti, P., Saraf, D., Singh, S., Mishra, K., Gupta, P., Mishra, A., Sharma, T., Gupta, S., Singh, P., Sood, S., Subramani, P., Dubey, A. K., Strych, U., ... Bottazzi, M. E. (2021). Identification of vaccine targets in pathogens and design of a vaccine using computational approaches. *Scientific Reports, 11*(1), 1–25. doi:10.103841598-021-96863-x PMID:34475453

Rayen, L. P., & Sreeranganachiyar, T. (2017). A Study on Problems Faced by the Customer in Relation to Customer Relationship Management Practices. *Sumedha Journal of Management, 6*(3), 24–38.

Redmon, J., Divvala, S., Girshick, R., & Farhadi, A. (2016). You Only Look Once: Unified, Real-Time Object Detection. *2016 IEEE Conference on Computer Vision and Pattern Recognition (CVPR)*. 10.1109/CVPR.2016.91

Ren, S., He, K., Girshick, R., & Sun, J. (2015). *Faster R-CNN: Towards Real-Time Object Detection with Region Proposal Networks*. Retrieved from arXiv.org website: https://arxiv.org/abs/1506.01497

RF Wireless World. (2012). *Automatic speech recognition system*. https://www.rfwireless-world.com/Terminology/automatic-speech-recognition-system.html

Riley, S., Fraser, C., Donnelly, C. A., Ghani, A. C., Laith, J., Hedley, A. J., Leung, G. M., Ho, L., Lam, T., Thuan, Q., Chau, P., Chan, K., Lo, S., Leung, P., Tsang, T., Ho, W., Lee, K., Lau, E. M. C., Ferguson, N. M., & Anderson, R. M. (2020). *Transmission Dynamics of the Etiological Agent of SARS in Hong Kong : Impact of Public Health Interventions*. American Association for the Advancement of Science. https://www.jstor.org/stable/3834535

Rong, G., Mendez, A., Bou Assi, E., Zhao, B., & Sawan, M. (2020). Artificial Intelligence in Healthcare: Review and Prediction Case Studies. *Engineering, 6*(3), 291–301. doi:10.1016/j.eng.2019.08.015

Rosenthal, D. I., Chew, F. S., Dupuy, D. E., Kattapuram, S. V., Palmer, W. E., Yap, R. M., & Levine, L. A. (1998). Computer-based speech recognition as a replacement for medical transcription. *AJR. American Journal of Roentgenology, 170*(1), 23–25. doi:10.2214/ajr.170.1.9423591 PMID:9423591

Roy, A., Kumbhar, F. H., Dhillon, H. S., Saxena, N., Shin, S. Y., & Singh, S. (2020). *Efficient Monitoring and Contact Tracing for COVID-19: A Smart IoT-Based Framework*. Academic Press.

Russakovsky, O., Deng, J., Su, H., Krause, J., Satheesh, S., Ma, S., Huang, Z., Karpathy, A., Khosla, A., Bernstein, M., Berg, A. C., & Fei-Fei, L. (2015). ImageNet Large Scale Visual Recognition Challenge. *International Journal of Computer Vision*, *115*(3), 211–252. doi:10.100711263-015-0816-y

Ruta, D., & Gabrys, B. (2005). Classifier selection for majority voting. *Information Fusion*, *6*(1), 63–81. doi:10.1016/j.inffus.2004.04.008

Rygielski, C., Wang, J., & Yen, D. C. (2002). Data mining techniques for customer relationship management. *Technology in Society*, *24*(4), 483–502. doi:10.1016/S0160-791X(02)00038-6

Sa'di, S., Maleki, A., Hashemi, R., Panbechi, Z., & Chalabi, K. (2015). Comparison of Data Mining Algorithms in the Diagnosis of Type Ii Diabetes. *International Journal on Computational Science & Applications*, *5*(5), 1–12. doi:10.5121/ijcsa.2015.5501

Sachdev, K., & Gupta, M. K. (2019). A comprehensive review of feature based methods for drug target interaction prediction. In *Journal of Biomedical Informatics* (Vol. 93, p. 103159). Academic Press. doi:10.1016/j.jbi.2019.103159

Saeed, Soobia, N. Z. Jhanjhi, Naqvi, M., Ponnusamy, V., & Humayun, M. (2020). Analysis of Climate Prediction and Climate Change in Pakistan Using Data Mining. *Industrial Internet of Things and Cyber-Physical Systems: Transforming the Conventional to Digital*, 321-338.

Sakalihasan, N., Limet, R., & Defawe, O. D. (2005). Abdominal aortic aneurysm. *Lancet*, *365*(9470), 1577–1589. doi:10.1016/S0140-6736(05)66459-8 PMID:15866312

Salah, M., & Abou-Shouk, M. (2019). *The effect of customer relationship management practices on airline customer loyalty*. Academic Press.

Salo, J. A., Soisalon-Soininen, S., Bondestam, S., & Mattila, P. S. (1999). Familial occurrence of abdominal aortic aneurysm. *Annals of Internal Medicine*, *130*(8), 637–642. doi:10.7326/0003-4819-130-8-199904200-00003 PMID:10215559

Sanakal, R., & Jayakumari, S. T. (2014). Prognosis of Diabetes Using Data mining Approach-Fuzzy C Means Clustering and Support Vector Machine. *International Journal of Computer Trends and Technology*, *11*(2), 94–98. doi:10.14445/22312803/IJCTT-V11P120

Sanjay, M. (2020). Model Performance boosting with Voting-Classifier - Analytics Vidhya. *Medium*. https://medium.com/analytics-vidhya/performance-boosting-with-voting-classifier-ea69313a367c

Saravananathan, K., & Velmurugan, T. (2016). Analyzing Diabetic Data using Classification Algorithms in Data Mining. *Indian Journal of Science and Technology*, *9*(43). Advance online publication. doi:10.17485/ijst/2016/v9i43/93874

Sarica, A., Cerasa, A., & Quattrone, A. (2017). Random Forest Algorithm for the Classification of Neuroimaging Data in Alzheimer's Disease: A Systematic Review. *Frontiers in Aging Neuroscience*, *0*(OCT), 329. doi:10.3389/fnagi.2017.00329 PMID:29056906

Satpathy, S., Mangla, M., Sharma, N., Deshmukh, H., & Mohanty, S. (2021). Predicting mortality rate and associated risks in COVID-19 patients. *Spatial Information Research*, *29*(4), 455–464. doi:10.100741324-021-00379-5

Saurabh, S., & Prateek, S. (2017). Role of contact tracing in containing the 2014 Ebola outbreak: A review. *African Health Sciences*, *17*(1), 225–236. doi:10.4314/ahs.v17i1.28 PMID:29026397

Scan, C. A. T. (n.d.). (CT) -. *Chest*.

Schapire, R. E. (2013). Explaining adaboost. In *Empirical inference* (pp. 37–52). Springer. doi:10.1007/978-3-642-41136-6_5

Selcuk, S. (2017). Predictive maintenance, its implementation and latest trends. *Proceedings of the Institution of Mechanical Engineers. Part B, Journal of Engineering Manufacture, 231*(9), 1670–1679. doi:10.1177/0954405415601640

Shafi, U., Mumtaz, R., Anwar, H., Qamar, A. M., & Khurshid, H. (2018). Surface Water Pollution Detection using Internet of Things. *15th International Conference on Smart Cities: Improving Quality of Life Using ICT & IoT (HONET-ICT),* 92-96. 10.1109/HONET.2018.8551341

Shankar, S., Chopra, A., Kanaparti, R., Kang, M., Singh, A., & Raskar, R. (2020). *Proximity sensing for contact tracing.* ArXiv.

Sharma, A. (2020). *Decision Tree vs. Random Forest – Which Algorithm Should you Use? Analytics Vidhya.* https://www.analyticsvidhya.com/blog/2020/05/decision-tree-vs-random-forest-algorithm/

Shehab, A., Ismail, A., Osman, L., Elhoseny, M., & El-Henawy, I. M. (2017). Quantified self using IoT wearable devices. In *International conference on advanced intelligent systems and informatics.* Springer.

Sheidaei, A., Hunley, S. C., Zeinali-Davarani, S., Raguin, L. G., & Baek, S. (2011). Simulation of abdominal aortic aneurysm growth with updating hemodynamic loads using a realistic geometry. *Medical Engineering & Physics, 33*(1), 80–88. doi:10.1016/j.medengphy.2010.09.012 PMID:20961796

Shulman, J. G., & Cervantes-Arslanian, A. M. (2019). Infectious etiologies of stroke. Seminars in Neurology, 39(4), 482-494. doi:10.1055-0039-1687915

Shuttleworth, J. (2019, January 7). *SAE J3016 automated-driving graphic.* Retrieved December 22, 2020, from Sae.org website: https://www.sae.org/news/2019/01/sae-updates-j3016-automated-driving-graphic

Sifakis, N. I. (1998). Quantitative mapping of air pollution density using Earth observations: A new processing method and application to an urban area. *International Journal of Remote Sensing, 19*(17), 3289–3300. doi:10.1080/014311698213975

Sifakis, N., & Deschamps, P. Y. (1992). Mapping of air pollution using SPOT satellite data. *Photogrammetric Engineering and Remote Sensing, 58*, 1433–1433.

Simon, S. T., Mandair, D., Tiwari, P., & Rosenberg, M. A. (2021). Prediction of Drug-Induced Long QT Syndrome Using Machine Learning Applied to Harmonized Electronic Health Record Data. *Journal of Cardiovascular Pharmacology and Therapeutics, 26*(4), 335–340. doi:10.1177/1074248421995348 PMID:33682475

Singh, P., Yadav, S., & Pall, L. (2018). Sales Trend Analysis of Products Using Customer Relationship Management Tool. *International Journal of Scientific Research in Computer Science, Engineering and Information Technology,* 289-294. doi:10.32628/CSEIT183888

Sipos, R., Fradkin, D., Moerchen, F., & Wang, Z. (2014). Log-based predictive maintenance. In *Proceedings of the 20th ACM SIGKDD international conference on Knowledge discovery and data mining* (pp. 1867-1876). 10.1145/2623330.2623340

Sirsat, M. S., Fermé, E., & Câmara, J. (2020). Machine learning for brain stroke: A review. *Journal of Stroke and Cerebrovascular Diseases, 29*(10), 105162. doi:10.1016/j.jstrokecerebrovasdis.2020.105162 PMID:32912543

Skurichina, M., & Duin, R. P. (1998). Bagging for linear classifiers. *Pattern Recognition, 31*(7), 909–930. doi:10.1016/S0031-3203(97)00110-6

Sneha, N., & Gangil, T. (2019). Analysis of diabetes mellitus for early prediction using optimal features selection. *Journal of Big Data, 6*(1), 13. Advance online publication. doi:10.118640537-019-0175-6

Sneha, R. V., & K, M. (2012). Water quality analysis of Bhadravathi taluk using GIS. *International Journal of Environmental Sciences,* 2443–2453.

Song, I., Yoon, J., Kang, J., Kim, M., Jang, W. S., Shin, N.-Y., & Yoo, Y. (2019). Design and implementation of a new wireless carotid neckband doppler system with wearable ultrasound sensors: Preliminary results. *Applied Sciences (Basel, Switzerland)*, *9*(11), 2202. doi:10.3390/app9112202

Sony, M., & Naik, S. S. (2019). Key ingredients for evaluating Industry 4.0 readiness for organizations: A literature review. *Benchmarking*, *27*(7), 2213–2232. doi:10.1108/BIJ-09-2018-0284

Soon, G. T. J., Zhi, P. K. L., Krishnan, S. M., & Meng, C. K. (2019). A review of aortic disease research in Malaysia. *The Medical Journal of Malaysia*, *74*(1), 67. PMID:30846666

Spence, J. D. (2020). Uses of ultrasound in stroke prevention. *Cardiovascular Diagnosis and Therapy*, *10*(4), 955–964. doi:10.21037/cdt.2019.12.12 PMID:32968653

Stodola, P., & Stodola, J. (2020). Model of Predictive Maintenance of Machines and Equipment. *Applied Sciences (Basel, Switzerland)*, *10*(1), 213. doi:10.3390/app10010213

Strozzi, F., Colicchia, C., Creazza, A., & Noé, C. (2017). Literature review on the 'Smart Factory' concept using bibliometric tools. *International Journal of Production Research*, *55*(22), 6572–6591. doi:10.1080/00207543.2017.1326643

Subramaniyam, M., Hong, S. H., Yu, J., & Park, S. J. (2017). Wake-Up Stroke Prediction through IoT and Its Possibilities. *2017 International Conference on Platform Technology and Service (PlatCon)*, 1-5. 10.1109/PlatCon.2017.7883738

Subramaniyam, M., Lee, K.-S., Park, S. J., & Min, S. N. (2020). Development of Mobile Application Program for Stroke Prediction Using Machine Learning with Voice Onset Time Data. In *International Conference on Human-Computer Interaction*. Springer. 10.1007/978-3-030-50726-8_87

Sun, P., Kretzschmar, H., Dotiwalla, X., Chouard, A., Patnaik, V., Tsui, P., … Caine, B. (2020). *Scalability in perception for autonomous driving: Waymo open dataset*. Academic Press.

Sun, F., Liu, H., Fu, H.-x., Li, C.-b., & Geng, X.-j. (2020). *Predictive factors of hemorrhage after thrombolysis in patients with acute ischemic stroke* (Vol. 11). Frontiers in Neurology.

Susto, G. A., Schirru, A., Pampuri, S., McLoone, S., & Beghi, A. (2015). Machine Learning for Predictive Maintenance: A multiple classifier approach. *IEEE Transactions on Industrial Informatics*, *11*(3), 812–820. doi:10.1109/TII.2014.2349359

Suzuki, K. (2011). *Artificial neural networks: Methodological advances and biomedical applications*. BoD–Books on Demand. doi:10.5772/644

Szegedy, C., Liu, W., Jia, Y., Sermanet, P., & Reed, S. (2015). *Going deeper with convolutions*. Retrieved from https://arxiv.org/abs/1409.4842

Szegedy, C., Ioffe, S., Vanhoucke, V., & Alemi, A. A. (2017). *Inception-v4, inception-resnet and the impact of residual connections on learning*. AAAI Press.

Tanre, D., Herman, M., Deschamps, P., & de Leffe, A. (1979). Atmospheric modeling for space measurements of ground reflectances, including bidirectional properties. *Applied Optics*, *18*(21), 3587. doi:10.1364/AO.18.003587 PMID:20216655

Taylor, P. (2009). *Text-to-speech Synthesis*. Cambridge University Press. doi:10.1017/CBO9780511816338

Taylor, S. J., & Letham, B. (2018). Forecasting at scale. *The American Statistician*, *72*(1), 37–45. doi:10.1080/00031305.2017.1380080

Tayyab, M., Marjani, M., Jhanjhi, N., & Hashem, I. A. T. (2021). A Light-weight Watermarking-Based Framework on Dataset Using Deep Learning Algorithms. *2021 National Computing Colleges Conference (NCCC)*, 1-6. 10.1109/NCCC49330.2021.9428845

Tayyab, M., Marjani, M., Jhanjhi, N., Hashim, I. A. T., & Almazroi, A. A (2021). Cryptographic Based Secure Model on Dataset for Deep Learning Algorithms. *Computers Materials & Continua*, *69*(1), 1183–1200. doi:10.32604/cmc.2021.017199

The State of the Self-Driving Car Race. (2020). *Bloomberg Hyperdrive*. Retrieved from https://www.bloomberg.com/features/2020-self-driving-car-race/

Topp, S. N., Pavelsky, T. M., Jensen, D., Simard, M., & Ross, M. R. V. (2020). Research Trends in the Use of Remote Sensing for Inland Water Quality Science: Moving Towards Multidisciplinary Applications. *Water (Basel)*, *12*(1), 169. doi:10.3390/w12010169

Toth, L., Hoffmann, I., Gosztolya, G., Vincze, V., Szatloczki, G., Banreti, Z., Pakaski, M., & Kalman, J. (2018). A Speech Recognition-based Solution for the Automatic Detection of Mild Cognitive Impairment from Spontaneous Speech. *Current Alzheimer Research*, *15*(2), 130–138. doi:10.2174/1567205014666171121114930 PMID:29165085

Toutiaee, M., Li, X., Chaudhari, Y., Sivaraja, S., Venkataraj, A., Javeri, I., Ke, Y., Arpinar, I., Lazar, N., & Miller, J. (2021). Improving COVID-19 Forecasting using eXogenous Variables. ArXiv Preprint ArXiv:2107.10397.

Troianovski, A., & Grundberg, S. (2012, July 19). Nokia's Bad Call on Smartphones. *Wall Street Journal*. Retrieved from https://www.wsj.com/articles/SB10001424052702304388004577531002591315494

Troncoso, C., Payer, M., Hubaux, J.-P., Salathé, M., Larus, J., Bugnion, E., Lueks, W., Stadler, T., Pyrgelis, A., Antonioli, D., Barman, L., Chatel, S., Paterson, K., Čapkun, S., Basin, D., Beutel, J., Jackson, D., Roeschlin, M., Leu, P., . . . Pereira, J. (2020). Decentralized Privacy-Preserving Proximity Tracing. https://arxiv.org/abs/2005.12273

Tulshyan, V., Sharma, D., & Mittal, M. (2020). An Eye on the Future of COVID'19: Prediction of Likely Positive Cases and Fatality in India over A 30 Days Horizon using Prophet Model. *Disaster Medicine and Public Health Preparedness*, 1–7. Advance online publication. doi:10.1017/dmp.2020.444 PMID:33203489

Tuzunkan, D. (2018). Customer relationship management in business-to-business marketing: Example of tourism sector. *Geo Journal of Tourism and Geosites*, *22*(1), 329. doi:10.30892/gtg.22204-291

Un, K.-C., Wong, C.-K., Lau, Y.-M., Lee, J. C.-Y., Tam, F. C.-C., Lai, W.-H., Lau, Y.-M., Chen, H., Wibowo, S., Zhang, X., Yan, M., Wu, E., Chan, S.-C., Lee, S.-M., Chow, A., Tong, R. C.-F., Majmudar, M. D., Rajput, K. S., Hung, I. F.-N., & Siu, C.-W. (2021). Observational study on wearable biosensors and machine learning-based remote monitoring of COVID-19 patients. *Scientific Reports*, *11*(1), 1–9. doi:10.103841598-021-82771-7 PMID:33623096

Unold, O., Nikodem, M., Piasecki, M., Szyc, K., Maciejewski, H., Bawiec, M., Dobrowolski, P., & Zdunek, M. (2020). IoT-Based Cow Health Monitoring System. *Computational Science – ICCS 2020: 20th International Conference, Amsterdam, The Netherlands, June 3–5, 2020. Proceedings*, *5*(12141), 344–356. doi:10.1007/978-3-030-50426-7_26

Uppal, S., Goel, S., Randhawa, B., & Maheshwary, A. (2020). Autoimmune-Associated Vasculitis Presenting as Ischemic Stroke With Hemorrhagic Transformation: A Case Report and Literature Review. *Cureus*, *12*(9). Advance online publication. doi:10.7759/cureus.10403 PMID:33062521

Usmani, R. S. A., Saeed, A., & Tayyab, M. (2021). Role of ICT for Community in Education During COVID-19. In *ICT Solutions for Improving Smart Communities in Asia* (pp. 125–150). IGI Global. doi:10.4018/978-1-7998-7114-9.ch006

Vamsi Badi, D. B. (2020). *Prediction of Brain Stroke Severity Using Machine Learning. International Information and Engineering Technology Association*. IIETA. doi:10.18280/ria.340609

Vashistha, R., Yadav, D., Chhabra, D., & Shukla, P. (2019). Artificial intelligence integration for neurodegenerative disorders. In *Leveraging Biomedical and Healthcare Data* (pp. 77–89). Elsevier. doi:10.1016/B978-0-12-809556-0.00005-8

Verma, A. K., & Aggarwal, R. (2021). Repurposing potential of FDA-approved and investigational drugs for COVID-19 targeting SARS-CoV-2 spike and main protease and validation by machine learning algorithm. *Chemical Biology & Drug Design*, *97*(4), 836–853. doi:10.1111/cbdd.13812 PMID:33289334

Verma, A., Jaiswal, S., & Sheikh, W. R. (2020). Acute thrombotic occlusion of subclavian artery presenting as a stroke mimic. *Journal of the American College of Emergency Physicians Open*, *1*(5), 932–934. doi:10.1002/emp2.12085 PMID:33145542

Vijayakumar, P., Khokhar, A., Pal, A., & Dhawan, M. (2020, July). Air Quality Index Monitoring and Mapping Using UAV. In *2020 International Conference on Communication and Signal Processing (ICCSP)* (pp. 1176-1179). IEEE. 10.1109/ICCSP48568.2020.9182374

Vos, T., Lim, S. S., Abbafati, C., Abbas, K. M., Abbasi, M., Abbasifard, M., Abbasi-Kangevari, M., Abbastabar, H., Abd-Allah, F., Abdelalim, A., Abdollahi, M., Abdollahpour, I., Abolhassani, H., Aboyans, V., Abrams, E. M., Abreu, L. G., Abrigo, M. R. M., Abu-Raddad, L. J., Abushouk, A. I., ... Murray, C. J. L. (2020). Global burden of 369 diseases and injuries in 204 countries and territories, 1990–2019: A systematic analysis for the Global Burden of Disease Study 2019. *Lancet*, *396*(10258), 1204–1222. doi:10.1016/S0140-6736(20)30925-9 PMID:33069326

Wagenseil, J. E., & Mecham, R. P. (2012). Elastin in Large Artery Stiffness and Hypertension. *Journal of Cardiovascular Translational Research*, *5*(3), 264–273. doi:10.100712265-012-9349-8 PMID:22290157

Wang, P., Zheng, X., Li, J., & Zhu, B. (2020). Prediction of epidemic trends in COVID-19 with logistic model and machine learning technics. *Chaos, Solitons, and Fractals*, *139*, 110058. Advance online publication. doi:10.1016/j.chaos.2020.110058 PMID:32834611

Wang, S., Li, Y., Tian, J., Peng, X., Yi, L., Du, C., Feng, C., Liu, C., Deng, R., & Liang, X. (2020). A randomized controlled trial of brain and heart health manager-led mHealth secondary stroke prevention. *Cardiovascular Diagnosis and Therapy*, *10*(5), 1192–1199. doi:10.21037/cdt-20-423 PMID:33224743

Waxter, T. M. (2014). *Analysis of Landsat Satellite Data to Monitor Water Quality Parameters in Tenmile Lake*. Civil and Environmental Engineering Master's Project Reports. doi:10.15760/CEEMP.35

Wei, C., & Yuan, H. M. (2014). An improved GA-SVM algorithm. *Proceedings of the 2014 9th IEEE Conference on Industrial Electronics and Applications, ICIEA 2014*, 2137–2141. 10.1109/ICIEA.2014.6931525

Weininger, D. (1988). SMILES, a chemical language and information system. 1. Introduction to methodology and encoding rules. *Journal of Chemical Information and Computer Sciences*, *28*(1), 31–36. doi:10.1021/ci00057a005

WHO/UNICEF Joint Monitoring Program (JMP) for Water Supply. (2015). Progress on sanitation and drinking water: 2015 update and MDG assessment. World Health Organization.

Wijeyaratne, S. M. (2011). Diagnosis of Aortic Aneurysm. *Diagnosis, Screening and Treatment of Abdominal, Thoracoabdominal and Thoracic Aortic Aneurysms*, 69.

Wilkinson, D. A., Daou, B. J., Nadel, J. L., Chaudhary, N., Gemmete, J. J., Thompson, B. G., & Pandey, A. S. (2021). Abdominal aortic aneurysm is associated with subarachnoid hemorrhage. *Journal of Neurointerventional Surgery*, *13*(8), 716–721. doi:10.1136/neurintsurg-2020-016757 PMID:33158992

Wilmink, T. B., Quick, C. R., & Day, N. E. (1999). The association between cigarette smoking and abdominal aortic aneurysms. *Journal of Vascular Surgery*, *30*(6), 1099–1105. doi:10.1016/S0741-5214(99)70049-2 PMID:10587395

Wise, E. S., Hocking, K. M., & Brophy, C. M. (2015). Prediction of in-hospital mortality after ruptured abdominal aortic aneurysm repair using an artificial neural network. *Journal of Vascular Surgery*, *62*(1), 8–15. doi:10.1016/j. jvs.2015.02.038 PMID:25953014

Witten, Frank, Hall, & Pal. (2005). Practical machine learning tools and techniques. *Data Mining*, *2*, 4.

Wong, H. Y. F., Lam, H. Y. S., Fong, A. H.-T., Leung, S. T., Chin, T. W.-Y., Lo, C. S. Y., Lui, M. M.-S., Lee, J. C. Y., Chiu, K. W.-H., Chung, T. W.-H., Lee, E. Y. P., Wan, E. Y. F., Hung, I. F. N., Lam, T. P. W., Kuo, M. D., & Ng, M.-Y. (2020). *Frequency and Distribution of Chest Radiographic Findings in Patients Positive for COVID-19*. doi:10.1148/ radiol.2020201160

Woo, P. C. Y., Lau, S. K. P., Wong, B. H. L., Tsoi, H., Fung, A. M. Y., Kao, R. Y. T., Chan, K., Peiris, J. S. M., & Yuen, K.PC. (2005). Differential sensitivities of severe acute respiratory syndrome (SARS) coronavirus spike polypeptide enzyme-linked immunosorbent assay (ELISA) and SARS coronavirus nucleocapsid protein ELISA for serodiagnosis of SARS coronavirus pneumonia. *Journal of Clinical Microbiology*, *43*(7), 3054–3058. doi:10.1128/JCM.43.7.3054-3058.2005 PMID:16000415

World Health Organization. (2004). *The atlas of heart disease and stroke*. Author.

Wu, A., Peng, Y., Huang, B., Ding, X., Wang, X., Niu, P., Meng, J., Zhu, Z., Zhang, Z., Wang, J., Sheng, J., Quan, L., Xia, Z., Tan, W., Cheng, G., & Jiang, T. (2020). Genome Composition and Divergence of the Novel Coronavirus (2019-nCoV) Originating in China. *Cell Host & Microbe*, *27*(3), 325–328. doi:10.1016/j.chom.2020.02.001 PMID:32035028

Wu, H., Yang, S., Huang, Z., He, J., & Wang, X. (2018). Type 2 diabetes mellitus prediction model based on data mining. *Informatics in Medicine Unlocked*, *10*, 100–107. doi:10.1016/j.imu.2017.12.006

Wurtz, K., Camerlink, I., D'Eath, R. B., Fernández, A. P., Norton, T., Steibel, J., & Siegford, J. (2019). Recording behaviour of indoor-housed farm animals automatically using machine vision technology: A systematic review. *PLoS One*, *14*(12), e0226669. doi:10.1371/journal.pone.0226669 PMID:31869364

Wu, S., Wu, B., Liu, M., Chen, Z., Wang, W., Anderson, C. S., Sandercock, P., Wang, Y., Huang, Y., Cui, L., Pu, C., Jia, J., Zhang, T., Liu, X., Zhang, S., Xie, P., Fan, D., Ji, X., Wong, K.-S. L., … Zhang, S. (2019). Stroke in China: Advances and challenges in epidemiology, prevention, and management. *Lancet Neurology*, *18*(4), 394–405. doi:10.1016/S1474-4422(18)30500-3 PMID:30878104

Yadaw, A. S., Li, Y., Bose, S., Iyengar, R., Bunyavanich, S., & Pandey, G. (2020). Clinical features of COVID-19 mortality: Development and validation of a clinical prediction model. *The Lancet. Digital Health*, *2*(10), e516–e525. doi:10.1016/ S2589-7500(20)30217-X PMID:32984797

Yamanishi, Y., Araki, M., Gutteridge, A., Honda, W., & Kanehisa, M. (2008). Prediction of drug–target interaction networks from the integration of chemical and genomic spaces. *Bioinformatics (Oxford, England)*, *24*(13), i232–i240. doi:10.1093/bioinformatics/btn162 PMID:18586719

Yazdani, Z., Rafiei, A., Yazdani, M., & Valadan, R. (2020). Design an efficient multi-epitope peptide vaccine candidate against SARS-CoV-2: An in silico analysis. *Infection and Drug Resistance*, *13*, 3007–3022. doi:10.2147/IDR.S264573 PMID:32943888

Yii, M. K. (2003). Epidemiology of abdominal aortic aneurysm in an Asian population. *ANZ Journal of Surgery*, *73*(6), 393–395. doi:10.1046/j.1445-2197.2003.t01-1-02657.x PMID:12801335

Yorio, Z., El-Tawab, S., & Heydari, M. H. (2021). Room-Level Localization and Automated Contact Tracing via Internet of Things (IoT) Nodes and Machine Learning Algorithm. *2021 IEEE Systems and Information Engineering Design Symposium, SIEDS 2021*, 46–51. 10.1109/SIEDS52267.2021.9483667

Young, N. E., Anderson, R. S., Chignell, S. M., Vorster, A. G., Lawrence, R., & Evangelista, P. H. (2017). A survival guide to Landsat preprocessing. *Ecology*, *98*(4), 920–932. doi:10.1002/ecy.1730 PMID:28072449

Young, S., & Woodland, P. (1994). State clustering in hidden Markov model-based continuous speech recognition. *Computer Speech & Language*, *8*(4), 369–383. doi:10.1006/csla.1994.1019

Yu, H. Q. (2019). *Experimental Disease Prediction Research on Combining Natural Language Processing and Machine Learning*. IEEE 7th International Conference on Computer Science and Network Technology, Dalian, China. 10.1109/ICCSNT47585.2019.8962507

Yu, D., & Deng, L. (2014). *Automatic Speech Recognition: A Deep Learning Approach*. Springer.

Yu, L., Wu, S., Hao, X., Li, X., Liu, X., Ye, S., Han, H., Dong, X., Li, X., Li, J., Liu, N., Liu, J., Zhang, W., Pelechano, V., Chen, W.-H., & Yin, X. (2020). *Rapid colorimetric detection of COVID-19 coronavirus using a reverse transcriptional loop-mediated isothermal amplification (RT-LAMP) diagnostic platform: iLACO*. MedRxiv., doi:10.1101/2020.02.20.20025874

Yusop, S. M., Abdullah, K., San, L. H., & Bakar, M. N. (2011). Monitoring Water Quality from Landsat TM Imagery in Penang, Malaysia. *Proceeding of the 2011 IEEE International Conference on Space Science and Communication (IconSpace)*, 249-253. 10.1109/IConSpace.2011.6015893

Yu, W., Dillon, T., Mostafa, F., Rahayu, W., & Liu, Y. (2020). A global manufacturing big data ecosystem for fault detection in predictive maintenance. *IEEE Transactions on Industrial Informatics*, *16*(1), 183–192. doi:10.1109/TII.2019.2915846

Z, L., Y, Y., X, L., N, X., Y, L., S, L., R, S., Y, W., B, H., W, C., Y, Z., J, W., B, H., Y, L., J, Y., W, C., X, W., J, C., Z, C., … F, Y. (2020). Development and clinical application of a rapid IgM-IgG combined antibody test for SARS-CoV-2 infection diagnosis. *Journal of Medical Virology, 92*(9), 1518–1524. doi:10.1002/jmv.25727 doi:10.1002/jmv.25727

Zainal, A., & Yusha, A. (1998). Profile of patients with abdominal aortic aneurysm referred to the Vascular Unit, Hospital Kuala Lumpur. *The Medical Journal of Malaysia*, *53*(4), 423–427. PMID:10971988

Zhang, C., & Ma, Y. (2012). *Ensemble machine learning: methods and applications*. Springer. doi:10.1007/978-1-4419-9326-7

Zhang, D., & Woo, S. S. (2020). Real time localized air quality monitoring and prediction through mobile and fixed IoT sensing network. *IEEE Access: Practical Innovations, Open Solutions*, *8*, 89584–89594. doi:10.1109/ACCESS.2020.2993547

Zhang, J., Tan, L., & Tao, X. (2019). On relational learning and discovery in social networks: A survey. *International Journal of Machine Learning and Cybernetics*, *10*(8), 2085–2102. doi:10.100713042-018-0823-8

Zheng, H., Paiva, A. R., & Gurciullo, C. (2020). *Advancing from Predictive Maintenance to Intelligent Maintenance with AI and IIoT*. ArXiv, abs/2009.00351.

Zhou, Q., Tao, W., Jiang, Y., & Cui, B. (2020). *A comparative study on the prediction model of COVID-19*. doi:10.1109/ITAIC49862.2020.9338466

Zhou, X., Bargshady, G., Abdar, M., Tao, X., Gururajan, R., & Chan, K. C. (2019). *A Case Study of Predicting Banking Customers Behaviour by Using Data Mining*. Paper presented at the 2019 6th International Conference on Behavioral, Economic and Socio-Cultural Computing (BESC).

Zhu, J. S., Ge, P., Jiang, C., Zhang, Y., Li, X., Zhao, Z., Zhang, L., & Duong, T. Q. (2020). Deep-learning artificial intelligence analysis of clinical variables predicts mortality in COVID-19 patients. *Journal of the American College of Emergency Physicians Open*, *1*(6), 1364–1373. doi:10.1002/emp2.12205 PMID:32838390

Zhuoyue, W., Xin, Z., & Xinge, Y. (2003). IFA in testing specific antibody of SARS coronavirus. *South China Journal of Preventive Medicine, 29*(3), 36–37.

Zitouni, I. (2007). Backoff hierarchical class n-gram language models: Effectiveness to model unseen events in speech recognition. *Computer Speech & Language, 21*(1), 88–104. doi:10.1016/j.csl.2006.01.001

Zou, Z., Shi, Z., Guo, Y., & Ye, J. (2019). *Object Detection in 20 Years: A Survey.* Retrieved from https://ui.adsabs.harvard.edu/abs/2019arXiv190505055Z/abstract

About the Contributors

Muneer Ahmad completed his Master's degrees in Mathematics (1999) and Computer Science (2002) from Punjab University and Islamia University, Pakistan, respectively. Later, he got his MPhil (MS, 2006) degree in Computer Science from COMSATS University, Pakistan. He completed his Ph.D. (2015) from Universiti Teknologi PETRONAS, Malaysia in Computer Science. Dr. Muneer Ahmad has authored numerous research articles in reputed journals indexed in WoS and Scopus. Dr. Muneer also secured a research grant of approximately 350, 000 SAR in numerous research projects. His research interests are in the discipline of data science that includes applied machine learning applications in pattern recognition, image processing, signal analysis, e-commerce, cybersecurity, edge computing, and cloud computing. He is investigating machine learning, deep learning, and data-driven approaches to propose optimized solutions in different areas of computing. The long-term goal of his research is to develop enhanced and novel solutions that add values to industry 4.0 applications.

Noor Zaman received the Ph.D. degree in IT from UTP, Malaysia. He has great international exposure in academia, research, administration, and academic quality accreditation. He was with ILMA University, KFU for a decade, and currently with Taylor's University, Malaysia. He has 19 years of teaching & administrative experience. He has an intensive background of academic quality accreditation in higher education besides scientific research activities, he had worked a decade for academic accreditation and earned ABET accreditation twice for three programs at CCSIT, King Faisal University, Saudi Arabia. Dr. Noor Zaman has awarded as top reviewer 1% globally by WoS/ISI (Publons) recently. He has edited/authored more than 11 research books with international reputed publishers, earned several research grants, and a great number of indexed research articles on his credit. He has supervised several postgraduate students including masters and Ph.D. Dr. Jhanjhi is an Associate Editor of IEEE ACCESS, Guest editor of several reputed journals, member of the editorial board of several research journals, and active TPC member of reputed conferences around the globe.

* * *

Javier M. Aguiar-Pérez (PhD, MSc Telecommunications Engineering) is an Associate Professor at University of Valladolid, and Head of the Data Engineering Research Unit. His research is focused on Big Data, Artificial Intelligence, and Internet of Things. He has managed international research projects and he has contributed to the standardisation field as expert at the European Telecommunications Standards Institute. He serves as editor, guest editor and reviewer, and author in several international journals, books, and conferences. Furthermore, he has been involved as reviewer and rapporteur in several international research initiatives.

Ashfaq Ahmad is currently a senior faculty member in the Department of Computer Science, College of CS and IT at Jazan University (JU) – Saudi Arabia. Before his recent appointment at the JU, he served PUCIT at Punjab University (PU) – Pakistan as an Assistant Professor. Dr. Ahmad received his Postgraduate Diploma (computer science) as well as his M.Sc. (computer science) degree from PU, his M.S. (computer science) degree from GC University, Lahore – Pakistan, and his Ph.D. (information technology) from IIUM – Malaysia. He has research interests in the semantic web, knowledge management, computer programming, algorithms, and machine learning. Dr. Ahmad has published several research papers in prestigious journals.

Hassan Ali is a Software Engineering graduate with hands-on experience in processing and analyzing satellite images data and to display it in a suitable form. Also experienced in building web applications in ReactJS using NodeJS and MongoDB on the backend.

Miguel Alonso-Felipe received his M.S. degrees in telecommunication engineering from the University of Valladolid, Spain. In addition, he is PhD Candidate at University of Valladolid and Researcher of the Data Engineering Research Unit. His research is mainly focused on Big Data, Artificial Intelligence, and Internet of Things. Besides, he is co-author of some publications in journals, dealing with topics related to his lines of research.

Sadia Aziz has over 14 years of academic experience in teaching, research, and supervision. She received her Ph.D. Degree in Computer Engineering in 2007 from Wuhan University of Technology, Wuhan, China, in 2007. She worked in the King Faisal University of Saudi Arabia as Assistant Professor for six years and played a significant role in curriculum development, examination management, and infrastructure management. She worked in CASE(Centre for Advanced Studies in Engineering) in Pakistan for almost five years, leading two wireless network research groups. Currently, she is working as a lecturer with the School of Information Technology and Engineering (SITE), Melbourne Institute of Technology (MIT), and Didasko-Latrobe, Melbourne, Australia. Her research interests are Wireless communications and networking, Mobile ad hoc networks, QoS and reliability in real-time systems, Distributed system design, and teaching and learning. She possesses some prestigious professional certifications and has authored numerous refereed publications.

Mikel Barrio-Conde is a PhD candidate at University of Valladolid, who received his M.S. degrees in telecommunication engineering from the University of Valladolid, Spain. He is researcher of the Data Engineering Research Unit, and his research is focused on Artificial Intelligence, and Internet of Things. Also, he is co-author of some publications in journals, dealing with topics related to his lines of his research.

Aymen Bashir is an Electrical Engineer graduate from National University and Sciences and Technology, Islamabad, Pakistan with Remote Sensing, Machine Learning and Deep Learning as area of research.

Javier del Pozo-Velázquez received his M.S. degrees in telecommunication engineering from the University of Valladolid, Spain. In addition, he is PhD Candidate at University of Valladolid and Researcher of the Data Engineering Research Unit. His research is mainly focused on Big Data, Artificial Intelligence, and Internet of Things. Besides, he is co-author of some publications in journals, dealing with topics related to his lines of research.

Qazi Mudassar Ilyas received his BS and MS degrees from the University of Agriculture, Faisalabad, Pakistan, in 1998 and 2000, respectively. He received his Ph.D. degree from Huazhong University of Science and Technology, China, in 2005. He has served in several universities in Pakistan, including Ghulam Ishaq Khan Institute, Topi, the University of the Punjab, Lahore, and the COMSATS Institute of Information Technology, Abbottabad. Currently, he is working with King Faisal University, Saudi Arabia, as an Associate Professor in the College of Computer Sciences and Information Technology. His research interests include machine learning knowledge management, and information retrieval.

Abid Mehmood received the M.Sc. degree in computer science from Quaid i Azam University, Islamabad, Pakistan, in 2001 and the Ph.D. degree in computer science from Universiti Teknologi Malaysia, Johor Bahru, Malaysia, in 2014. His research interests include neural networks and deep learning, Internet of Things, dynamic and self-adaptive systems, model-driven engineering, and aspect-orientation. Previously, from 2001 to 2009, he worked in the software development industry in different roles and contributed to the design and development of various high-performance enterprise applications.

Rafia Mumtaz (Senior Member, IEEE) received the Ph.D. degree in remote sensing and satellite image processing from the University of Surrey, U.K., in 2010. She is currently working as Tenured Associate Professor and the Director of the Internet of Things (IoT) Lab, NUST-SEECS. She was a recipient of several national and international research grants worth PKR 43 million. She received the NUST-SEECS Best Researcher Award, in 2019.

Rafia Mumtaz received her PhD in 2010 from University of Surrey, UK, in the area of Remote Sensing and Satellite image processing. She is currently working as a Tenured Associate Professor and the Director Internet of Things (IoT) Lab, at NUST-SEECS. She is the recipient of several national and international research grants worth PKR 43 million. She has been awarded NUST-SEECS Best Researcher Award in the year 2019.

Mitra Muralitharan is an Executive of Internal Client Services, SEA Clients & Markets at Deloitte. Her work focuses specifically on the management of content updates for all Deloitte's website and intranet using content management system and responding to and troubleshooting all website and intranet issues. Her favourite thing to do during her free time is to read fictional books.

María A. Pérez-Juárez received her M.S. and Ph.D. degrees in telecommunication engineering from the University of Valladolid, Spain, in 1996 and 1999, respectively. She is presently Associate Professor at University of Valladolid, and member of the Data Engineering Research Unit. Her research is focused on Big Data, Artificial Intelligence, Internet of Things, and the application of technology to the learning process. She has managed or participated in numerous international research projects. She is author or co-author of many publications in journals, books, and conferences. In addition, she has been involved as reviewer in several international research initiatives.

Saúl Rozada-Raneros is a PhD candidate at University of Valladolid, who received his M.S. degrees in telecommunication engineering from the University of Valladolid, Spain. He is researcher of the Data Engineering Research Unit, and his research is focused on Internet of Things and Virtual Reality. Also, he is co-author of some publications in journals, dealing with topics related to his lines of his research.

Muhammad Ali Tahir has completed his Ph.D. in Automatic speech recognition in 2015 and M.Sc. Media Informatics from RWTH Aachen University, Germany. During his 11 year stay in Germany he has worked on different EU-funded speech recognition and AI projects namely QUAERO, EU-Bridge and TransLectures; primarily aiming to bridge the language divide in a multilingual Europe. Since 2016, during his time at NUST he has been teaching software engineering courses, while also collaborating with local industry to create Urdu voice based solutions for TV analytics and vehicle navigation. He has also launched an Urdu website UrduAsr which has been used by thousands of people for Urdu voice based typing.

Muhammad Tayyab has completed all the requirements for PhD in Computer Science for Taylor's University, Lake-side Campus, Subang Jaya, Malaysia. He has received M.S. degree in Information security MS(IS) from COMSATS University Islamabad, Islamabad Campus, Pakistan in 2016 and Bachelors of Computer engineering (BCE) from Bahira University Islamabad, Pakistan in 2012.. Currently, he is working as an assistant professor at the University of Sialkot, Sialkot campus. From 2013 to 2015 he was a high school teacher in the public govt education department, in Rawalpindi, Pakistan. After that from 2015 to 2018, he worked as a lecturer in one of the leading private universities named the University of Management and Technology (UMT) Sialkot Campus, Punjab, Pakistan. His research area is Big data, Cryptography, Applied cryptography and Data Science.

Muhammad Usama is a Software Engineer by profession and has been a part of a lot of research teams. He has vast knowledge in the field of climate change and its impacts. He has been working on the practical applications of machine learning towards environmental sciences.

Hafsa Zubair is a Software Engineer with hands-on experience in web development using MERN stack, machine learning, satellite image processing, and data analytics. Passionate about software research and development, and consultancy.

Index

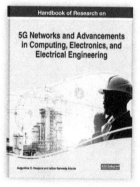

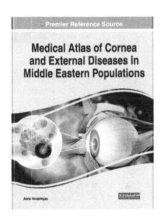

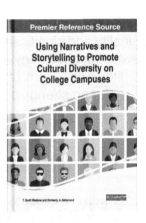

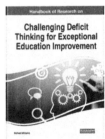